# Studying Distant Galaxies

## A Handbook of Methods and Analyses

## Advanced Textbooks in Physics

ISSN: 2059-7711

---

*Published*

Trapped Charged Particles: A Graduate Textbook with Problems and Solutions
*edited by Richard C Thompson, Niels Madsen & Martina Knoop*

Studying Distant Galaxies: A Handbook of Methods and Analyses
*François Hammer, Mathieu Puech, Hector Flores & Myriam Rodrigues*

*Forthcoming*

An Introduction to Particle Dark Matter
*Stefano Profumo*

An Introducion to String Theory and D-Brane Dynamics: With Problems and Solutions (3rd Edition)
*by Richard J Szabo*

Advanced Textbooks in Physics

# Studying Distant Galaxies

## A Handbook of Methods and Analyses


**François Hammer**
**Mathieu Puech**
**Hector Flores**
**Myriam Rodrigues**

*Paris Observatory, France*



World Scientific

NEW JERSEY • LONDON • SINGAPORE • BEIJING • SHANGHAI • HONG KONG • TAIPEI • CHENNAI • TOKYO

*Published by*

World Scientific Publishing Europe Ltd.

57 Shelton Street, Covent Garden, London WC2H 9HE

*Head office:* 5 Toh Tuck Link, Singapore 596224

*USA office:* 27 Warren Street, Suite 401-402, Hackensack, NJ 07601

**Library of Congress Cataloging-in-Publication Data**
Names: Hammer, François.
Title: Studying distant galaxies : a handbook of methods and analyses /
François Hammer (Paris Observatory, France) [and three others].
Description: New Jersey : World Scientific, 2016. | Series: Advanced textbooks in physics
Identifiers: LCCN 2016015279 | ISBN 9781786340542 (hc : alk. paper)
Subjects: LCSH: Astronomy--Observers' manuals. | Galaxies--Observers' manuals.
Classification: LCC QB64 .S78 2016 | DDC 523.1/12--dc23
LC record available at https://lccn.loc.gov/2016015279

**British Library Cataloguing-in-Publication Data**
A catalogue record for this book is available from the British Library.



Desk Editors: Herbert Moses/Mary Simpson

Typeset by Stallion Press
Email: enquiries@stallionpress.com

Printed in Singapore

# Foreword

During the 1990s, two of us had been deeply involved in the Canada France Redshift Survey (CFRS), the first systematic effort to understand how galaxies evolved during the past 8 billion years. It has led us to design, implement, and gather from the literature a multiplicity of techniques to study their statistics, morphologies, and chemistry. During the past 25 years, our group has devoted its efforts to pioneering and refining many techniques, including those elaborated to interpret the challenging spatially-resolved kinematics of distant galaxies. These efforts have lead us to identify the gradual paucity of star formation and heavy element production, and the disordering of galaxy internal motions. In turn, they contributed to our discovery of the evolution of the Hubble Sequence.

Since the CFRS, myriads of papers on galaxy evolution have been published. This has lead to mixed results. On one hand, increasingly more distant galaxies have been identified and observed with instruments and telescopes which number and power has grown rapidly. On the other hand, besides improved technicalities, methods of analysis have progressed less. According to many pioneers in the field, this has affected too large a fraction of the literature, independently of their impact or citation rate. A consensus on the most important processes driving galaxy evolution and formation has not been reached yet, and this is perhaps due to a lack of sufficiently-defined methodology, instead of observational limits. This is especially problematic before the advent of extremely large telescopes.

This book was especially designed to tackle the issues stemming from this lack of well-defined methodology. It strives to provide an observation-oriented handbook with clear and common sense guidelines for each critical step: from sample selection, preparation, and analysis of observations, up to interpretation. The goal is precision, i.e., to reduce or better control the numerous sources of uncertainties and systematics that limit the understanding of observations of distant galaxies. Master and PhD students as well as academics may find here a simple way to prepare, handle, or evaluate a research project, an article, or a proposal for observations.

Writing and correcting the content of this book has been a long and sometimes difficult process with many lively discussions, questions, and interrogations. Given the number of articles in the field, we have had to compromise exhaustiveness in favor of comprehensibility. We deeply apologize to the colleagues we may have unfairly neglected and trust they will forgive us. We are very grateful to all our colleagues who have kindly agreed to share their scientific materials with us. This project could not have been achieved without the many encouragements and discussions from which it has spawn, and which, stone after stone, have nurtured it: namely with Lia Athanassoula, Albert Bosma, David Crampton, Pavel Kroupa, Olivier Le Févre, Yanchun Liang, Simon Lilly, Benoit Neichel, Jim Peebles, Didier Pelat, Daniel Schaerer, and Sidney van den Bergh. We are further indebted to Yanbin Yang, Karen Disseau, and Susanna Vergani for their considerable help in elaborating a thorough methodology and for their reading of preliminary versions. We are very grateful to John Carter and Nor Pirzkal for their careful proof reading and numerous useful suggestions. We apologize for any errors the reader might find, which remain our responsibility, and we will be very pleased to receive a correction.

F. Hammer, M. Puech, H. Flores, and M. Rodrigues

*Meudon, June 2016*

GEPI, Observatoire de Paris
PSL Research University, CNRS
Univ. Paris Diderot, Sorbonne Paris Cité
Place Jules Janssen
92195 Meudon, France

# List of Figures

# List of Tables

# List of Examples

# Glossary

**ACS** Advanced Camera for Surveys onboard the HST

**ADC** Atmospheric Diffraction Corrector — Optical device used to correct from atmospheric diffraction in ground base instruments

**ADU** Analog–Digital Units; Digital unit in which the counts produced by a light detector are recorded. The conversion between the number of electrons acquired by the CCD and the ADUs is given by the detector gain

**AGB** Asymptotic Giant Branch; Region of the Hertzsprung–Russel diagram corresponding to the final evolution stage of low- and intermediate-mass stars. This phase of evolution is characterized by nuclear burning of hydrogen and helium in thin shells on top of the electron-degenerate core of carbon and oxygen

**AGN** Active Galactic Nucleus

**ALMA** Atacama Large Millimeter/submillimeter Array

**AO** Adaptive Optics; Technology used to improve the performances of optical systems by correcting the distortion of wavefronts due to the atmospheric turbulence

**BAO** Baryon Acoustic Oscillations

**BzK** Distant galaxy selected based on $B - z$ and $z - K$ colors

**CANDELS** Cosmic Assembly Near-infrared Deep Extragalactic Legacy Survey; Imaging survey conducted with the HST ACS and WFC3 cameras on several cosmological fields

**CCD** Charge-Coupled Device; Digital image sensor widely used in optical astronomy

**CDFS** Chandra Deep Field South

**CFHT** Canada France Hawaii Telescope

**CFRS** Canada France Redshift Survey

**CGM** Circum Galactic Medium

**CHeB** Core Helium Burning phase; phase that marks the end of the RGB phase in stellar evolution, in which a star burns Helium in its core

**CMOS** Complementary Metal-Oxide-Semiconductor; technology for constructing integrated circuits, including image NIR sensors

**COG** Curve-of-Growth. Method to measure the column density of a specie from the absorption equivalent width

**COSMOS** Cosmic Evolution Survey

**CSP** Composite Stellar Population; Theoretical population of stars formed at different time and having different metallicities used by Stellar Population Synthesis models; Both ages and metallicities are usually obtained from the convolution of a Simple Stellar Population with a Star Formation History law

**CTE** Charge Transfer Efficiency; fraction of electrons that are successfully transferred from one pixel to another during one CCD read-out cycle

**DEEP2** Deep Extragalactic Evolutionary Probe project, phase 2

**DLA** Damped Lyman-alpha system; absorption line saturating the spectra of distant sources and caused by HI column density above $2 \times 10^{20}\text{cm}^{-2}$

**DRG** Distant Red Galaxy

**DS9** Widely used Astronomical imaging and data visualization application, originally named SAOimage DS9

**E-ELT** European Extremely Large Telescope; Future ESO 39-meter telescope

**ELT** Extremely Large Telescope; Next-generation telescope with diameter larger than ~20m

**ETC** Exposure Time Calculator

**FIR** Far Infrared; wavelengths in the range ~25/40–200/350 $\mu$m

**FLAMES/GIRAFFE** Fiber-fed multi-object and multi-IFU medium resolution, optical spectrograph at the VLT

**FORS2** FOcal Reducer/low dispersion Spectrograph 2 at the VLT

**FoV** Field of View; solid angle through which a detector is sensitive to light

**FRD** Focal Ratio Degradation

**FUV** Far-UV wavelength domain Wavelengths in the range ~100–200 nm

**FWHM** Full Width at Half Maximum; Often used to define the width of a distribution

**GALEX** 0.5m space telescope dedicated to the UV (NUV and FUV windows) operated from 2003 to 2012

**GALFIT** Program that fits 2D analytic functions to galaxies and point sources

**GAMA** Galaxy And Mass Assembly survey; Spectroscopic survey of ~300,000 galaxies carried out at the Anglo-Australian Telescope over 2008–2014

**GLAO** Ground-Layer Adaptive Optics; Adaptive Optics system in which only the lowest altitude turbulente layer, which usually dominates the overall atmospheric turbulence strength, is corrected

**GMASS** Galaxy Mass Assembly ultra-deep Spectroscopic Survey

**GOODS** Great Observatories Origins Deep Survey

**GRB** Gamma Ray Burst; extremely energetic emission of gamma rays in galaxies

**GSC** Guide Star Catalog; also known as the Hubble Space Telescope, Guide Catalog

**Herschel** Herschel space telescope; IR and sub-mm 3.5m space telescope operated between 2009 and 2013

**HST** Hubble Space Telescope; 2.4m space telescope working at UV-optical-NIR wavelengths

**HUDF** Hubble Ultra Deep Field; $2.4 \times 2.4$ arcmin$^2$ region of the southern sky on which the deepest HST observations have been conducted

**IFU** Integral Field Unit

**IGM** Inter-Galactic Medium

**IMAGES** Intermediate Mass Galaxy Evolution Sequence

**IMF** Initial Mass Function; distribution of stellar mass for a population of newly borned stars

**IR** Infrared; Wavelengths in the range ~1–800 $\mu$m

**IRAC** The Infrared Array Camera onboard the Spitzer telescope with four channels centered at 3.6, 4.5, 5.8, and 8.0 $\mu$m

**IRAF** Image Reduction and Analysis Facility; general purpose software system for the reduction and analysis of astronomical data, written and supported by the National Optical Astronomy Observatories (NOAO) in Tucson, Arizona

**IRAS** The Infrared AStronomical Satellite was a 0.57m space telescope operated during 10 months in 1983. It performed the first all-sky survey in the IR at 12, 25, 60, and 100 $\mu$m

**ISM** Inter-Stellar Medium

**ISO** Infrared Space Observatory; 0.6m telescope working between 2 and 200 $\mu$m, which was operated by ESA between 1995 and 1998

**JWST** James Webb Space Telescope; 6.5m space telescope working between 0.6 and 28 $\mu$m to be launched in 2018

**KMOS** K band Multi-Object Spectrographat the VLT

**L*** Characteristic luminosity of the luminosity function at a given redshift; Luminosity describing the knee of the Luminosity Function, which corresponds to galaxies dominating the luminosity density at a given redshift

**LAE** Lyman Alpha Emitter

**LBG** Lyman Break Galaxy

**LF** Luminosity Function

**LINER** Low Ionization Narrow Emission Line. Active Galactic nuclei galaxies characterized by a optical line ratios compatible with low ionization

**LIRG** Luminous IR Galaxy; galaxy with $L_{IR} \geq 10^{11}\,\mathrm{L}_{\odot}$

**LSF** Line Spread Function; Spectral response of a system to a point-like source

**MIPS** Multiband Imaging Photometer for Spitzer, is a IR camera onboard the Spitzer space telescope with two channels centered at 24 and 70 $\mu$m

**MIR** Mid Infrared; wavelengths in the range ~5–25/40 $\mu$m

**MOAO** Multi-Object Adaptive Optics; Adaptive Optics system in which the correction is optimized along a limited number of directions in the FoV

**MOS** Multi-Object Spectrograph

**MOSAIC** E-ELT MOS for Astrophysics, IGM and Cosmology

**MS** Main Sequence stars; stars located along a continuous region of the Hertzsprung-Russel diagram that characterizes stars burning hydrogen into helium in a dense core. Generally, only low mass stars have lifetimes long enough to populate significatively the MS

**MUSE** Optical integral field spectrograph at the VLT; MUSE is currently the IFU with the most extended FoV (i.e., up to 1×1 arcmin$^2$)

**MW** Milky Way

**NICMOS** Near Infrared Camera and Multi-Object Spectrometer

**NIR** Near-InfraRed; wavelengths in the range ~0.7/1.0–2.5 $\mu$m

**NUV** Near-UV wavelength domain Wavelengths in the range ∼200–300 nm

**OH** HydrOxyl radical; molecules that are responsible for the numerous sky emission lines in the NIR

**OSIRIS** Near-infrared integral field spectrograph designed for the Keck Adaptive Optics System

**P.A.** Position Angle; on-sky orientation of a galaxy photometric major axis, counted anti-clockwise from North to East

**PSF** Point Spread Function; optical response of a system to a point-like source

**PWV** Precipitable water vapor. Amount of water vapor in the atmosphere, express in mm

**QE** Quantum Efficiency; percentage of photons that are converted into charges by a detector

**QSO** Quasi Stellar Object or quasar

**RGB** Red Giant Branch; Region of the Hertzsprung–Russel diagram corresponding to a stage in evolution of main sequence stars that have burnt all their hydrogen leading to a readjustement of the internal structure of the star, with expansion and subsequent reaction of the envelope. RGB stars are giant by their volume and luminosity but not by their masses

**S/N** Signal-to-Noise ratio

**SDSS** Sloan Digital Sky Survey

**SED** Spectral Energy Distribution; spectral distribution of flux/luminosity of a galaxy as a function of wavelength and at very low spectral resolution

**SEXTRACTOR** Program that builds a catalog of objects from an astronomical image, particularly oriented towards reduction of large scale galaxy-survey data or moderately crowded star fields

**SFH** Star-formation History

**SFR** Star Formation Rate; rate at which a galaxy forms stars, in $M_\odot \cdot yr^{-1}$

**SGB** Sub Giant Branch; Region of the Hertzsprung–Russel diagram that characterizes stars that have stopped burning hydrogen into helium in their core but have yet started doing so in a surrounding shell

**SINFONI** Spectrograph for INtegral Field Observations in the Near Infrared at the VLT

**SINS** Spectroscopic Imaging survey in the Near-infrared with SINFONI

**SKA** Skare Kilometer Array; project of radio telescope with a total collecting area of approximately one square kilometer

**SMC** Small Magellanic Cloud

**Spitzer** 0.85m space telescope launched in 2003 an dedicated to the IR with the IRAC and MIPS camera, and the IRS spectrograph

**SPS** Stellar Population Synthesis; set of methods and models used to compute the spectrophotometric properties of galaxies from those of the underlying stellar populations

**SSP** Simple Stellar Population; theoretical population of stars used by Stellar Population Synthesis models assumed to be born at the same time from a single gas could according to a given IMF and having the same metallicity

**TFR** Tully–Fisher Relation; tight relation between the velocity width or rotation velocity of galaxies and their luminosity

**TP-AGB** Thermally Pulsing Asymptotic Giant Branch

**UCAC** USNO CCD Astrograph Catalog

**ULIRG** Ultra Luminous IR Galaxy; Galaxy with $L_{IR} \geq 10^{12} L_{\odot}$

**USNO** United States Naval Observatory

**UV** Ultraviolet; wavelengths in the range ∼10–400nm

**VIMOS** VIsible Multi-object Spectrograph at the VLT

**VLT** European Southern Observatory Very Large Telescope

**VVDS** VIMOS VLT Deep Survey

**WFC3** Wide Field Camera 3 onboard the HST

**WFPC2** Wide Field and PLanetary Camera 2; Camera onboard the HST installed in 1993 and replaced in 2009 by the WFC3. This camera, which is sensitive to wavelengths from 120 to 1000 nm, was used to image the Hubble Deep Field in 1995

**WiggleZ** WiggleZ Dark Energy Survey; large-scale galaxy redshift survey conducted between 2006 and 2011n which measured 240,000 galaxy redshifts over 1000 square degrees up to $z = 1$

**WR** Wolf–Rayet star. Massive young stars with ages <10 Myr

**X-SHOOTER** UV to NIR single-target, intermediate-resolution spectrograph at the VLT

# Introduction

Galaxies are complex objects comprising hundreds of billions of stars, multiple gas phases, and dust. Present-day galaxies are organized into a well-defined sequence, the Hubble Sequence. Galaxy complexity naturally increases with redshift, since distant galaxies are likely the subject of transformation mechanisms. The study of distant galaxies is particularly enthralling since the Cosmological Principle infers that they are causally linked to the present-day ones.

There are considerable difficulties in studying distant galaxies. The more distant they are, the more they bear witness to the earliest stages of galaxy formation, but this is accompanied by faintness, smallness, and redshifted emissions, all challenges for interpreting their observation. Theoretical interpretations require the determination of stellar and gas mass, star formation, and heavy element abundances, all of which are quantities obtained indirectly, and which are afflicted by uncertainties and systematics both large and numerous. Galaxies are also not homogeneously distributed in space, and their spectral energy distributions are affected by redshift, stellar ages, abundances and dust, thus complicating the rationale of their selection.

This handbook includes a state-of-the-art description of the sampling and selection effects, based on observations with current instruments at large telescopes and using complementary approaches including photometry, imagery, as well as integrated and spatially-resolved spectroscopy. Such methods can be found in a myriad of scientific papers or reviews, synthetized for the first time in a single book. It aims to be a vital tool for teaching final-year undergraduate and postgraduate courses on the topics of "Galaxies" and Observational Cosmology. A number of illustrated examples will assist the reader in deriving various galaxy physical quantities from observables. The book shows or proposes numerous methods to derive physical quantities from observations, but also expresses reasonable doubts about far-reaching conclusions made on minimal or insufficient data. Masters and PhD students are encouraged to use it while preparing research projects, including telescope time proposals or articles. Academics may inquire about the numerous aspects surrounding their field of research, which they can better evaluate.

The first chapter establishes how a galaxy sample can be gathered and the many associated caveats which may hamper theoretical interpretations. It addresses the statistical completeness issue, the redshift measurement, the luminosity function analysis, and finally the possibility to causally link galaxy samples from different epochs. The second chapter provides an overview of the basics of imagery and photometry and their uncertainties, as well as the heightened complexity of morphological analyses. A method to derive the latter analyses is proposed, which minimizes systematics while allowing to derive stellar mass and star formation from the energy distribution. The third chapter provides a comprehensive basis for integrated spectroscopy, showing how one can derive the heavy element abundances of their gas phases and star formation rates (SFRs), as well as determine the metal abundances from UV absorption lines and constrain the stellar populations. In principle, velocity is the most directly determined parameter for distant galaxies, and it is the role of Chapter 4 to introduce the considerable limitations caused by galaxy faintness and smallness. It then includes a full description of existing methods for reducing data cubes with a comprehensive treatment of sky residuals, and proposes a method to derive then classify the dynamical state of galaxies, called "morpho-kinematics". Chapter 5 describes how to synthesize the numerous observations provided for a single, distant galaxy, to determine its physical properties and allow its modeling. It addresses the important question of stellar mass evaluation as well as that of the gas in its various phases. It finally shows the subsequent strengths and weaknesses of scaling relations between mass, velocity, abundance, and star formation.

# Contents

# Chapter 1

# Samples and Selection Effects

## 1.1 Introduction: which sample for which science goal?

### 1.1.1 *General rules to optimize the scientific returns from a survey of distant galaxies*

The present-day generation of instruments allows us to study distant galaxies in considerable details. We can now determine the mass, dynamics, abundances, stellar population, and star formation efficiency of galaxies. While one could hope to learn a lot from previous work done in the local Universe, distant galaxies are unfortunately not like those in the local Universe both observationally (they are faint and small) and intrinsically (they are in principle less evolved).

To study how galaxies form and evolve, to test the curvature of the Universe or to investigate the epoch of reionization, astronomers rely on samples of galaxies for which they measure their redshifts, i.e., the so-called redshift surveys. While the redshift is assumed to be a fair indicator of the distance of a galaxy, there are many different criteria that can be used to select galaxies. One must therefore ensure that the selection criteria that were used are appropriate for the scientific objectives. For example, to study galaxy evolution one can study galaxy populations that are at different redshifts within the same redshift survey, or, alternatively, gather a comparison sample of galaxies at lower redshifts or in the Local Universe.

As was learned from the numerous redshift surveys, a simple, single, selection criterion such as apparent luminosity is preferred.

This avoids having to deal with multiple selection effects that can cause some results to be very difficult to interpret, particularly with evolutionary studies. Even when using a single selection criterion for a redshift survey, one should carefully investigate possible biases that may prevent a representative analysis (see Ex. *Introduction to the k-correction effects*).

We will also discuss other means to select galaxies, using a color selection or photometric redshift techniques (see Sec. 1.5). These two selection techniques do not, however, provide true measurements of redshift[1] and we therefore discuss these later. These methods are increasingly popular, especially when estimating the redshifts of distant galaxies which are often selected using similar spectral energy distribution based techniques (see Sec. 1.6).

**Example 1.1: *Introduction to the k-correction effects***

It would be hard to test the evolution of elliptical galaxies up to $z = 1.5$ using a redshift survey based on the observed luminosity in a filter bluer than 800nm. This is because the luminosity of these galaxies drops abruptly below the 400nm (so-called 4000 Å) break. At $z \geq 1$, such a selection will be principally driven by episodic star formation events dominated by UV luminous hot stars, and not by the bulk of the stellar mass. A very large $k$-correction is required when selecting $z \geq 1$ galaxies in the rest-frame UV. This effect can result in an apparent lack of elliptical galaxies in the redshift survey since their UV emission is so faint. It is therefore essential to select objects at a rest-frame wavelength greater than 400nm (see Sec. 1.3.6).

Once a consistent method to select galaxies at a chosen redshift has been established, one should also prepare for additional observations that are needed to estimate additional physical quantities such

[1] A measurement comes with an uncertainty distribution; this is discussed in Sec. 1.5.

as mass, abundances, stellar population, star formation. These additional measurements often need to be made at various wavelengths which bring additional selection effects, often caused by insufficient depth or resolution. For example, far-IR (FIR) observations of distant galaxies are often more limited in depth and spatial resolution than optical observations, i.e., detecting only the lower number density sub-sample of FIR-bright galaxies.

Table 1.1 illustrates how complementary observations can be used. It also shows how duplicating estimates at two different wavelengths can product a more robust estimate of a given physical quantity.

The way a galaxy sample is defined has considerable impact on the science that can be done with. It is a complex problem that must be handled carefully before any observations are attempted.

This chapter provides elementary guidelines that are based on common sense but it also tries to identify most of the usual pitfalls that can seriously affect studies of distant galaxies. It presents a logical sequence of problems encountered by scientific investigators aiming to assemble a sample of distant galaxies.

Section 1.2 describes the magnitude systems used along the book and presents general rules for estimating magnitudes. Section 1.3 shows how a galaxy sample can be obtained using imaging data, describing and providing solutions to the numerous problems linked to the photometric depth, to selection effects and biases, and to $k$-corrections. Section 1.4 describes how to prepare a galaxy redshift survey and how to derive the luminosity function (LF), and outlines the limitations of existing surveys and what can and cannot be derived from them. Section 1.5 presents an alternative way to estimate redshifts using photometry, and also explains why these photometric redshifts are not as robust distance measurements as spectroscopic redshifts. Section 1.6 deals specifically with very distant galaxies redshift surveys. Selecting these objects using photometry is difficult since they are overwhelmingly dominated in numbers by foreground galaxies. Section 1.7 describes techniques to identify a sample of distant galaxies that can be causally linked to nearby galaxies.

Table 1.1: Combination of measurements required for a robust determination of physical quantities of distant galaxies.

| Physical quantities | Required measurement 1 | Required measurement 2 | Errors |
|---|---|---|---|
| Galaxy mass | Dynamics (3D spectrosc.)<br>$R \geq 3000$ | Stellar mass<br>K band + $\geq$2 vis. bands | Error 0.3–0.5 dex on stellar mass |
| Gas extinction | Balmer line ratio corrected for absorption<br>$R \geq 1000$ | IR/visible ratio aperture corrected | Error <0.3 dex |
| SFR | FIR<br>or mid-IR (MIR) | $H\alpha$ or $H\beta$<br>extinction corrected | Error $\ll$0.3 dex |
| Gas: metal abundance | Extinction corr. $R_{23}$<br>$R \geq 1000$ | $[N\,II]/H\alpha$ | Error $\sim$0.1 dex |
| Morphology | "Eyeball" classification | Automatic software to decompose light profile | $\geq$2 indep. analyses<br>$k$ correction indep. |

## 1.2 General rules for estimating magnitudes

Before one can begin to understand the complexity of the selection effects present in distant galaxy samples, important steps needed to define the astronomical magnitude system must be clarified. The reader is encouraged to read Chapter 2 to learn the basic of photometric measurements and their uncertainties.

Several fundamental galaxy properties are directly related with luminosities in various bands. This includes the star formation rate (SFR) (combination of UV and mid to far IR luminosities), the stellar mass (which can be indirectly related to the near-infrared (NIR) luminosity), and bolometric luminosity. Special care must therefore be taken to measure magnitudes that are independent of the distance of a galaxy.

### 1.2.1 *Monochromatic and integrated magnitudes*

Galaxies emit light over a wide range of order of magnitudes. It is therefore useful to use a logarithmic scale to quantify observed fluxes. Apparent magnitudes and fluxes are related following the classic Pogson relation:

$$m = -2.5 \times \log (f_\nu) + ZP,$$

where $f_\nu$ is the flux spectral density in erg/s/cm$^2$/Hz, and $ZP$ is the zero point used to set the origin of the logarithmic scale. In optical wavelengths, it is conventional to deal with spectral densities per wavelength unit rather than per frequency unit. Both quantities are related as follows:

$$f_\lambda[\text{erg/s/cm}^2/\text{Å}] = f_\nu[\text{erg/s/cm}^2/\text{Hz}] \times \frac{c}{\lambda^2}.$$

When $f_\nu$ is the average flux density over a spectral filter $T$, the magnitude is often written as $m_T$, and is derived as follows:

$$m_T = -2.5 \times \log \left( \frac{\int_0^\infty T(\nu) f_\nu(\nu) d\nu}{\int_0^\infty T(\nu) r(\nu) d\nu} \right),$$

where $r(\nu)$ is the monochromatic response of a reference source taken as zero point, and $T(\nu)$ is the filter transmission as a function of frequency (see, e.g., Sterken and Manfroid, 1992).

### 1.2.2 *Photometric systems*

A photometric system is defined by a set of zero points defined at different wavelengths (or filters).

There are two commonly used photometric systems. Historically, people have used the Vega system, which uses the Vega star as a zero point in all filters. However, detectors have now become too sensitive to use Vega itself for calibration, and observing secondary standard stars is necessary. These stars are themselves calibrated against Vega, which has the drawback of introducing "closure errors" during the calibration. The zero point uncertainties end up being dominated by measurement errors on the standard star itself, which must be observed every time a new photometric system is used or the observation conditions have changed.

More recently, the AB system (Oke, 1974) was defined and it is independent of these measurement errors. Instead of a standard star, the AB system assumes a constant monochromatic flux distribution as a function of frequency, with $f_\nu(\nu) = 3631$ Jy. In this system, an apparent magnitude is given by:

$$m_{AB} = -2.5 \times \log f_\nu - 48.6 \tag{1.1}$$

at any wavelengths, with $f_\nu$ in erg/s/cm$^2$/Hz. When $f_\nu$ is given in nJy (1 nJy $= 10^{-32}$ erg/s/cm$^2$/Hz), then $m_{AB} = -2.5 \times \log f_\nu - 31.43$. The uncertainty on the zero point is dominated by uncertainties on the global throughput of the system (atmosphere included).

While the Vega system is no longer used in the distant Universe, it is still used in the local Universe because the stellar physics community uses this system widely. This has led to the confusing situation where the reader must often guess what system was used. We strongly recommend the use of the AB system in extragalactic works, since it allows direct comparison between spectroscopy and photometry without having to apply a color correction. The exact relation between AB and Vega magnitudes is filter dependent. Usually, both systems give the same apparent magnitudes in the V band. Table 1.2 lists some approximate conversion factors between the Vega and AB magnitude systems. Definition of the numerous filters used at many observatories can be found in Table 9 of Fukugita *et al.* (1995), see also the definition of zero points in their Tables 3 and 4

Table 1.2: Conversion factors between the AB and Vega systems.

| Filter | U | B | V | R | I | J | H | K |
|---|---|---|---|---|---|---|---|---|
| $\lambda_{sel}$ (nm) | 357.1 | 434.4 | 545.6 | 644.2 | 799.4 | 1235.5 | 1645.8 | 2160.3 |
| $AB_{\rm Vega}$ | 0.79 | −0.09 | 0.0 | 0.21 | 0.45 | 0.91 | 1.39 | 1.85 |

This Table assumes Bessel filters, with $m_{AB} = m_{\rm Vega} + AB_{\rm Vega}$ (from Blanton and Roweis, 2007).

(see Table 1.2 for conversion). A full description and explanation of the different magnitude systems can be found in Bessell (2005). In the following, all magnitudes will be given in the AB system, except if noticed otherwise.

### 1.2.3 *Total magnitudes and proxies*

In distant galaxies, the effective size of an observed galaxy disk is set by the image sensitivity, i.e., where the object light distribution falls below the detection level. It is therefore useful to characterize total object magnitudes using isophotal magnitudes, which integrates the object flux down to a given surface brightness limit. The 25th isophotal magnitude is often used as a proxy for estimating the total light from the optical disk of a local galaxy because their optical radius is close to $R_{25}$, i.e., the 25th isophotal radius.

In the distant Universe, the detection threshold is often selected to be the lowest reliable isophotal level. Isophotal magnitudes must be corrected for the fraction of light that is lost outside the lowest isophote. This is often done using a functional fit to the object light distribution which can be extrapolated to estimate the total object flux. This correction depends heavily on the shape of the PSF and the actual galaxy light profile and it can be poorly defined for galaxies with large light profiles such as elliptical galaxies.

When comparing galaxies at different redshifts, isophotal magnitudes can be problematic, since they suffer from a $(1+z)^5$ dimming factor[2] as a function of redshift.

[2] A factor $(1+z)^4$ comes from the cosmological dimming affecting fluxes in a nonEuclidean geometry, while an additional $(1+z)$ factor is due to the $k$-correction (see Sec. 1.3.6).

The light profile of distant galaxies is observed to be more irregular than nearby galaxies that simply sample the Hubble sequence (see Fig. 1.9). This is why large apertures are often used to estimate the total flux of distant galaxies. However using large aperture is not always optimal in terms of S/N, since a significant sky contribution can be present in the aperture, as well as light from other objects that are close to the same line of sight. A good compromise is to use an aperture size slightly larger than the galaxy diameter (see Chapter 2). Simulations using artificial sources can be very useful to understand the uncertainties in these measurements as well as to optimize the flux extraction method.

There are other popular techniques that were developed to estimate the total amount of galaxy light, such as the Kron and Petrosian magnitudes (Kron, 1980; Petrosian, 1976; Blanton *et al.*, 2001). However, these should not be applied to images of distant galaxies since they require a good sampling of the galaxy flux distribution and because they assume a varying fraction of the total flux as a function of Hubble type (Bertin and Arnouts, 1996; Graham *et al.*, 2005; Graham and Driver, 2005), while many distant galaxies cannot be easily classified this way.

### 1.2.4 *Absolute magnitudes*

Absolute magnitudes are a measure of the intrinsic luminosity of an object. By convention, they assume that objects are located at a standard luminosity distance $D_L = 10$ pc from the Sun. An absolute magnitude $M_T$ in a filter $T$ can be derived from the corresponding apparent magnitudes $m_T$:

$$M_T = m_T - 5(\log D_L - 1).$$

Absolute magnitudes must be corrected for atmospheric extinction (e.g., Sterken and Manfroid, 1992), extinction by our own galaxy (see Chapter 2), and for $k$-correction (see Sec. 1.3.6). The more complex corrections for internal extinction are discussed in Chapter 3. Absolute magnitudes $M$ and intrinsic luminosities $L$ in a given filter are

Table 1.3: Solar absolute magnitudes in different bands for the Vega and AB systems.

| Filter | U | B | V | R | I | J | H | K |
|---|---|---|---|---|---|---|---|---|
| Vega | 5.61 | 5.48 | 4.83 | 4.42 | 4.08 | 3.64 | 3.32 | 3.28 |
| AB | 6.36 | 5.36 | 4.83 | 4.65 | 4.55 | 4.57 | 4.71 | 5.19 |

This table assumes Bessel filters from Binney and Tremaine (1987).

related by:

$$\frac{L}{L_{\odot}} = 10^{-\left(\frac{M-M_{\odot}}{2.5}\right)},$$

where $L_{\odot}$ and $M_{\odot}$ are solar luminosities and absolute magnitudes respectively (see Table 1.3).

## 1.3 Pre-selection of a sample prior to redshift measurement

### 1.3.1 *Photometric catalogs*

The selection of a sample begins with a photometric catalog, providing positions and magnitudes of the sources to be observed spectroscopically. In principle, comparing galaxies at different distances should allow one to evaluate how galaxies evolve, and ultimately how they form. This is however a complex problem because there are numerous biases that are related to how far objects are. In the following subsections we describe selection effects related to statistics, to the photometric depth and to the choice of the wavelength of the limiting magnitude ($k$-correction).

### 1.3.2 *Statistics: which size for a redshift survey?*

The importance of number statistics can be summarized by the simple Poisson uncertainty of $\sqrt{N}/N$, e.g., a 10% nonreducible error is related to each sub-sample of $N = 100$ galaxies in a given redshift bin $z$ (for evolution), luminosity bin $L$ (to characterize a given range in stellar mass, star formation etc.) and/or color $C$ bin (for a given spectral type).

This has led observers to undertake very large surveys reaching several 10,000s of galaxies.

However, very large surveys are not always free of selection effects or systematics and these can become important when the selection criteria are multiple or too complex.

For example, using an additional criteria (e.g., color or photometric redshift) to eliminate low redshift galaxies can bias a sample considerably (see Ex. *Statistical bias*). Such pre-selection criteria should be kept in mind with every use of such galaxy sample since they can dramatically affect the representativeness of the sample in ($z$, $L$, $C$) bins.

**Example 1.2: *Statistical bias***

Consider a magnitude limited survey for which the investigators have added an additional condition to enhance their efficiency in selecting a larger fraction of high redshift galaxies. To do so, they used photometric redshifts based on the BRI colors, a criterion that is known to misidentify ∼10% of the high redshift galaxies as low redshift galaxies. Assuming a survey of 10,000 high redshift galaxies, the above systematic error is greater than the Poisson error and the survey might not be better than a survey of ... 100 high redshift galaxies extracted from a complete survey.

Moreover, redshift surveys are affected by the cone effect limiting the number of galaxies having similar absolute luminosity at the low and high redshift ends of the survey. This has to be carefully accounted for when defining a redshift survey as it also constrains the effective solid angle that should be used for the survey (see Ex. *The cone effect*).

**Example 1.3: *The cone effect***

Let us consider preparing a survey including $z = 0.3$ to $z = 1$ galaxies, aiming at selecting sources within angular-diameter distances from 1.54 Gpc to 6.6 Gpc. Assuming a limiting magnitude allowing to detect several 100s of bright, Milky Way-like (or so-called $L*$, i.e., the knee of the LF) luminosity galaxies

(*Continued*)

**Example 1.3: *Continued***

at $z = 1$, there could be only few galaxies similarly luminous at $z = 0.3$ simply because the solid angle and hence the volume[3] are too small at low redshift.

### 1.3.3 *Cosmological variance*

Cosmological variance plays an important role in studies of distant galaxies.

Variations of the galaxy density are expected at clustering scales of 5 Mpc to 10 Mpc, implying a minimal size of a deep pencil beam of ~100 arcmin$^2$. Nevertheless, large scale structures like filaments are larger than that and some well-studied fields (e.g., the GOODS-South field and the scientific results derived from them may be significantly affected, see Gilli *et al.*, 2003; Vanzella *et al.*, 2005; Ravikumar *et al.*, 2007).

Typical variations in luminosity density can be up to a factor 2 from one field to another and large scale structures may have a considerably effect on the LF that one derives (see Sec. 1.4.2 and example therein). This is a strong reason for one to consider more than one field when studying distant galaxies. Generally an uneven number of fields, from 3 to 7 is ideal to perform a redshift survey, allowing a statistically valid check of cosmological variance effects.

### 1.3.4 *Depth of the photometry and selection effects related to surface-brightness*

Galaxies should be selected from a sufficiently deep photometric catalog. The photometric catalog can be tested by the photometric completeness to the deepest counts in the literature. It is generally accepted that the catalog should be complete down to ~1–2 magnitudes fainter than the limit used to perform the galaxy selection in a redshift survey.

[3]For example, the volume between $z = 0.2$ and 0.4 is five times smaller than the volume between $z = 0.9$ and 1.1, assuming a common solid angle.

Doing otherwise can cause an important fraction of extended sources with relatively low surface brightness to be missed, especially at high redshifts where cosmological dimming effects become very important. Additionally, selecting more distant galaxies is increasingly difficult as they are many more intrinsically faint foreground galaxies (see Sec. 1.6). These distant galaxies often need photometric and spectroscopic measurements in NIR, which are further affected by surface brightness dimming. The light from extended distant galaxies is affected significantly by cosmological dimming with a factor of $(1+z)^{-4}$ that applies to the surface brightness (bolometric) luminosity of these objects. The strong redshift dependence of this effect can significantly affect the fraction of light detected in the outer isophotes of a galaxy, and can lead to galaxies being excluded from the sample. Cosmological dimming strongly affects extended sources as well as the faint sources at large redshifts. Deep photometry is therefore required to pre-select galaxies in an unbiased way before attempting to determine their redshifts from spectroscopy (see Ex. *Photometric depth to recover* $z = 0.9$ *spiral galaxies*).

**Example 1.4: *Photometric depth to recover* $z = 0.9$ *spiral galaxies***

Let us assume that most spiral-like galaxies at $z = 0.9$ have a central surface brightness similar to that of local spiral galaxies ($\mu_0(B) = 21.5$). Figure 1.1 shows that to recover most spiral-like galaxies up to $z \sim 0.9$ in a $I = 22.5$ limited sample, it requires a photometry with a completeness up to $I = 23.5$. This also implies that detecting lower surface brightness galaxies requires increasingly deeper photometry.

Distant galaxies are forming more stars than present day galaxies and their intrinsic surface-brightness can be expected to be enhanced, making these objects easier to detect. However, the fraction of low surface brightness galaxies in the distant universe is unknown and these objects are most likely to be excluded from a sample. It is therefore crucial to begin by requiring deep photometry as a starting

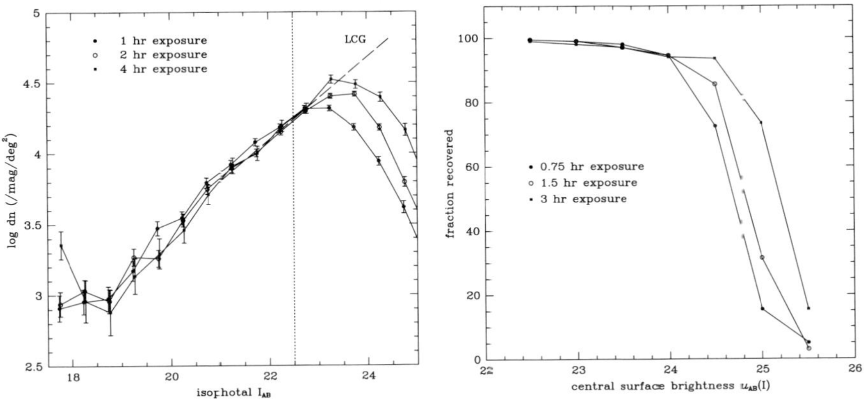


Fig. 1.1: Photometry of distant galaxies and related surface brightness selection effects (from the CFRS project, Lilly *et al.* (1995), ©AAS, reproduced with permission). *Left*: Number counts of objects with different exposure times. The CFRS reached down to $I_{AB} = 22.5$ (assumed to be $I_{AB}$, vertical line) and all source counts are complete to this limit, although only 4 h exposure (with CFHT) allow a completeness to $I_{AB} = 23.5$ (compare it to the dashed line that represents the deepest counts). *Right*: Fraction of $I_{AB} = 22.5$ galaxies that are recovered by the CFRS image detection algorithm, as a function of the central surface brightness. A large exposure time (3–4 h) is necessary to recover $\geq$90% of objects with $\mu(I) = 24.5$ that correspond to a face-on, $L*$exponential disk with a scale-length $h = 6.6$kpc at $z = 0.9$ and with a rest-frame central surface brightness of $\mu_0(B) = 21.5$.

point if one wants to avoid severe biases while selecting a sample of distant galaxies.

### 1.3.5 *Other selection effects: Malmquist bias, Eddington bias, and lensing effects*

There are several gotchas that can negatively affect the quality of a sample of distant galaxies. Even when attempting to select a sample of distant galaxies with a simple criterion, there are several selection effects that can impact scientific results derived from that sample. It is therefore necessary to carefully select the field, the limiting magnitude and the sample size of a survey.

This also implies that one needs to avoid several selection criteria, as the combination of the many related selection effects can be complex and can very severely affect the representativeness and then the whole scientific outputs derived from such a sample.

#### 1.3.5.1 *Malmquist bias*

The Malmquist bias is probably the best known selection effect, although it affects all average properties of galaxies, such as the luminosity density, and therefore star formation or stellar density. A survey that is limited by a simple apparent magnitude criterion would be unable to detect intrinsically faint galaxies in its highest redshift bin, while those would be included in the sample in the lower redshift bins. This is illustrated in Fig. 1.2, which also indicates (see the dotted lines) how to prepare a sample not affected by

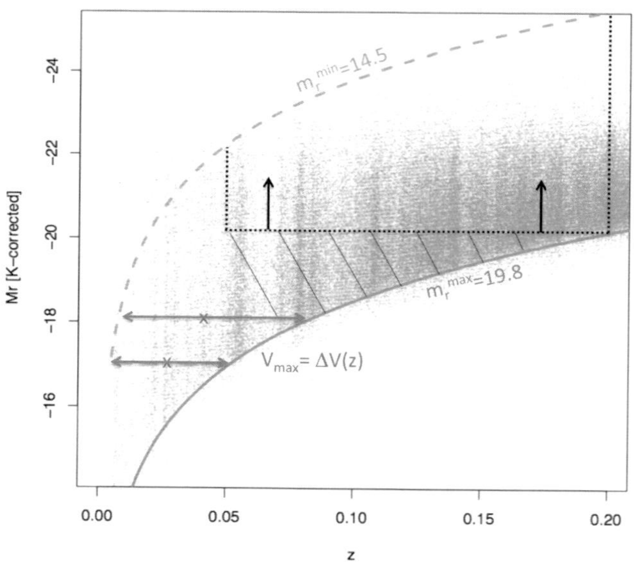


Fig. 1.2: Absolute, $k$-corrected $M_r$ SDSS magnitudes of galaxies (black points) selected from the GAMA survey as a function of redshift (adapted from Rodrigues *et al.*, 2016). These absolute magnitudes are relatively good proxies for stellar mass. The solid and dashed lines represent the GAMA selection with $m_r \leq 19.8$ and $\geq 14.5$, respectively. Selecting $0.05 < z < 0.2$ galaxies between these two lines would cause an important Malmquist bias, since for example, low luminosity galaxies with $M_r > -20$ are not detected at $z = 0.2$, while they represent most of the sample at $z = 0.05$. Avoiding such a bias requires selecting galaxies in the area delineated by the dotted lines (see also the two vertical arrows), which eliminates the low luminosity galaxies lying within the dashed area. The crosses represent two galaxies for which the horizontal, double arrows indicate how $V_{\max}$ can be calculated for establishing the LF (see Sec. 1.4.2 and Eq. (1.5)).

the Malmquist bias. Neglecting this may lead to significant errors in interpreting, for example, scaling relations (see Chapter 5).

Naturally, calculations of the luminosity density evolution need to be corrected for this effect. At a given redshift bin, one can use the immediately lower redshift bin to evaluate the contribution of the galaxies that have absolute luminosity just below the luminosity cut-off (see, e.g., Lilly *et al.*, 1996). This approach however requires a significant number of galaxies in relatively narrow redshift bins to allow corrections that do not depend too much on evolution.

#### 1.3.5.2 *Eddington bias: magnitude errors near the magnitude limit*

Even the most simple and secure case of a magnitude limited sample can be affected by subtle selection effects. For example, consider a sample limited to the magnitude of $m$, and for which the error in measuring the magnitude is $\delta m$. It is often true that the number of fainter galaxies (those with $m + \delta m$) is larger than the number of brighter galaxies with $m - \delta m$. For a random distribution of errors between $\pm\delta m$, there will be more sources intrinsically fainter than the limiting magnitude included in the sample than sources intrinsically brighter: this is the Eddington bias. It may seriously affect a sample or a part of a sample where the number count density is very steep.[4] This is the case for example with rare luminous ($L > L*$) galaxies at the highest redshift bin of a survey, or with any bright sources that are barely detected at large distances.

The Eddington bias can become important when combined with significant photometric errors.

To recover the actual number count, Monte Carlo simulations accounting for the distribution of the noise signal can be compared to the deepest observed counts (see Ex. *The Eddington bias for intermediate redshift LIRGs observed by ISO in 1998*). This again shows the need for accurate magnitude measurements, especially near the magnitude limit of the galaxy sample.

[4]This happens when $p > 1$ for number count density following $n \sim f_\nu^{-p}$.

### Example 1.5: *The Eddington bias for intermediate redshift LIRGs observed by ISO in 1998*

The satellite ISO detected for the first time the Luminous InfraRed Galaxies (LIRGs) at intermediate redshifts, providing the first estimate of the density evolution of the star formation including both UV and IR wavelengths (Flores *et al.*, 1999). However ISO was significantly less sensitive than Spitzer and these LIRGs were detected with S/N $>$ 4 and $>$ 3 for the complementary list. Although LIRGs are much more abundant at intermediate $z$ than locally, they are still rare sources and the slope of their number counts, as derived using Spitzer, is $p = 1.7$. However number counts from ISO (see Fig. 1.3) result in a steeper slope of $p = 2.3$ because objects with flux below 250 microJy are far more numerous than the brighter ones. This is therefore a good demonstration of the effect of Eddington bias.

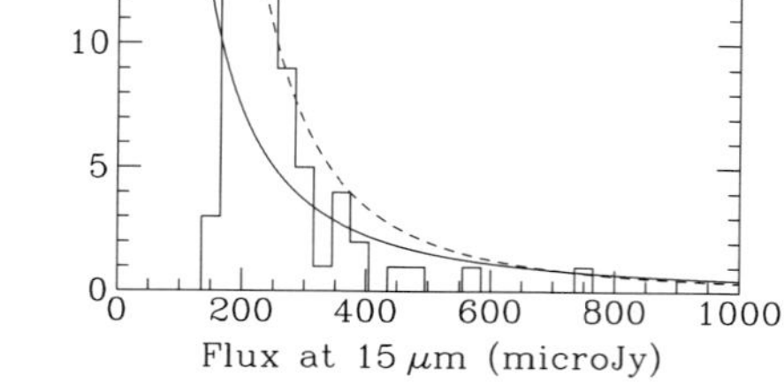


Fig. 1.3: The number counts of ISO sources with S/N $>$ 3 are shown by the histogram. The dashed line corresponds to a slope $p = 2.3$ ($n \sim f_\nu^{-p}$) fitting approximately the observed counts. However, deeper observations showed that the slope should be $p = 1.7$ (solid line). This effect is due to the Eddington bias, and approximately 25% of the sources in this histogram are almost certainly below the ISO delection limit (250 microJy) and have included into the sample as a result of their large flux calculations errors (from Flores, 1999).

#### 1.3.5.3 *Effect of gravitational lensing near the magnitude limit*

Gravitational lensing can also cause a somewhat similar selection bias when a galaxy sample is obtained near the line-of-sight of a mass density such as a galaxy cluster or a super-cluster.

Gravitational lensing affects the observed sources density since it causes a brightening of background sources by $\delta\log(f_\nu) = \log(\text{magnification})$ (or $\delta m = -2.5 \times \log(\text{magnification})$) while reducing the area behind the lens due to the convergence of the light beam by a factor equals to $(\text{magnification})^{-1}$. This results in a factor of $\log(n/n_0) = (p-1) \times \log(\text{magnification})$, where $n$ is the observed galaxy number density, $n_0$ is the source density if there were no lensing, and $p$ is the actual slope of the number density counts. As it is the case with the Eddington bias, gravitational lensing especially affects the brightest sources in a sample (or sub-sample) since the latter often shows steep number counts.

### 1.3.6 *K-correction and choice of the selecting magnitude*

Last but not least, the preparation of a redshift survey requires a careful choice of the wavelength at which apparent magnitudes are used to perform the sample selection. As was discussed in Sec. 1.1.1, the choice of the selecting wavelength $\lambda_{\rm sel}$ is crucial to (1) warrant a fair selection of galaxies independently of their rest-frame color properties; (2) to accurately calculate the rest-frame luminosities.

In fact, most redshift surveys have the goal of comparing the properties of distant galaxies at different redshifts within the survey as well as with the properties of nearby galaxies, where optical filters are defined from the $B$ to the $R$ band and also more recently up to the $K$ band.

Let us consider a redshift survey that is limited by a magnitude limit defined by an observed filter centered at $\lambda_{\rm sel}$, i.e., including all galaxies with $m_{\lambda_{\rm sel}} \leq m_{\lambda_{\rm sel}}^{\rm max}$. For each galaxy in the survey, a common rest-frame absolute magnitude at a reference wavelength, $\lambda_{\rm ref}$, can be computed as:

$$M_{\lambda_{\rm ref}} = m_{\lambda_{\rm sel}} - 5 \times \log(d_L(z)) - \kappa, \tag{1.2}$$

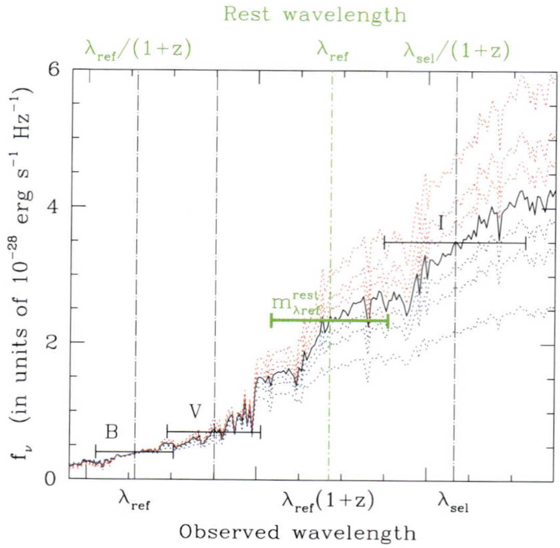


Fig. 1.4: Spectral energy distribution $f_\nu(\lambda)$ of a CFRS galaxy at $z = 0.5$, observed in $B$, V, $I$ broad-band filters (see horizontal solid bars). Fluxes ($f_\nu$) can be derived from magnitudes using Eq. (1.1) and vice-versa. To derive $m^{\rm rest}_{\lambda_{\rm ref}}$ (here the rest-frame $B$ magnitude at $\lambda_{\rm ref} = 457$nm, see the green, bold horizontal bar), one needs to adjust the broad-band measurements using a spectrum (here a Sc type at $z = 0.5$, see the solid line) extracted from a grid of templates (here with types ranging from Sb to Sd, see red and blue dotted lines). Rest-frame (observed) characteristic wavelengths are indicated on the top (bottom) of the figure, respectively (see also the vertical dot-dash lines). Equation (1.4) provides directly the $k$-corrected absolute magnitude from $m^{\rm rest}_{\lambda_{\rm ref}}$.

where $d_L(z)$ is the cosmological luminosity distance and $k$ is the $k$-correction. When a source is observed at a redshift $z$ in a filter centered at $\lambda_{\rm sel}$, this provides a direct estimate of the rest-frame absolute magnitude at $\lambda_{\rm sel}/(1+z)$, a wavelength that is generally different than the desired $\lambda_{\rm ref}$ (see Fig. 1.4).

The $k$-correction $k$ has to be computed to account for this discrepancy. If a galaxy happens to be at a redshift that is exactly $z = \lambda_{\rm sel}/\lambda_{\rm ref} - 1$, one can compute $M_{\lambda_{\rm ref}}$ and the $k$-correction is simply $k = -2.5 \times \log(1+z)$ to account for the fact that the luminosity $L_\nu$ (or the flux $f_\nu$ that is related to AB magnitude, see Eq. (1.1)) is not bolometric but a density per unit frequency[5] (Hogg *et al.*, 2002).

[5]This assuming the energy conservation ($E = \nu \times f_\nu = \nu_0 \times f_{\nu_0}$), which leads to a flux at rest $f_{\nu_0}$ that is equal to $f_\nu/(1+z)$.

In most cases, however, *a priori* knowledge of the galaxy spectral energy distribution (SED) is required to compute the $k$-correction. The exact, and complex definition of the $k$-correction can be found in Hogg *et al.* (2002).

In practice, an estimate of the $k$-correction can be obtained with a limited knowledge of the SED that only samples the rest-frame wavelength range from $\lambda_{\rm ref}$ to $\lambda_{\rm sel}/(1+z)$. Estimating $k$-corrections is simpler if all of the galaxies have actually been observed at $\lambda_{\rm ref}$ in addition to $\lambda_{\rm sel}$, providing the very useful $(m_{\lambda_{\rm ref}} - m_{\lambda_{\rm sel}})$ observed color (see Fig. 1.4). The $k$-correction is then:

$$k = -2.5 \times \log(1+z) - (m^{\rm rest}_{\lambda_{\rm ref}} - m_{\lambda_{\rm sel}}) \tag{1.3}$$

and Eq. (1.2) simplifies to:

$$M_{\lambda_{\rm ref}} = m^{\rm rest}_{\lambda_{\rm ref}} - 5 \times \log(d_L(z)) + 2.5 \times \log(1+z), \tag{1.4}$$

where $m^{\rm rest}_{\lambda_{\rm ref}}$ is the rest-frame magnitude that the galaxy would have at $\lambda_{\rm ref}$ in the rest frame. Figure 1.4 shows how it can be interpolated from the $(m_{\lambda_{\rm ref}} - m_{\lambda_{\rm sel}})$ color, which is used to assign an SED type to each galaxy, using a grid of colors obtained from redshifted galaxy synthesis models.

Other colors can be used to interpolate $m^{\rm rest}_{\lambda_{\rm ref}}$ and the $k$-correction, but they should be defined over a wavelength range that includes[6] $\lambda_{\rm ref} \times (1+z)$. It is however strongly recommended to avoid any extrapolation that can lead to considerable errors in estimating $k$-corrections.

For samples of galaxies spanning a small redshift range, the associated errors can be very small (∼0.05 mag) by setting $\lambda_{\rm ref}$ accordingly to the median redshift ($z_{\rm med}$), i.e., $\lambda_{\rm ref} \sim \lambda_{\rm sel}/(1+z_{\rm med})$ (Hammer *et al.*, 2001). The calculations of magnitudes, colors and the use of SED libraries are described in Chapter 2.

Old stellar populations (F, G, and K) have a strong signature at 400nm (the 400nm break) causing a strong decrease in luminosity at $\lambda < 400$nm (see Chapter 2). This signature has to be accounted for when defining $\lambda_{\rm sel}$ in redshift surveys that aim at comparing distant galaxy properties with the properties of local galaxies. This allows

[6]For example, the $V - I$ color may be used to interpolate the (reference) $B$ mag from $z = 0.2$ to 0.8 (see Fig. 1.4).

for a better determination of the stellar mass of a galaxy as the galaxy mass can be dominated by old stars emitting most of their light above 400nm rather than by young and blue stars with low $M_{\rm stellar}/L$ ratio (see Ex. *Selecting wavelength for comparing* $z = 0$ *to* $z = 0.8$ *galaxies*).

**Example 1.6: *Selecting wavelength for comparing*** $\boldsymbol{z = 0}$ ***to*** $\boldsymbol{z = 0.8}$ ***galaxies***

Figure 1.5 compares the $k$-corrections of a survey aiming at characterizing the rest-frame $B$ band properties of galaxies from $z = 0$ to 0.8. If galaxies are selected in the observed $B$ band, with increasing redshift this band progressively samples the rest-frame ultraviolet light (see, e.g., Fig. 1.4). This leads a very wide range of $k$-corrections within the different SED types, that are up to 4 mag at $z = 0.75$. This results in large uncertainties, especially since the properties of the different galaxy types are not very understood in the UV.

In contrast, a selection in the $I$ band is an exact match to the rest-frame $B$ band at $z = 0.82(\lambda_{\rm ref} = \lambda_B = \lambda_I/(1+z) = \lambda_{\rm sel}/(1+z))$. The $(B-I)$ color (equivalent to the $(m_{\lambda_{\rm ref}} - m_{\lambda_{\rm sel}})$ color) allows us to interpolate with a far better accuracy and results in much smaller $k$-corrections all the way from $z = 0$ to $z = 0.82$, because $\lambda_B \le \lambda_I/(1+z) \le \lambda_I$. Figure 1.5 illustrates that by selecting distant galaxies using the apparent $B$ band magnitude, one would exclude elliptical galaxies at high redshift, since these galaxies are made of old and red stellar populations with a low continuum emission in the UV. The $I$ band selection allows for a much more representative sample of galaxies up to $z = 0.82$.

This further demonstrates how a selection in the $I$ band ($\lambda_{\rm sel} =$ 832nm) is the best way to characterize the rest-frame $B$-band properties for a survey of $z \le 0.8$ galaxies.

Surveys at $z \le 2, 3$, and 4 require a selection done in the $J$ ($\lambda_{\rm sel} = 1200$nm), $H$ ($\lambda_{\rm sel} = 1600$nm) and $K$ ($\lambda_{\rm sel} = 2100$nm) band, respectively.

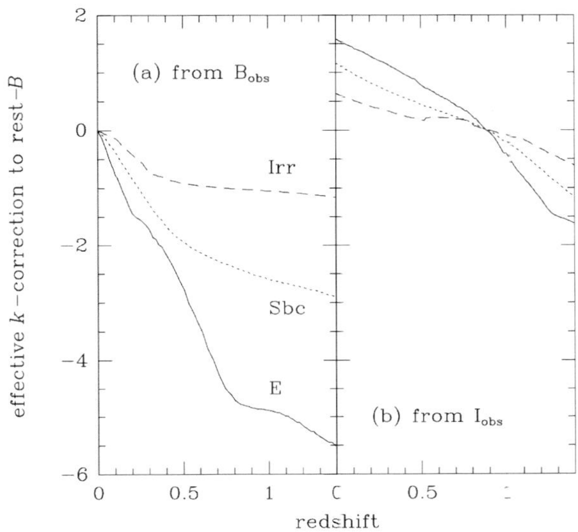


Fig. 1.5: Variations of the $k$-corrections in magnitude (from Lilly *et al.*, 1995, ©AAS, reproduced with permission) for three types of galaxies, E, Sbc, and Irr, assuming the LF is computed in absolute magnitude in $B$ band ($\lambda_{\rm ref} = 457$nm). *Left*: Assuming that galaxies have been selected from their observed B band magnitudes ($\lambda_{\rm sel} = 457$nm). *Right*: Assuming that galaxies have been selected from their observed I band magnitudes ($\lambda_{\rm sel} = 832$nm).

When possible, it is even better to select objects in even redder filters to sample the rest-frame properties of galaxies at wavelengths that are not strongly affected by star formation and dust extinction, such as the rest-frame $R, I$, or $K$ bands.

## 1.4 Redshift surveys

In this section, we assume that a redshift survey is undertaken on the basis of a photometric catalog, as was described in the previous section. The methods related to spectroscopy are described in Chapter 3.

### 1.4.1 *Redshift measurements and spectroscopic selection effects*

There are currently many multi-object spectrographs (MOSs) at large telescope facilities that can be used to perform redshift surveys.

Table 1.4: Comparison of multi-spectrograph capabilities for redshift surveys.

| Spectrograph type | Multi-slits | Multi-fibers |
|---|---|---|
| Instrumental needs and biases | | |
| Sky subtraction requirements: | 10 arcseconds slit length (1) | significant number of sky-fibers (2) |
| Wavelength range: | depends on the vertical position of the slit in the mask | same for all fibers and all targets |
| Biases (aperture effects) against: | extended sources elongated perpendicular to the slit (3) | extended sources size > fiber diameter |
| Measurements | | |
| Redshift determination: | different $\lambda$ range from the top to the bottom of the mask | |
| Extinction, SFR and metallicity | requires $R \geq 1000$ (4) | flux calibration could be an issue |
| Velocity | depends on the object position in the slit | requires IFU or large aperture (for $\sigma$) |

References are: (1) Le Fevre *et al.* (1995); (2) Rodrigues *et al.* (2012); (3) Hammer *et al.* (1995); (4) Liang *et al.* (2004).

Some are multi-slits spectrographs with generally low spectral resolution (e.g., from $R = 150$ to $R = 1500$) and some are multi-fibers spectrographs with moderate to high resolution (e.g., from $R = 1500$ to $R = 30000$). Table 1.4 lists some of their respective advantages and disadvantages (see also Chapters 3 and 4).

To perform efficient redshift measurements in a consistent manner for all targets in different $(z, L, C)$ bins, where $L$ and $C$ are luminosity and color bins, respectively, requires a robust methodology.

In the CFRS, all the 1,000 spectra were reduced and examined by three independent researchers, and the final redshift was determined after careful comparisons. This method has the noticeable advantage of being robust as well as to provide a well-defined quality standard attached to each redshift estimate. But this simple approach becomes difficult when performing surveys of several 10,000s spectra.

Several automated procedures (see e.g., EZ software from Garilli *et al.*, 2010, or RUNZ software described in Drinkwater *et al.*, 2010) have been developed, and they perform very well, especially for high S/N and/or emission line spectra. Complete automatization of the redshift measurements has however proven to be difficult because of the noisy nature of many of the spectra and the presence of artifacts such as residuals from imperfect cosmic ray or sky removal, fringing, and crosstalk.

This often makes it necessary to have software that automatically generate quality flags, as well as perform numerous tests, so that only spectra with low quality flags can be examined by a human. Usually, this implies that about a third to a half of all spectra still need to be examined by researchers, making large surveys a daunting task.

Having photometric measurements in the largest possible number of photometric bands is particularly useful when trying to constrain the spectroscopic nature of an object. This is particularly important when attempting to estimate the redshift of quiescent galaxies devoid of emission lines and the redshift has to be determined using continuum breaks.

There are several effects that can negatively impact redshift estimates. For example, low resolution spectroscopy of a galaxy can be affected by bright OH lines at $\lambda \geq 720$nm, causing the redshift to be wrongly estimated to be $z \sim 0.8$. Similarly, M stars can be misidentified to be distant galaxies.

In addition to possible instrumental problems (see Table 1.4), the following lists some of the most important selected effects related to spectroscopy, as well as some advice on how to deal with them:

- *Confusion between stars and galaxies*: In deep surveys, galaxies and stars can have similar spectral characteristics. For example, low S/N spectra of K stars populating the Milky Way halo often produce featureless spectra that can easily be confused with spectra of moderate redshift early-type galaxies. This can however be overcome by checking whether the source is a point source but this requires a very good image quality and imaging information that is significantly deeper than the spectroscopic limit. Separating galaxies from stars before attempting spectroscopy is a significant source of error. Although color–color diagrams and deep imaging can help separate stars from galaxies, there are many galaxies that have similar colors than stars in visible filters (see Crampton *et al.*, 1995) and there are also very compact galaxies that are not even resolved by the HST.

The only robust way to distinguish between these two populations is using spectroscopy. This is because spectra of distant compact galaxies often show emission lines (see e.g., Hammer *et al.*, 2001).

- *Single emission line spectra*: This is a problem that can especially affect surveys conducted over narrow wavelength ranges and surveys done using multi-slit spectrographs (see Table 1.4, wavelength range line). Confusion between moderate and low redshift objects occurs when a single emission line is misidentified. For example, a single [O II]$\lambda$372.7nm emission line, or a single $H\alpha$ emission line, identified to be Ly$\alpha$($\lambda$121.5nm) would result in a much higher redshift estimate. Some, but not always all of the degeneracy can be mitigated by using photometric classification. Performing spectroscopic observations at resolution high enough, i.e., $R \sim 3000$–5000, to resolve the [O II]$\lambda\lambda$372.6;372.9nm doublet is a more robust approach. Confusion between [O II]$\lambda$372.7nm and Ly$\alpha$ lines in some of the most distant galaxies has been the source of a significant amount of literature.
- *Position biases in fields with high numerical density of sources*: In most surveys, the number density of sources within the magnitude limit is significantly greater than the number of sources for which a spectrum can be obtained using a spectrograph. This can result in an artificial two-point correlation since the positioning of both slit and fiber spectrographs is restricted by instrumental constraints (see e.g., Le Fevre *et al.*, 1995). The only way to address this problem is to use several different observations of the same field, ultimately resulting in spectroscopic observations of all the sources in the field, down to a chosen magnitude limit.

### 1.4.2 *Determination of the LF, uncertainties and completeness tests*

The LF, characterized by $\phi(L)$ (in units of erg$^{-1}$ s Mpc$^{-3}$) or $\phi(M)$ (in units of mag$^{-1}$ Mpc$^{-3}$), is a fundamental tool to probe the distribution of galaxies over cosmological time. It allows one to compare the number density of several sets of galaxies at different redshifts, for different galaxy types, or environments, providing crucial statistical tests for galaxy evolution studies. Significant progresses have been made since the establishment of the first LFs in the 1930s, and a complete summary of the different methods used to derive the LF can be found in Johnston (2011) and references therein. Here we specifically address aspects of the LF determination as it relates to distant galaxies. It is affected by all the selection effects described in the previous sections, either from the pre-selection using photometric measurements or from redshift measurements. To properly derive the LF and its uncertainties, it is necessary to pay close attention to all the sources of statistical errors and biases. A table is often helpful to clearly present the results of such investigations.

Let us consider a pencil beam redshift survey, limited by two apparent magnitudes, $m^{\min}_{\lambda_{\rm sel}}$ and $m^{\max}_{\lambda_{\rm sel}}$, where $\lambda_{\rm sel}$ is the central wavelength of the (broad-band) filter used to select galaxies as illustrated in Fig. 1.2, for which $\lambda_{\rm sel}$ is defined by the SDSS $r$ filter. To build the LF a specific weight must be assigned to each galaxy in the sample.

For example, a dwarf galaxy could be only observed in a small volume of the pencil beam, because it is too faint to be part of the highest redshift bins of the survey (see horizontal, double arrows in Fig. 1.2). It does not mean that there is no dwarf galaxies in the highest redshift bins, simply that they are not likely to be included in the sample since they are faint with $m_{\lambda_{\rm sel}} > m^{\max}_{\lambda_{\rm sel}}$. To compensate for this effect, one calculates the maximum comoving volume,[7] $V_{\max}$, over which this galaxy could be observed, with $m^{\min}_{\lambda_{\rm sel}} < m_{\lambda_{\rm sel}} < m^{\max}_{\lambda_{\rm sel}}$ (see horizontal, double arrows in Fig. 1.2). An estimate of the LF can then be computed using (see e.g., Felten, 1976):

$$\phi(M)dM = \sum_{i=1}^{n} 1/V_{\max}(i), \tag{1.5}$$

where $dM$ is the chosen interval of absolute magnitude to sample the LF and $n$ is the number of galaxies. For studying the LF in different

[7]The comoving volume is the volume measure in which number densities of non-evolving objects locked into the Hubble flow are constant with redshift (see, e.g., Peebles, 1993). There are many available routines to calculate the comoving volume (e.g., Astro:Cosmology at `http://hea-www.cfa.harvard.edu/~dburke/PDL/Cosmology-0.90.html`) or one can use `http://www.icosmos.co.uk/index.html` for an individual redshift.

bins of redshift, absolute luminosity or color, this may be expanded to a multivariate LF, as:

$$\phi(M, z, \text{color})dM = \sum_{i=1}^{n} 1/V_{\max}(i). \tag{1.6}$$

For a galaxy $i$ with observed $m_{\lambda_{\text{sel}}}$ and a redshift $z$, the $k$-correction allows to calculate its absolute magnitude, $M_{\lambda_{\text{ref}}}$, given an estimate of its spectral type (see Sec. 1.3.6). To calculate $V_{\max}(i)$ one estimates the redshift range over which a galaxy with this spectral type and with $M_{\lambda_{\text{ref}}}$ would be part of the sample, i.e., have an observed magnitude within $m_{\lambda_{\text{sel}}}^{\min}$ and $m_{\lambda_{\text{sel}}}^{\max}$ (see an illustration of this in Fig. 1.2, with the two horizontal lines with crosses and the horizontal, long dashed lines that match the limiting apparent magnitudes). To calculate the minimum (and maximum) redshift, one needs to solve Eq. (1.2) after substituting $m_{\lambda_{\text{sel}}}$ by $m_{\lambda_{\text{sel}}}^{\min}$ ($m_{\lambda_{\text{sel}}}^{\max}$) and $z$ by $z_{\min}$ ($z_{\max}$), respectively. The $k$-correction is obtained following Eq. (1.3) in a similar manner (substituting $z$) and by computing $m_{\lambda_{\text{ref}}}^{\text{rest}}$ for the spectral type of the galaxy (see Sec. 1.3.6 and Fig. 1.2). Again, these calculations are much more accurate when $\lambda_{\text{sel}}$ is carefully chosen to minimize the amount of the $k$-correction.

**Example 1.7: *Calculation of the rest-frame B band LF of the CFRS-like survey***

The CFRS gathered a statistically complete sample of 730 galaxies limited in I band ($\lambda_{\text{sel}} = 832$nm) with $m_{\lambda_{\text{sel}}}^{\min} = 17.5$ and $m_{\lambda_{\text{sel}}}^{\max} = 22.5$. The best achieved photometry includes V and I filter and the $V - I$ color straddles the rest-frame B band ($\lambda_{\text{ref}} = 457$nm) for $0.2 < \text{z} < 0.8$. For each $0.2 < z < 0.8$ galaxy one may:

- assign a template spectrum by matching the observed $(V-I)$ color to a library of redshifted templates (see Fig. 1.4);
- interpolate $m_{\lambda_{\text{ref}}}^{\text{rest}} = m_B^{\text{rest}}$ and derive the rest-frame absolute magnitude $M_{\lambda_{\text{ref}}} = M_B$;

*(Continued)*

**Example 1.7: *Continued***

- using the template spectrum and $M_{\lambda_{\rm ref}}$, calculate the minimal redshift $z_{\rm min}$ for which the galaxy apparent magnitude ($m_{\lambda_{\rm sel}} = I$) reaches $m^{\rm min}_{\lambda_{\rm sel}}$;
- same as above for the maximal redshift $z_{\rm max}$ for which the galaxy I band magnitude reaches $m^{\rm max}_{\lambda_{\rm sel}}$;
- derive the maximal comoving volume for which the galaxy can be observed $V_{\rm max} = V_{z_{\rm max}} - V_{z_{\rm min}}$, knowing that CFRS comoving volumes are based on an effective solid angle of 112 arcmin$^2$.

Then the LF can be calculated by summing the contribution of each galaxy through Eqs. (1.5) or (1.6). It can be described by using a Schechter function:

$$\phi(M_B)dM_B = 0.4\ \ln 10\ \phi^*[10^{0.4(M_B^* - M_B)}]^{\alpha+1} \times \exp[-10^{0.4(M_B^* - M_B)}]\ dM_B, \qquad (1.7)$$

where $\phi^*$, $M_B^*$, and $\alpha$ are the characteristic density, absolute magnitude and faint end slope, respectively. Besides an accurate determination of these parameters, the CFRS LF has been derived (by two independent groups) to characterize its changes with epochs. It has to be verified whether or not the comparison using different redshift bins is affected by the cone effect (see Ex. *The cone effect*). It is recommended to list the expected errors and systematics in a table, accounting for the impact of photometric errors, non or misidentified galaxies, $k$-correction errors, large and small scale clustering (see below).

The main effects affecting the LF of distant galaxies are:

(1) Errors in the sample selection process when photometric measurements are used (see Sec. 1.3). In order to minimize these errors, the photometric completeness should to be at least 1 magnitude deeper than the limiting magnitude for the redshift survey.

(2) Uncertainties in redshift measurements (see Sec. 1.4.1) causing some galaxies to have uncertain redshift estimates, or no redshift estimates at all. To quantify the impact of nonidentified galaxies, it is advised to repeat the spectroscopic observations of a representative subset of the sample, and to then re-integrate them under the assumption of their representativeness. Without this procedure, one needs to weigh each galaxy with a redshift estimate by the inverse of the fractional identification rate of the sample as a function of the magnitude.
(3) Correction of the chosen rest-frame magnitude to the observed magnitude used to select the galaxies, i.e., accuracy of the $k$-corrections as described in Sec. 1.3.6.
(4) Large scale structure (seen as redshift peaks) affecting narrow pencil beam surveys and which can impact estimates of the galaxy number density evolution. This problem is best resolved by observing galaxies in several pencil beams pointed at different locations of the sky (see Sec. 1.3.3 and Ex. *Effect of large scale structures on the LF and photometric redshifts*). Lacking this, the reader is referenced to Johnston (2011) for LF determinations less affected by large scale structures, although applying this to narrow pencil beams is not beyond doubt.
(5) The small scale clustering of galaxies. This decreases the number of independent galaxies by a factor of approximately 2.5 (Lilly *et al.*, 1996). This increases the uncertainty of each bin of the LF, as it effectively reduces the number $N$ of galaxies with an uncertainty proportional to $(N/2.5)^{-0.5}$.

**Example 1.8: *Effect of large scale structures on the LF and photometric redshifts***

The GOODS-S field is amongst the most studied field for galaxy evolution. However it shows (see Fig. 1.6) three large scale structures that dominate the number counts. Conversely the CFRS is far less affected because it is based on five different fields. Comparison of the luminosity densities between GOODS-S and CFRS (see Fig. 1.6) shows that the evolution of the luminosity

(*Continued*)

**Example 1.8: *Continued***

density is hardly affected by structures in the rest-frame $U$ band. However, large scale structures have a considerable impact in the $K$ band, mimicking an additional, unphysical evolution of 0.45 dex between $z = 0.35$ and $z = 0.675$! This is caused by the larger number of massive and red galaxies present in large scale structures than in the field. The effect is less prominent (0.25 dex) when adopting the photometric redshift technique (see Sec. 1.5) because it smoothes out large scale structures in the redshift distribution. However this renders the analysis more complex, since the smoothed large scale structures become difficult to detect and the unphysical evolution could be mistaken as a real one.

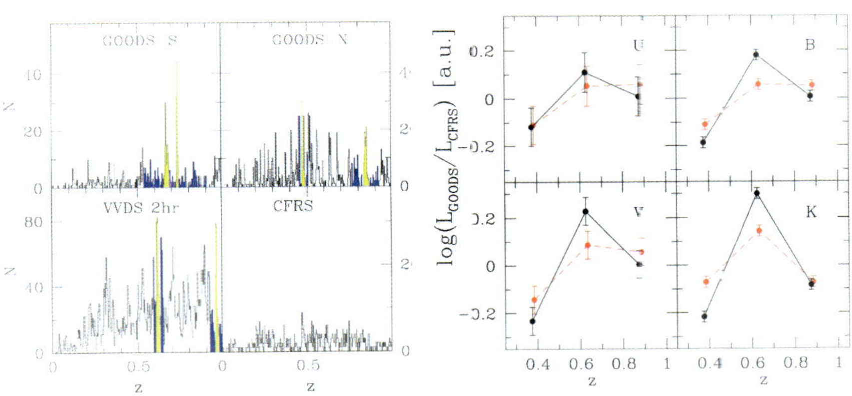


Fig. 1.6: Effect of large scale structures in the GOODS-S field and how this could affect the apparent evolution (Credit: Ravikumar *et al.*, 2007, A&A, 465, 1099, reproduced with permission ©ESO). *Left*: Comparison between the redshift distribution of GOODS-S ($I_{AB} < 23.5$) redshift distribution to that of GOODS-N ($R_{AB} < 24.5$, Wirth *et al.*, 2004), VVDS-02 h ($I_{AB} < 24$, Le Fèvre *et al.*, 2005), and CFRS ($I_{AB} < 22.5$, five fields of view, Crampton *et al.*, 1995).The two most prominent structures in GOODS-S, GOODS-N, and VVDS-02 h are shown in yellow. The redshift area surrounding the structure, which contains the same number of galaxies as the two largest structures is shown in blue. *Right*: Evolution in the ratios of luminosity densities of observed galaxies from the GOODS-S spectroscopic sample and of observed galaxies from CFRS as a function of redshift, for 3 redshift bins (0.25–0.5; 0.5–0.75, and 0.75–1). Values calculated using spectroscopic and photometric redshifts are shown using black and red-dashed lines, respectively.

Still, the uncertainties in deriving the LF from most redshift surveys are dominated by the incompleteness effects described in Secs. 1.3 and 1.4.

There are several tests that have been developed to test *a posteriori* the apparent magnitude completeness of a redshift survey. For example the $V/V_{\rm max}$ test (V being the actual comoving volume delineated by the galaxy at its observed redshift) should result in an average value of 1/2 if the survey is complete. However it is in practice difficult to interpret the results of the $V/V_{\rm max}$ test with surveys of distant galaxies because significant departure from 1/2 could also be caused by clustering and evolutionary effects. More modern methods have been developed (see Johnston *et al.*, 2007) but these have yet to be applied to narrow pencil beam surveys.

### 1.4.3 *Examples of surveys (CFRS, DEEP2, VVDS, zCOSMOS, and WIGGLEZ)*

Table 1.5 presents some of the properties of several redshift surveys, although it is not an exhaustive list. It does not include redshift surveys that have been done on specific areas (e.g., those observed by the HST), since in many cases those are made by using a plethora of criteria rendering them difficult to interpret. The first redshift survey of galaxies with redshift up to $z = 1$ was the CFRS, at the time the state of the art in galaxy selection prior to spectroscopy, and in redshift identifications. It was however limited by its size as well as by its low spectral resolution. Former redshift surveys relied on blue band object selection and thus provided mostly low redshift galaxies. The Hawaii Deep Survey (Cowie *et al.*, 1996) identified 393 $K < 20$ objects in a smaller area (26.2 arcmin$^2$) but reached a similar depth than the CFRS.

The main goal of the CFRS, DEEP2, and VVDS redshift surveys was to determine and study the evolution of the galaxy LF. The zCOSMOS survey aimed at delineating the cosmic web by characterizing the environments of galaxies on the scale of galaxy groups. Finally, BAO surveys such as WiggleZ or SDSS-III attempted to determine the cosmological constants using baryonic acoustic oscillations.

Studying the galaxy LF requires at least the control of cosmic variance effects, explaining the need of several different fields of view. Characterizing the environments of galaxies requires that a large fraction of galaxies be spectroscopically observed. BAO surveys observe a very large numbers of galaxies although often at the cost of numerous selection and incompleteness effects.

Most redshift surveys have additional secondary goals such as studying sources detected at other wavelengths, or the measurements of galaxy physical properties, such as SFRs, gas phase metal abundance or kinematic measurements. Many of these measurements require sufficient spectral resolution, as described in Table 1.1 (see also Chapter 3).

As Table 1.5 shows, the limiting magnitude of redshift surveys imposes a limit on their redshift range. Even in the case of deep surveys, the number density of high redshift galaxies is much lower than that of low luminosity lower redshift galaxies.

The VVDS derived galaxy LFs from $z = 0.2$ to $z = 4.5$, but only observed a small fraction of $z > 1.5$ galaxies. This is an example of what was discussed in Sec. 1.3.6: selecting galaxies in the $I$ band results in an increasing number of UV dominated galaxies as redshift increases (see also the example below). Selecting $z \sim 2$ galaxies would better be done at a wavelength larger than 1200nm ($J$ band) rather than 835nm ($I$ band).

Some redshift surveys were designed with additional selection criteria (e.g., DEEP2, WiggleZ) in order to avoid the numerous galaxies at lower redshift. However whatever the technique ($z_{\text{phot}} > 0.7$ for DEEP2, FUV-NUV vs. NUV-$r$ for WiggleZ), additional requirements also imply additional uncertainties since none of them techniques can ensure that a specific population of high-$z$ galaxies has not also been excluded. This might results in significant biases in evolutionary studies (see Ex. *Statistical bias* in Sec. 1.3), while it impacts BAO determinations less.

Specific selections of very high-$z$ galaxies are presented in Sec. 1.6.

Table 1.5: Comparison of several redshift surveys.

| Survey<br>References<br>(a) | $R_{\text{Spectral}}$<br>$N_{\text{fields}}$ (size)<br>(b) | $\lambda_C$<br>$m^{\max}_{\lambda_{\text{sel}}}$ ($m^{\min}_{\lambda_{\text{sel}}}$)<br>(c) | Phot<br>$m^{\text{phot}}_{\lambda_{\text{sel}}}$<br>(d) | $z_{\text{median}}$<br>$z_{\text{range}}$<br>(e) | $\lambda_{\text{ref}}$<br>color<br>(f) | $N_{\text{gal}}$<br>$N_{\text{unident.}}$<br>(g) |
|---|---|---|---|---|---|---|
| CFRS | 160 | 832 ($I_{AB}$) | Y | 0.65 | 457 ($B_{AB}$) | 743 |
| (1, 2, 3, 4) | 5 (0.14) | 22.5 (17.5) | 23.5 | 0.2–1 | $(V-I)_{AB}$ | 146 (20%) |
| DEEP2 | 3200 | 650 ($R_{AB}$) | Y | $\sim 1$ | 435 ($B_{AB}$) | 47700 |
| (5, 6, 7) | 3 (0.86) | 24.1 (18.5) | 24.75 | 0.7–1.4 | $(R-I)_{AB}$ | $\sim$ 14000 (30%) |
| VVDS(deep) | 210 | 835 ($I_{AB}$) | Y | 0.76 | $B_{AB}$ to $I_{AB}$ | 10212 |
| (8, 9) | 2 (0.61) | 24 (17.5) | 25.3 | 0–5 | $B_{AB}$ to $I_{AB}$ | 2372 (23%) |
| zCOSMOS(bright) | 600 | F814W ($I_{AB}$) | Y | 0.65 | 457 ($B_{AB}$) | 9600 |
| (10, 11, 12) | 1 (1.69) | 22.5 | ∼23.0 | 0.2–1 | $U_{AB}$ to $K_{AB}$ | 1150 (12%) |
| Wiggle Z | 600 | 225 (NUV) | Y | 0.6 | 457 ($B_{AB}$) | 10000 |
| (13, 14) | 7 (1000) | ∼22.8 | 23.25 | 0.2–1 | FUV to i (SDSS) | 40000 (40%) |

References are: (1) Lilly *et al.* (1995); (2) Le Fevre *et al.* (1995); (3) Hammer *et al.* (1995); (4) Crampton *et al.* (1995); (5) Coil *et al.* (2004); (6) Willmer *et al.* (2006); (7) Faber *et al.* (2007); (8) Le Fèvre *et al.* (2005); (9) Ilbert *et al.* (2005); (10) Lilly *et al.* (2007); (11) Lilly *et al.* (2009); (12) Pozzetti *et al.* (2010); (13) Drinkwater *et al.* (2010); (14) Blake *et al.* (2010). *Notes*: (b) total field size is in square degree; (c) central wavelength are given in nm; (d) $Y$ means prior photometry to select galaxies and $m^{\text{phot}}_{\lambda_{\text{sel}}}$ indicates the completeness magnitude of the photometric counts when compared to the deepest counts (e) $z_{\text{range}}$ is the applicable range to determine galaxy LF; (f) $\lambda_{\text{ref}}$ and color are the central wavelength used to establish the LF and the color used to determine the $k$-correction, respectively; (g) $N_{\text{gal}}$ is the number of galaxies with spectroscopy and $N_{\text{unident.}}$ of galaxies without a secure redshift (fraction in parenthesis). Comments on selection: (CFRS) stars observed, unidentified galaxies re-observed; (DEEP2) *a priori* elimination of stars, selection effects due to $z_{\text{phot}} > 0.7$; (VVDS-deep) stars observed, unidentified sources with $z_{\text{phot}}$; (zCOSMOS-bright) some stars observed, unidentified galaxies re-observed; (WiggleZ) stars observed, numerous selection effects (see text).

Representative samples of galaxies could be extracted from the existing samples of distant galaxies listed in Table 1.5. For example one could identify galaxies in a given range of absolute magnitude (or stellar mass), that could be studied kinematically or using different wavelengths.

This should be done after verifying that the desired range of absolute magnitude (or stellar mass) is well represented in the different redshift bins of the LF (see Ex. *Two sub-samples of distant galaxies from the VVDS*).

**Example 1.9: *Two sub-samples of distant galaxies from the VVDS***

Let us consider the VVDS for preparing a survey aiming at studying the kinematics of $z = 0.6$ and $z = 1.5$ galaxies. One can assemble two separate sub-samples of such galaxies to perform spatially resolved spectroscopic observations. However because the VVDS is limited by the $I_{AB}$ selecting magnitude at 835nm, the corresponding rest-frame magnitudes are at $\lambda = 521$ and 334nm, respectively, i.e., on both sides of the 400nm break. The consequence of this is that at $z = 0.6$ one will select galaxies that are somewhat representative of the stellar mass function, but at $z = 1.5$ one will be selecting a mixture of massive galaxies with moderate amount of extinction and of very strong star-forming dwarves that have very small amount of dust. A direct comparison of these two sub-samples in an attempt to, for example, study galaxy evolution requires to account for such a large selection effect.

In such studies, one should use redshift surveys that have the lowest fraction possible of unidentified galaxies, have not additional section criteria, and finally have sufficient spectral resolution to ensure quality redshift estimates. The latter is particularly important when redshift estimates are based on lines redshifted beyond 720nm, where OH sky lines become increasingly dominant.

## 1.5 Photometric redshifts

It is tempting to estimate redshifts using multiple band photometry since spectroscopic surveys are often limited in depth and leave a significant fraction of objects with undetermined redshifts.

This approach can be used with a large number of objects at the relatively modest cost of only performing photometry in multiple bands and it has become very attractive. Several projects are even based on photometric redshifts to estimate cosmological parameters through BAO or weak lensing studies. In addition, as photometric redshifts circumvent some of the limitations of magnitude-limit redshift surveys, they have played an important part in identifying high redshift galaxies.

Photometric redshift techniques can be divided into two main groups,[8] as described in Dahlen *et al.* (2010), who also provides references to many existing photometric redshift codes (see also Hildebrandt *et al.*, 2010).

The first group is based on template fitting. The spectral energy distribution (SED, see Chapter 2) of a galaxy is compared to a set of template SEDs and $\chi^2$ minimization is used to determine the best fitting template and to provide an estimate of the photometric redshift.

The second group requires a sample of existing spectroscopic redshifts that is then used to train an artificial neural networks. This approach is particularly well suited to estimate the redshifts of all the galaxies in a field where only a sub-sample of galaxies have known spectroscopic redshifts.

Photometric redshift estimates have typically small r.m.s. errors, generally ranging from 0.02 to 0.08. This is generally determined after comparing these estimates with known spectroscopic redshifts. There are however cases of catastrophic photo-$z$ failures, which are defined as $\Delta z/(1+z) > 0.1$–$0.2$ or at $>3$ times the r.m.s. These failures usually have very large associated errors, since they are often

[8]There is now a third category (Morris *et al.*, 2015) that combines open grism observations at $R = 130$ (3D-HST) to photometric redshifts. It considerably improves the redshift estimation, although it is unclear whether it can resolve the issue of catastrophic photo-$z$ failures. A description of the open grism technique can be found in Chapter 3.

caused by the confusion between the Lyman break at 91.2nm and the 400nm break or the Balmer break at 364.6nm. For example, dwarves at $z \sim 0.2$ can be confused with luminous starbursts at $z \sim 3$.

The quality of photometric redshifts can be characterized by the fraction of catastrophic failures, which can be as low as a few percent and as high as a few tens of percent.

The rate of catastrophic failure can be reduced with the use of very deep photometry in bluer bands, since the Lyman break ensures that there should be no emission below the rest-frame wavelength of 91.2nm, unless an AGN is present.

Additional ways to reduce the rate of catastrophic failures are:

- ensure a high S/N for each photometric band used to determine the redshift (see Margoniner and Wittman, 2008);
- rely on excellent aperture photometry that is consistent from one band to another (see Chapter 2);
- use deep photometry that covers the largest wavelength range possible, i.e., including deep UV, NIR, and MIR photometry in addition to the visible band photometry (see Dahlen *et al.*, 2010).

A fundamental concern about photometric redshift estimates is that these depend on what spectroscopic templates were adopted. These are typically either based on population synthesis models or empirical SEDs observed at different redshifts.

The choice of template to use should be made carefully since this essentially places *a priori* assumptions on the physical properties of the very galaxies that are to be studied. The use of Bayesian techniques to weigh the probability occurrence of a photometric redshift only exacerbate the problem: how do we know the SED distribution of a population that is under study? In general, SED-template selection effects dominate over magnitude prior errors (Abrahamse *et al.*, 2011).

Different population of galaxies can be affected differently by the accuracy and catastrophic failure rate of photometric redshift techniques. This is because:

- population synthesis models are limited in their ability to reproduce an observed SED. This is particularly true over some wavelength ranges such as the NIR;

- the choice of the dust extinction models can impact the results for some galaxies, e.g., extremely dusty galaxies; it is recommended to use a wide range of extinction laws for while assembling a set of spectroscopic templates;
- the use of templates with solar or near solar metallicity can severely bias the photometric redshift estimates of distant galaxies with low metallicity (Kotulla, 2011);
- AGN light can be present in some galaxies, causing the assumed depth of the Lyman break (MacDonald and Bernstein, 2010) to be wrong;
- the actual wavelength of the Lyman break can be different than the one that is assumed by the models since it depends on the exact distribution of intervening matter along the line of sight, which is a redshift dependent effect;
- emission lines, as well as nebular continuum emission generated in strong and low-abundance starbursts can affect broad-band photometry (see Chapter 2).

It is therefore difficult to homogeneously determine the photometric redshifts of a population of distant galaxies that may have properties that are different that the ones assumed by the SED templates. This also implies that one should be extremely careful with using a photometric redshift estimate for an individual galaxy.

The neural network method, which relies on pre-existing spectroscopic redshift, is also limited in that there is no reason why the SEDs of fainter galaxies should be similar to those of brighter galaxies for which spectroscopic observations were obtained.

As a solution to this problem, some have proposed adopting probabilities weighted by the observed magnitude (i.e., assuming that faint galaxies are more likely to be distant), by the surface brightness or by assuming that galaxies in pair have better chances to be at same redshift.

It is very tempting to use the photometric redshift for deriving galaxy rest-frame properties, as well as to calculate cosmic density of various quantities such as stellar mass or star formation (see Chapter 2). One could estimate the stellar mass, extinction, SFR directly from the best SED template selected during the photometric redshift determination. These are however severely inaccurate when these are applied to a wide range of galaxy populations.

**Example 1.10: *Stellar mass and star formation densities estimated through photo-z***

Let us consider a sample of galaxies with a 10% failure rate which is caused by the confusion between low and high redshift galaxies. The inclusion of these low-$z$ galaxies, misidentified as high-$z$ galaxies, could severely affect the cosmic stellar mass density estimates that are derived in the highest redshift bins. The intrusion of just a few low-$z$ dwarves assumed to be high-$z$ very massive galaxies can dominate the calculation of the cosmic stellar mass density.

It is even more problematic to estimate the extinction law through SED fitting, since the resulting SFR may be easily mismatched, by factors of 10 or larger, and since it is also heavily dependent on the *a priori* adopted extinction law. The reader is referenced to Chapter 3 for proper estimates of the extinction.

A large effort is currently being undertaken to improve the results from various existing photometric redshift codes, by adopting a common method to compare them (photo-$z$ Accuracy Testing programme, see Hildebrandt *et al.*, 2010). At high redshift the main limitation appears to be related to the specific treatment of the Lyman break and of the Ly$\alpha$ forest.

Meanwhile, while the complexity of photo-$z$ techniques increases, their use for galaxy evolution studies remain unreliable. This is particularly true in cases lacking a comparative sample of galaxies with spectroscopic redshifts over a similar magnitude range.

## 1.6 Selection of very distant galaxies at $z \gg 1$

### 1.6.1 *Lyman break galaxies (LBGs)*

One of the most remarkable success of the Keck telescope has been the discovery of an abundant populations of $z \geq 3$ galaxies

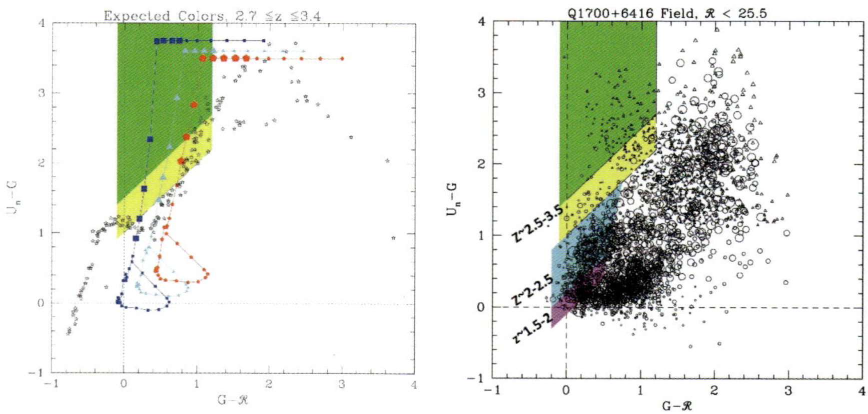


Fig. 1.7: Selection of LBGs (from Steidel *et al.*, 2003, 2004, ©AAS, reproduced with permission). *Left*: $U-G$ vs. $G-R$ color: U band luminosity decreases when the redshifted Lyman break passes through it, causing a large $U-G$ color change, thus allowing high-$z$ galaxies to be selected ($2.5 < z < 3.5$, green and yellow shaded areas). Curves show the results of three SED modeling, and stars represent the stellar sequence. The large points on each curve correspond to galaxies in the redshift interval $2.7 < z < 3.4$. *Right*: same diagram with $R < 25.5$ galaxies and with the addition of two additional categories of LBGs, at $z = 1.5$–2 and $z = 2$–2.5 (magenta and cyan shaded areas, respectively).

(Steidel *et al.*, 1996). Besides being the most powerful telescope in the early 90s, this success was made possible by a specific selection of distant galaxy candidates using color criteria that are sensitive to the presence of a Lyman continuum break superimposed onto an otherwise very blue far-UV continuum. This is illustrated in Fig. 1.7 which shows the efficiency in identifying LBGs from $z = 1.5$ and all the way to $z = 3.5$. The surface density of LBG at $R \leq 25.5$ reaches 3.8, 5.2, and 2 arcmin$^{-2}$ for $z = 1.5$–2, $z = 2$–2.5, and 2.5–3.5 LBGs, respectively. Altogether LBGs represent approximately 25% of the whole galaxy population with $R \leq 25.5$. Spectroscopic observations of these $R < 25.5$ sources can routinely be done using blue-spectrographs that are mounted at 8m telescopes. The redshift of these sources can be be determined using the numerous and prominent high-ionization stellar lines of He II, C IV, Si IV, and N V, and strong interstellar absorption lines due to low-ionization stages of C, O, Si, and Al.

The LBG galaxies selected by Steidel *et al.* (2003, 2004) can be referred to as U-drop galaxies. Detecting more distant galaxies requires the assumption that the Lyman break enters into redder broad-band filters, allowing for the selection of $B$, $V$, and $I$ drop galaxies, with redshift ranges of $z = 2.8$–$4.4$ (Papovich *et al.*, 2004), $z = 4.4$–$5.5$ (Giavalisco *et al.*, 2004) and $z = 5.5$–$6.8$ (Dickinson *et al.*, 2004), respectively. At redshifts larger than 6, this technique relies on the opacity of the Ly$\alpha$ forest of the intergalactic medium, i.e., providing a photometric break at $\lambda \sim 120$nm instead of at 91.2nm.

In principle, $J$ drop galaxies would be at $z \sim 10$, but extremely deep photometry is required, up to $m_{AB} = 32$[9] in visible bands, in order to avoid contamination by foreground interlopers at $z \sim 2$ (Hayes *et al.*, 2012). Gravitational lensing by foreground galaxy clusters may help the detection of distant lensed sources, although they have to form large gravitational arcs to be efficiently magnified, and this increases the chance of being contaminated by foreground light emissions, which are not unexpected in the cluster cores. JWST and ELTs are certainly the best way to detect and study the most distant galaxies.

While this is a powerful technique to select high-z galaxies from a very large population of foreground objects, it does not select the overall galaxy population at high-$z$.

First, the reddest broad-band $R$ filter used by Steidel *et al.* (2003) corresponds to 162nm at rest for $z = 3$ LBGs, meaning it samples mostly nonextinguished hot stars in the rest-frame FUV. Relatively massive galaxies with nonnegligible extinction are probably missed by such a criterion. Second, the adopted cut in $G - R$ color ($\leq 1.2$) corresponds to a very blue far-UV continuum, which excludes galaxies with significant dust extinction or galaxies that have relatively old stellar populations. This is well illustrated by Marchesini *et al.* (2010) as the most massive galaxies they identified at $z = 3$–$4$ generally show $G - R$ color in excess of 1.2.

[9]Deepest photometry with the HST reaches $m_{AB} = 29.45$ (see Chapter 2, Sec. 2.2.2).

Third, there might be some galaxies with a nonnegligible AGN contribution that may not obey to the Lyman drop criterion in $U - G$ color. While the drop-out technique used by the Steidel team identified a very significant population of high-$z$ blue starbursts, it is far from providing a complete sample of galaxies at high-$z$.

### 1.6.2 *Balmer and 400nm break drop-out galaxies*

To complement the starburst population that is selected using the Lyman drop-out techniques, one can use redder filters that are more sensitive to older stars, and hence to the overall stellar population content of galaxies. The Balmer break at 364.6nm is caused by intermediate-age stars (~0.2–1 Gyr old, A stars) while the 400nm break is caused by absorption from ionized metals in the atmospheres of older (F, G, and K) stars. Two techniques have been used, based on the detection of the Balmer break and of the 400nm break using optical to NIR bands.

Distant red galaxies (DRGs) at $z \geq 2$ have been identified on the basis of a $(J - K)_{Vega}$ color that is greater than 2.3 (e.g., van Dokkum *et al.*, 2003).

BzK galaxies at $z = 1.4$–$2.5$ can be divided into two categories: star forming galaxies with $\text{BzK} = (z - K)_{AB} - (B - z)_{AB} \geq -0.2$ and quiescent galaxies with $\text{BzK} < -0.2$ and $(z - K)_{AB} > 2.5$ (e.g., Daddi *et al.*, 2004).

Down to $K_{AB} = 22.8$, star forming BzK, quiescent BzKs and DRGs have a surface density of 3, 0.24, and 1 arcmin$^{-2}$, respectively. There is considerable overlap between DRGs and BzKs (>95%), UV or $B$-drop galaxies and star forming BzK (~80%), while the overlap between DRGs and UV or $B$-drop galaxies is at the level of 12% (Reddy *et al.*, 2005; Grazian *et al.*, 2007).

The main problem with NIR selected galaxies is that they are very faint in the optical broad-band (generally fainter than $R = 25.5$), and only the very brightest ones with emission lines can be spectroscopically observed with NIR spectrographs at a reasonable S/N (see e.g., Förster Schreiber *et al.*, 2009; Onodera *et al.*, 2010, and references therein). The end result is that most DRGs and BzK galaxies are identified through photo-$z$ techniques and these samples are made up of a mixture of extinguished star forming, post-starburst, and quiescent galaxies. The stellar content of the most passive galaxies at $z \sim 2$ appears to be significantly larger than that of LBGs, resulting in large uncertainties on the actual stellar content of galaxies at high-$z$.

### 1.6.3 *Lyα emitters (LAEs)*

Narrow-band filters at wavelengths adjusted to avoid OH sky lines are a powerful tool to detect LAE up to very high redshifts. The main difficulty is in avoiding possible confusion between LAE and other emission lines at lower redshifts, such as [O II]372.7, [O III]500.7, or H$\alpha$. This usually leads to additional criteria, such as the presence of a Lyman break in a bluer filter. For example Ouchi *et al.* (2008) identified LAEs at $z = 3.1$ by looking for a significant excess between broad and narrow band filter, i.e., $V - NB_{503\mathrm{nm}} > 1.2$ and $B - V > 0.5$. At increasing redshifts this method requires very deep broad-band imagery since LAE spectra exhibit strong emission line generally associated with very faint continua. Several thousands of LAEs from $z = 2$ to $z = 7$ have been identified up to now. But the way LAEs have been selected may result in them being a significantly different populations of distant galaxies than those gathered using the selection methods described above.

In fact, at $z \sim 3$ approximately 20–25% of the LBGs of Shapley *et al.* (2003) are LAEs and reciprocally 23% of the LAEs are LBGs. The overlap between the LBG and LAE populations is influenced the rather complex process of LAE, as well as N(H I) and dust (Verhamme *et al.*, 2008), and perhaps even on the orientation of the galaxy relatively to the observer (Dijkstra and Wyithe, 2010).

### 1.6.4 *Indirect probes: damped Lyα systems (DLAs) and gamma ray burst hosts*

Absorption line spectra of bright QSOs have revealed the presence of DLAs probing gas column densities of N(H I) $\geq 2 \times 10^{20}\,\mathrm{cm}^{-2}$ (see also Chapter 3). Several hundreds DLAs have been detected from $z = 0$ to 5 (e.g., Prochaska *et al.*, 2003), providing a census of the HI gas content. In principle DLAs are not affected by the biases affecting magnitude-limited samples of galaxies. On the other hand little is known about the galactic counterparts of DLAs, which could be galactic halos, outer gas disks or gas clouds. Still, DLAs indubitably represent the gas-richest distant galaxies and can probe the galaxy evolution up to $z \sim 5$.

Another way to probe distant galaxies is provided by gamma-ray bursts that still hold the record of the most distant galaxy (GRB 090423 at $z = 8.4$, Tanvir *et al.*, 2009) at the time of writing this manuscript. At $z < 1$, GRB host galaxies are known to have lower metal abundances than galaxies with similar stellar masses (Han *et al.*, 2010). Thus GRB hosts likely sample a relatively premature and gas-rich medium when compared to other galaxies. It is however very difficult to observe high-$z$, GRB host galaxies following the GRB burst, because of their intrinsic faintness. Many detections of very distant GRB-hosts are expected in the near future, providing an interesting and possibly unique way to sample the low-mass end and possibly the most gas-rich part of high-$z$ galaxies.

### 1.6.5 *Dust-enshrouded galaxies*

Sub-mm, MIR, and FIR observations allow to study high-$z$ dust-enshrouded galaxies such as luminous or ultra luminous galaxies (LIRGs and ULIRGs with SFR $> 19$ and $190\ M_{\odot} \cdot yr^{-1}$, respectively). Although their number density is several times smaller than that of LBGs, they contribute to a significant fraction of the star formation density at $z > 1$ (Le Floc'h *et al.*, 2005) and up to 90% of the IR luminosity density at $z \sim 2$ (Caputi *et al.*, 2007). Current sub-mm interferometers allow to detect their cold dust emission to higher redshifts, and a far improved position accuracy is expected from ALMA.

### 1.6.6 *Towards an overview of the high-z galaxy population and a fair determination of the LFs*

As described in the previous sections, considerable effort was made to identify very high-$z$ galaxies using a wide variety of techniques.

It remains unclear whether or not the overall galactic population of these objects have been identified. Reaching a census at high-$z$ is far more complex than at $z \leq 1$ (see Ex. *A sub-sample of $z \sim 2$ galaxies for determining kinematics*). LFs can be computed for each categories of high-$z$ selected galaxies, as described in Sec. 1.4.2, while the galaxy selection criteria are kept in mind. Taking into account the overlaps between different galaxy selection allows to determine a more general overview of their LF, at least for the most massive galaxies. In principle, photometric redshift studies may provide an even more complete picture if they remained reliable in the case of faint galaxies at high-$z$.

A possible breakthrough could be a spectroscopic redshift survey of galaxies that are selected in near or MIR wavelengths and using NIR, MOS. This also underlines the need for MOS at the future ELTs.

GMASS (Cimatti *et al.*, 2008) provides a first glimpse of this, selecting $m_{4.5\mu m}(AB) < 23$ galaxies. Although it also imposes additional selection with $z_{phot} > 1.4$, and additional magnitude limits due to the use of optical spectrographs, it convincingly shows that most of the $z = 1.5$–2 passive galaxies are recovered with the BzK criterion.

It is likely that a reasonable census of the most massive galaxies (typically $M_{stellar} > 10^{11}$ $M_\odot$ and of the most actively star forming galaxies (at least ULIRGs) has now been achieved up to $z \sim 2.5$. However, this is probably not the case for lower luminosity galaxies, and strongly affects the overall census of stellar mass at high-$z$. The situation is most certainly worse with any attempt to evaluate the baryonic content at high-$z$: it is almost certain that the fraction of gaseous-rich galaxies increases with redshift, and there are very few chances that these can be observed in the near future.

**Example 1.11: *A sub-sample of $z \sim 2$ galaxies for determining kinematics***

Spatially resolved spectroscopy of very distant galaxies can be obtained using NIR 3D spectrographs (see Chapter 4). However,

(*Continued*)

**Example 1.11: *Continued***

to get a representative sub-sample of e.g., $z = 2$ galaxies, one may chose to aggregate samples from different origins. Let us consider a selection sample composed of 1/3 LBGs (see Fig. 1.7) and 2/3 of sources obeying the BzK criterion. What is the significance of the resulting internal motions of these galaxies? How can their properties be compared to those of the overall galaxy population at $z \sim 2$? It is necessary to compare the luminosity distribution of the observed galaxies relatively to galaxies from a more complete catalog that is known to be complete within a (red) magnitude limit, e.g., a catalog of K band limited galaxies. If within a given luminosity range, the luminosity distribution matches that of the complete catalog with similar number density, then this luminosity range was sufficiently probed by the sub-sample. If not, one has to understand whether or not some specific sources are rejected from the selection, and why. It is a very difficult exercise to aggregate samples that are from different origin, particularly since some of the selection criteria might actually be excluding some important galaxy population.

## 1.7 Relating distant and nearby galaxies

An important goal of redshift surveys is to probe galaxy evolution, comparing galaxies at different redshifts, ultimately aiming to identify the progenitors of present-day galaxies. The Cosmological Principle implies that progenitors of present-day galaxies should be present within the population of high redshift galaxies. This is somewhat of an hazardous task, because:

- during an Hubble time galaxies cannot be considered as "isolated islands": they may be colliding with other galaxies and may exchange gas with the intergalactic medium;
- even without major exchange with their environment, galaxies evolve secularly: their stellar populations are changing, forming

and dying, and their bulge and disk are experiencing internal dynamical processes;

- comparing distant and nearby galaxies requires the use of some common physical parameter, for example the mass; unfortunately the stellar mass cannot be determined very accurately (0.3 dex uncertainty, see Chapters 2 and 3). Moreover, only an indirect estimate of their gas mass is available (through inversion of the Kennicutt–Schmidt law), without large radio interferometers allowing detection of neutral hydrogen at $21 \times (1 + z)$ cm. While one could use the circular velocity of spiral galaxies or the intrinsic dispersion of ellipticals to characterize their dynamical masses, this incorrectly assumes that all of their progenitors were at equilibrium in the past;
- one needs to ensure that the selection parameter is robust enough to include all the distant galaxies within the given mass range that would allow them to be the progenitors of a given set of nearby galaxies: this is further complicated by undetected low surface brightness galaxies at large distances;
- to compare distant and nearby galaxies requires one have a reasonably complete knowledge of the properties of both distant and nearby galaxies: this includes a knowledge of fundamental quantities such as mass, "size", shape, and kinematics.

The above could be considered as a definitive show stopper for studying the progenitor–descendent relation between distant and nearby galaxies. However, under certain circumstances, it remains possible to infer some relations between distant galaxies and their descendants. For example, it is likely that dust-enshrouded, $z = 2$–3 galaxies detected at sub-mm wavelengths, have baryonic masses in excess of $10^{11}$ $M_\odot$, and could therefore be progenitors of present-day, massive elliptical galaxies. Another possible approach is to compare[10] the number density of a distant galaxy population to those of lower redshift galaxy populations. For example LIRGs at $z = 0.5$–1 are so numerous that they can only be related to the dominant population

[10]This technique has to be used with a lot of precautions, because luminosity evolution as well as mergers may strongly affect the galaxy number density at very different redshifts.

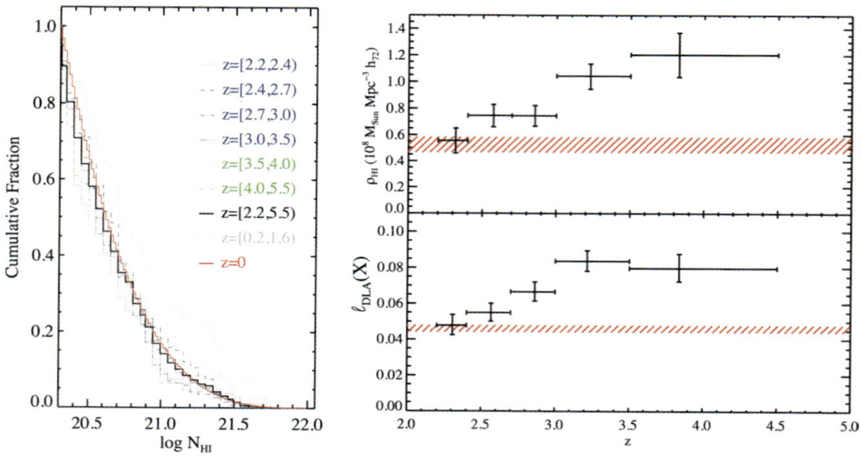


Fig. 1.8: Evolution of Damped Lyman Alpha systems (from Prochaska and Wolfe, 2009, ©AAS, reproduced with permission). *Left*: cumulative distribution functions of $\log(N_{HI})$ of galaxies in a series of redshift intervals. For $z > 2$, the shape is invariant and, remarkably, is a good match to the observed distribution of local HI disks. *Upper-right*: comoving HI mass density $\rho_{HI}$ of galaxies at $z > 2$. The HI mass density declines by $\sim$50% from $z = 4$ to 2.2, an interval spanning less than 2 Gyr. The red band shows the estimate of $\rho_{HI}$ at $z \sim 0$ from 21 cm surveys of local HI galaxies. *Bottom-right*: line density of DLAs per comoving absorption length $dX$. This quantity can be visualized as the integrated covering fraction per comoving path length for galactic HI gas. Following the mass density $\rho_{HI}$, the covering fraction decreases by 50% from $z = 4$ to 2 where it reaches the present-day value (red band) as estimated from 21 cm observations.

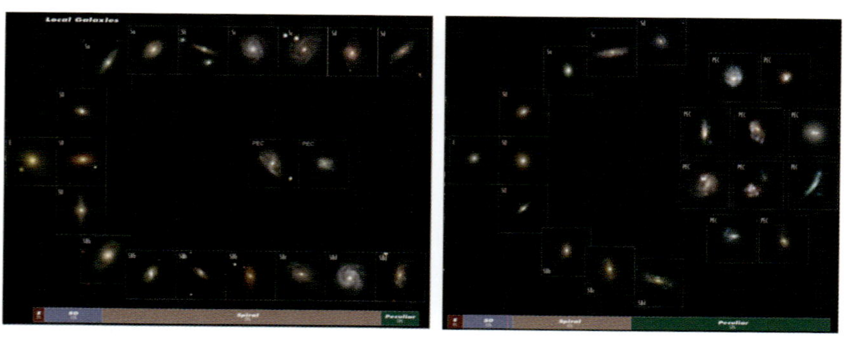

Fig. 1.9: Evolution of the Hubble Sequence of massive galaxies (Delgado-Serrano *et al.*, 2010). *Left*: morphological classification of nearby galaxies including from the left to the right, elliptical, lenticular, and spiral galaxies with decreasing bulge to disk ratio. The fork distinguishes galaxies with bar (bottom fork) and those without (upper fork). Each stamps represents 5% of the several hundred galaxies selected from the SDSS.

of galaxies at lower redshifts, e.g., spiral galaxies at $z = 0$ (Hammer *et al.*, 2005).

Many of the selection effects applying to magnitude limited surveys can be avoided using an unbiased selection of galaxies that is based on the presence of intervening absorption lines in QSO spectra. They allows the HI content of the Universe to be probed from $z = 0$ to $z = 5$.

Figure 1.8 shows that the distribution of HI column density is similar at all redshifts, which remains unexplained in current galaxy formation models. Studies of HI column density at high redshift from DLAs can be compared to radio observations at low redshift using the 21 cm HI line. Figure 1.8 shows that the HI galactic gas has not evolved significantly over the past 10 Gyr (Prochaska and Wolfe, 2009). Processes that remove the neutral gas from galaxies (either gas consumption to form stars, or gas expelled by feedback or perhaps by mergers) must be balanced by gas accretion during that period.

An important step in comparing nearby galaxies to their ancestor can be found in Delgado-Serrano *et al.* (2010) who compared samples of galaxies from two different epochs: 6 billion years ago and at present-day. At the moderate redshift of $z \sim 0.65$, the properties of galaxies can be studied nearly as well as those of local galaxies and surface brightness selection effects remain very modest. Moreover, the fraction of $z \sim 0.65$ galaxies expected to undergo additional merging during the next 6 billions years is expected to be small. Both present day and $z \sim 0.65$ galaxy samples can be selected using an unbiased absolute magnitude in the $J$ band ($M_J(AB) < -20.3$, calculated from observed $J$ and $K$ band, see Sec. 1.3.6) that correlates particularly well with stellar and baryonic mass ($M_{\rm stellar} > 1.5 \times 10^{10}\,{\rm M}_\odot$).

Fig. 1.9: (*continued*) Up to 10% of local galaxies have morphologies sufficiently distorted to prevent them from being included in the Hubble Sequence. *Right*: same but for $z \sim 0.65$ galaxies, i.e., galaxies 6 billion years ago. The fraction of galaxies that cannot be classified using the Hubble Sequence increases considerably, reaching 50%. Credit: NASA, ESA, Sloan Digital Sky Survey, R. Delgado-Serrano, and F. Hammer (Observatoire de Paris).

HST imaging using the ACS camera probes galaxy features on the scale of a few hundreds of parsecs at $z \sim 0.65$. Rest-frame spatial resolution and depth provided by ACS in the right panel of Fig. 1.9 are comparable to those of the SDSS (Fig. 1.9, left panel). These observations show that bulge-dominated galaxies (elliptical and lenticular galaxies) are not evolving in fraction, while more than 50% of the ancestors of disk-dominated galaxies have peculiar morphologies (see Chapter 2 for details about the morphological classification).

Spatially-resolved spectroscopy of $z \sim 0.65$ galaxies is now within the reach of large telescopes (see Chapter 4) and has been used to show that peculiar morphologies are associated to anomalous velocities (Neichel *et al.*, 2008).

We conclude this chapter by stating that, as long as selection effects are fully taken into account, samples of high redshift galaxies can be used to provide strong constraints on the formation and evolution of galaxies.

## Bibliography

Abrahamse A., Knox L., Schmidt S. *et al.* (2011). Characterizing and propagating modeling uncertainties in photometrically derived redshift distributions, *ApJ* **734**, p. 36.

Bertin E., Arnouts S. (1996). Sextractor: Software for source extraction, *A&AS* **117**, pp. 393–404.

Bessell M. S. (2005). Standard photometric systems, *ARA&A* **43**, pp. 293–336.

Binney J., Tremaine S. (1987). *Galactic Dynamics*. 1st edn. Princeton Series in Astrophysics. Princeton University Press.

Blake C., Brough S., Colless M. *et al.* (2010). The wigglez dark energy survey: The selection function and z = 0.6 galaxy power spectrum, *MNRAS* **406**, pp. 803–821.

Blanton M. R., Dalcanton J., Eisenstein D. *et al.* (2001). The luminosity function of galaxies in sdss commissioning data, *AJ* **121**, pp. 2358–2380.

Blanton M. R., Roweis S. (2007). $K$-corrections and filter transformations in the ultraviolet, optical, and near-infrared, *AJ* **133**, pp. 734–754.

Caputi K. I., Lagache G., Yan L. *et al.* (2007). The infrared luminosity function of galaxies at redshifts $z = 1$ and $z \sim 2$ in the goods fields, *ApJ* **660**, pp. 97–116.

Cimatti A., Cassata P., Pozzetti L. *et al.* (2008). Gmass ultradeep spectroscopy of galaxies at $z \sim 2$. ii. superdense passive galaxies: How did they form and evolve? *A&A* **482**, pp. 21–42.

Coil A. L., Davis M., Madgwick D. S. *et al.* (2004). The DEEP2 galaxy redshift survey: Clustering of galaxies in early data, *ApJ* **609**, pp. 525–538.

Cowie L. L., Songaila A., Hu E. M. *et al.* (1996). New insight on galaxy formation and evolution from keck spectroscopy of the hawaii deep fields, *AJ* **112**, p. 839.

Crampton D., Le Fevre O., Lilly S. J. *et al.* (1995). The Canada–France redshift survey. v. global properties of the sample, *ApJ* **455**, p. 96.

Daddi E., Cimatti A., Renzini A. *et al.* (2004). A new photometric technique for the joint selection of star-forming and passive galaxies at 1.4z $\leq$ zz $\leq$ 2.5, *ApJ* **617**, pp. 746–764.

Dahlen T., Mobasher B., Dickinson M. *et al.* (2010). A detailed study of photometric redshifts for goods-south galaxies, *ApJ* **724**, pp. 425–447.

Delgado-Serrano R., Hammer F., Yang Y. B. *et al.* (2010). How was the Hubble sequence 6 Gyr ago? *A&A* **509**, A78.

Dickinson M., Stern D., Giavalisco M. *et al* (2004). Color-selected galaxies at $z \sim 6$ in the great observatories origins deep survey, *ApJ* **600**, pp. L99–L102.

Dijkstra M., Wyithe J. S. B. (2010). Seeing through the trough: Outflows and the detectability of Ly$\alpha$ emission from the first galaxies, *MNRAS* **408**, pp. 352–361.

Drinkwater M. J., Jurek R. J., Blake C. *et al.* (2010). The Wiggle$Z$ dark energy survey: Survey design and first data release, *MNRAS* **401**, pp. 1429–1452.

Faber S. M., Willmer C. N. A., Wolf C. *et al.* (2007). Galaxy luminosity functions to $z \sim 1$ from DEEP2 and COMBO-17: Implications for red galaxy formation, *ApJ* **665**, pp. 265–294.

Felten J. E. (1976). On schmidt's $vm$ estimator and other estimators of luminosity functions, *ApJ* **207**, pp. 700–709.

Flores H. (1999). Distribution global d'énergie et spectrophotométrie des galaxies de champ, Ph.D. thesis, Université Paris.

Flores H., Hammer F., Thuan T. X. *et al.* (1999). 15 micron infrared space observatory observations of the 1415 + 52 canada–france redshift survey field: The cosmic star formation rate as derived from deep ultraviolet, optical, mid-infrared, and radio photometry, *ApJ* **517**, pp. 148–167.

Förster Schreiber N. M., Genzel R., Bouché N. *et al.* (2009). The SINS Survey: SINFONI integral field spectroscopy of $z \sim 2$ star-forming Galaxies, *ApJ* **706**, pp. 1364–1428.

Fukugita M., Shimasaku K., Ichikawa T. (1995). Galaxy colors in various photometric band systems, *PASP* **107**, p. 945.

Garilli B., Fumana M., Franzetti P. *et al.* (2010). Ez: A tool for automatic redshift measurement, *PASP* **122**, pp. 827–838.

Giavalisco M., Dickinson M., Ferguson H. C. *et al.* (2004). The rest-frame ultraviolet luminosity density of star-forming galaxies at redshifts $z > 3.5$, *ApJ* **600**, pp. L103–L106.

Gilli R., Cimatti A., Daddi E. *et al.* (2003). Tracing the large-scale structure in the chandra deep field south, *ApJ* **592**, pp. 721–727.

Graham A. W., Driver S. P. (2005). A concise reference to (projected) sérsic $r^{1/n}$ quantities, including concentration, profile slopes, petrosian indices, and kron magnitudes, *PASA* **22**, pp. 118–127.

Graham A. W., Driver S. P., Petrosian V. *et al.* (2005). Total galaxy magnitudes and effective radii from petrosian magnitudes and radii, *AJ* **130**, pp. 1535–1544.

Grazian A., Salimbeni S., Pentericci L. *et al.* (2007). A comparison of lbgs, drgs, and bzk galaxies: Their contribution to the stellar mass density in the goods-music sample, *A&A* **465**, pp. 393–404.

Hammer F., Crampton D., Le Fevre O. *et al.* (1995). The Canada–France redshift survey. IV. Spectroscopic selection effects and $0300 + 00$ field spectroscopic data, *ApJ* **455**, p. 88.

Hammer F., Flores H., Elbaz D. *et al.* (2005). Did most present-day spirals form during the last 8 gyr? A formation history with violent episodes revealed by panchromatic observations, *A&A* **430**, pp. 115–128.

Hammer F., Gruel N., Thuan T. X. *et al.* (2001). Luminous compact galaxies at intermediate redshifts: Progenitors of bulges of massive spirals? *ApJ* **550**, pp. 570–584.

Han X. H., Hammer F., Liang Y. C. *et al.* (2010). The wolf-rayet features and mass-metallicity relation of long-duration gamma-ray burst host galaxies, *A&A* **514**, A24.

Hayes M., Laporte N., Pelló R. *et al.* (2012). A peculiar galaxy appears at redshift 11: Properties of a moderate-redshift interloper, *MNRAS* **425**, pp. L19–L23.

Hildebrandt H., Arnouts S., Capak P. *et al.* (2010). Phat: Photo-$z$ accuracy testing, *A&A* **523**, A31.

Hogg D. W., Baldry I. K., Blanton M. R. *et al.* (2002). The K correction, *ArXiv Astrophysics e-prints*. astro-ph/0210394, Astrophysics, available at: http://cdsads.u-strasbg.fr/abs/2002astro.ph.10394H, Provided by the SAO/NASA Astrophysics Data System.

Ilbert O., Tresse L., Zucca E. *et al.* (2005). The VIMOS-VLT deep survey. Evolution of the galaxy luminosity function up to $z = 2$ in first epoch data, *A&A* **439**, pp. 863–876.

Johnston R. (2011). Shedding light on the galaxy luminosity function, *A&A Rev.* **19**, p. 41.

Johnston R., Teodoro L., Hendry M. (2007). Completeness — i. revisited, reviewed and revived, *MNRAS* **376**, pp. 1757–1766.

Kotulla R. (2011). *AAS 217th Meeting, Bulletin of the American Astronomical Society*, Vol. 43, p. 211.06.

Kron R. G. (1980). Photometry of a complete sample of faint galaxies, *ApJS* **43**, pp. 305–325.

Le Fevre O., Crampton D., Lilly S. J. *et al.* (1995). The Canada–France redshift survey. II. Spectroscopic program: Data for the 0000-00 and 1000+25 Fields, *ApJ* **455**, p. 60.

Le Fèvre O., Vettolani G., Garilli B. *et al.* (2005). The VIMOS VLT deep survey. First epoch VVDS-deep survey: 11 564 spectra with $17.5 \leq \text{IAB} \leq 24$, and the redshift distribution over $0 \leq z \leq 5$, *A&A* **439**, pp. 845–862.

Le Floc'h E., Papovich C., Dole H. *et al.* (2005). Infrared luminosity functions from the chandra deep field-south: The spitzer view on the history of dusty star formation at $0 \leq z \leq 1$, *ApJ* **632**, pp. 169–190.

Liang Y. C., Hammer F., Flores H. *et al.* (2004). Misleading results from low-resolution spectroscopy: From galaxy interstellar medium chemistry to cosmic star formation density, *A&A* **417**, pp. 905–918.

Lilly S. J., Le Brun V., Maier C. *et al.* (2009). The zCOSMOS 10k-bright spectroscopic sample, *ApJS* **184**, pp. 218–229.

Lilly S. J., Le Fevre O., Crampton D. *et al.* (1995). The Canada–France redshift survey. i. Introduction to the survey, photometric catalogs, and surface brightness selection effects, *ApJ* **455**, p. 50.

Lilly S. J., Le Fevre O., Hammer F. *et al.* (1996). The Canada–France redshift survey: The luminosity density and star formation history of the universe to $z$ approximately 1, *ApJ* **460**, p. L1.

Lilly S. J., Le Fèvre O., Renzini A. *et al.* (2007). zCOSMOS: A large VLT/VIMOS redshift survey covering $0 < z < 3$ in the COSMOS field, *ApJS* **172**, pp. 70–85.

MacDonald C. J., Bernstein G. (2010). Photometric redshift biases from galaxy evolution, *PASP* **122**, pp. 485–489.

Marchesini D., Whitaker K. E., Brammer G. *et al.* (2010). The most massive galaxies at $3.0 < z < 4.0$ in the newfirm medium-band survey: Properties and improved constraints on the stellar mass function, *ApJ* **725**, pp. 1277–1295.

Margoniner V. E., Wittman D. M. (2008). Photometric redshifts and signal-to-noise ratios, *ApJ* **679**, pp. 31–51.

Morris A. M., Kocevski D. D., Trump J. R. *et al.* (2015). A WFC3 grism emission line redshift catalog in the GOODS-South field, *AJ*, **149**, p. 178, doi:10.1088/0004-6256/149/6/178.

Neichel B., Hammer F., Puech M. *et al.* (2008). Images. ii. a surprisingly low fraction of undisturbed rotating spiral disks at $z \sim 0.6$ the morpho-kinematical relation 6 gyr ago, *A&A* **484**, pp. 159–172.

Oke J. B. (1974). Absolute spectral energy distributions for white dwarfs, *ApJS* **27**, p. 21.

Onodera M., Arimoto N., Daddi E. *et al.* (2010). A wide area survey for high-redshift massive galaxies. II. Near-infrared spectroscopy of BZK-selected massive star-forming galaxies, *ApJ* **715**, pp. 385–405.

Ouchi M., Shimasaku K., Akiyama M. *et al.* (2008). The subaru/xmm-newton deep survey (sxds). iv. evolution of ly$\alpha$ emitters from $z = 3.1$ to 5.7 in the 1 deg$^2$ field: Luminosity functions and agn, *ApJS* **176**, pp. 301–330.

Papovich C., Dickinson M., Ferguson H. C. *et al.* (2004). Evolution in the colors of lyman break galaxies from $z \sim 4$ to $z \sim 3$, *ApJ* **600**, pp. L111–L114.

Peebles P. J. E. (1993). *Principles of Physical Cosmology*. Princeton University Press.

Petrosian V. (1976). Surface brightness and evolution of galaxies, *ApJ* **209**, pp. L1–L5.

Pozzetti L., Bolzonella M., Zucca E. *et al.* (2010). zCOSMOS — 10k-bright spectroscopic sample. The bimodality in the galaxy stellar mass function: Exploring its evolution with redshift, *A&A* **523**, A13.

Prochaska J. X., Gawiser E., Wolfe A. M. *et al.* (2003). The age-metallicity relation of the universe in neutral gas: The first 100 damped ly$\alpha$ systems, *ApJ* **595**, pp. L9–L12.

Prochaska J. X., Wolfe A. M. (2009). On the (non)evolution of h i gas in galaxies over cosmic time, *ApJ* **696**, pp. 1543–1547.

Ravikumar C. D., Puech M., Flores H. *et al.* (2007). New spectroscopic redshifts from the cdfs and a test of the cosmological relevance of the goods-south field, *A&A* **465**, pp. 1099–1108.

Reddy N. A., Erb D. K., Steidel C. C. *et al.* (2005). A census of optical and near-infrared selected star-forming and passively evolving galaxies at redshift $z \sim 2$, *ApJ* **633**, pp. 748–767.

Rodrigues M., Foster C., Taylor E. *et al.* (2016). Galaxy And Mass Assembly (GAMA): Improved emission lines measurements in four representative samples at 0.07 < z < 0.3, *A&A*, **590**, A18, doi:10.1051/0004-6361/201527836.

Rodrigues M., Puech M., Hammer F. *et al.* (2012). A decrease of the gas exchanges between galaxies and the intergalactic medium, from 12 to 6 billion years ago, *MNRAS* **421**, pp. 2888–2903.

Shapley A. E., Steidel C. C., Pettini M. *et al.* (2003). Rest-frame ultraviolet spectra of $z \sim 3$ lyman break galaxies, *ApJ* **588**, pp. 65–89.

Steidel C. C., Adelberger K. L., Shapley A. E. *et al.* (2003). Lyman break galaxies at redshift $z \sim 3$: Survey description and full data set, *ApJ* **592**, pp. 728–754.

Steidel C. C., Giavalisco M., Pettini M. *et al.* (1996). Spectroscopic confirmation of a population of normal star-forming galaxies at redshifts $z > 3$, *ApJ* **462**, p. L17.

Steidel C. C., Shapley A. E., Pettini M. *et al.* (2004). A survey of star-forming galaxies in the $1.4 \leq z \leq 2.5$ redshift desert: Overview, *ApJ* **604**, pp. 534–550.

Sterken, C., Manfroid, J. (eds.) (1992). *Astronomical Photometry, A Guide, Astrophysics and Space Science Library*, Springer Netherlands, ISBN 9780792316534, available at: https://books.google.fr/books?id=ojzepXhy9DIC.

Tanvir N. R., Fox D. B., Levan A. J. *et al.* (2009). A $\gamma$-ray burst at a redshift of $z \sim 8.2$, *Nature* **461**, pp. 1254–1257.

van Dokkum P. G., Förster Schreiber N. M., Franx M. *et al.* (2003). Spectroscopic confirmation of a substantial population of luminous red galaxies at redshifts $z \geq 2$, *ApJ* **587**, pp. L83–L87.

Vanzella E., Cristiani S., Dickinson M. *et al.* (2005). The great observatories origins deep survey. vlt/fors2 spectroscopy in the goods-south field, *A&A* **434**, pp. 53–65.

Verhamme A., Schaerer D., Atek H. *et al.* (2008). 3d ly$\alpha$ radiation transfer. iii. constraints on gas and stellar properties of $z \sim 3$ lyman break galaxies (lbg) and implications for high-z lbgs and ly$\alpha$ emitters, *A&A* **491**, pp. 89–111.

Willmer C. N. A., Faber S. M., Koo D. C. *et al.* (2006). The deep evolutionary exploratory probe 2 galaxy redshift survey: The galaxy luminosity function to $z \sim 1$, *ApJ* **647**, pp. 853–873.

Wirth G. D., Willmer C. N. A., Amico P. *et al.* (2004). The team keck treasury redshift survey of the GOODS-north field, *AJ* **127**, pp. 3121–3136.

# Chapter 2

# Imaging and Photometry

## 2.1 Introduction

One of the most basic data that can be obtained in astronomy are images and their related photometry and astrometry. Astronomical images were first obtained using photometric plates in the middle of the 19th century but astronomers now use digital images obtained with CCDs in the optical (see review by Mackay, 1986), and CMOS-like detector arrays in the near-infrared NIR (see review by Rieke, 2007). These new devices allow the accurate numerical subtraction of unwanted background signals from the detector and from the sky.

The first step of extra-galactic surveys is obtaining images from which accurate flux measurements can be obtained (see Sec. 1.2). Crucially, measuring the morphology of distant galaxies is possible using space based images. It is however essential to make these measurements at rest-frame wavelengths that probe the bulk of the stellar content of these objects and to also account for the surface brightness dimming effect, which can be large. For example, while the optical disk of the Milky Way can be detected up to $z = 0.4$–$0.5$ (redshifted $R_{25}$ with signal-to-noise (S/N) $= 4$ in 2 resolution elements) using HST/ACS images from GOODS (3 orbits), only Ultra Deep Field like exposures (120 orbits) are sufficiently long to reach the same depth at $z = 1.3$. Large space telescopes such as the JWST are therefore needed to study the morphology of distant galaxies.

Important properties of distant galaxies can be estimated using photometry obtained using images taken at different wavelengths, as summarized in Table 2.1. These estimates are however strongly dependent on the many assumptions made about the

Table 2.1: Physical quantities that can be inferred from imaging data.

| Quantity | Technique | Requirements and/or main limitations |
|---|---|---|
| Stellar Mass | Aperture photometry | Rest-frame red filter<br>+ optical $(B - V)$ color<br>Model dependant |
| Stellar populations<br>and Metallicity $Z$ | Aperture photometry<br>+ SED fitting | UV to IR<br>Age-$Z$-dust degeneracies |
| Extinction | Aperture photometry<br>+ IR/UV luminosity ratio | UV + IR<br>Global estimate only |
| Star Formation Rate (SFR) | Aperture photometry<br>+ IR-UV luminosities | UV + IR |
| Morphology | Structure extraction | At least one<br>rest-frame red filter |

The last column indicates how the considered technique can lead to a reasonable estimate of the associated physical quantity.

spectral energy distribution (SED) of these objects and are heavily impacted by the impossibility for photometry to disentangle effects due to stellar ages, metallicity, and dust extinction.

Imaging can also be used to morphologically identify galaxy interactions and mergers (e.g., van den Bergh, 2002; Delgado-Serrano *et al.*, 2010), although additional spatially-resolved kinematic measurements are often required to guarantee such identifications (Neichel *et al.*, 2008). Spectroscopic follow-up (see Chapter 3) of these galaxies, including three-dimensional (3D) spectroscopy (see Chapter 4), are required to get accurate estimates of most quantities of interest.

In this chapter, we list some of the fundamental properties of an image and how one can select an imager to achieve a given scientific program (Sec. 2.2).

The main steps in the data reduction process of images are presented in Sec. 2.3, while the basis of photometry is presented in Sec. 1.2. High resolution images and color maps are crucial to provide a visual representation of the objects, i.e., of their morphology, and therefore, of their nature (e.g., star, galaxy, AGN). These images are therefore a central ingredient of any morphological classification of galaxies (see Sec. 2.4). Aperture photometry and SEDs techniques are presented in Secs. 2.5 and 2.6, respectively.

## 2.2 Basic Elements of imaging and image analysis

Fundamentally, imaging can be understood through the three following properties of an image: its spatial resolution (see Sec. 2.2.1), its depth (see Sec. 2.2.2), and its spectral bandpass (see Sec. 2.2.3). These properties are the most important factors to consider when choosing an instrument to obtain useful images (see Sec. 2.2.4).

### 2.2.1 *Spatial resolution*

Spatial resolution refers to the smallest spatial detail that can be identified in an image. The spatial resolution of ground-based imagers are generally limited by atmospheric turbulence ("seeing"), unless an adaptive optics (AOs) system is used. Space-borne imagers are however generally limited by the overall optical quality of the instrument. Optical systems can be characterized by their PSF, i.e., the response of the system to a point-like source. The size of this PSF defines the spatial resolution of the image. The diameter of the telescope, the observational conditions (e.g., seeing), and the spatial and temporal stability of the PSF are important drivers of the final image quality. It is essential to characterize the resolution of an image before attempting to analyze spatial structures contained in that image. The image resolution can be empirically determined by measuring the size of bright (unsaturated) stars in the field (see Ex. *How to estimate the spatial resolution of an image*?).

**Example 2.1: *How to estimate the spatial resolution of an image*?**

Visually inspect the image and identify stars. Plot and check their profiles to discard saturated stars. You can for example use the *imexam* task within IRAF to interactively measure the FWHM of suitable stars to derive an average value for your image.

(*Continued*)

**Example 2.1: *Continued***

If you are working with ground-based images with no suitable stars, you can interpolate between seeing measurements $\Delta\theta$ that were performed, for example, at the beginning and at the end of the night.

Be careful about wavelength: the seeing is always defined at 500nm; First, convert to other wavelengths using $\Delta\theta \propto \lambda^{-1/5}$. Then, convert the seeing value into the actual image FWHM using Eq. (19) of Tokovinin (2002). Note that at any wavelength, the properties of atmospheric turbulence result in FWHM $\leq \Delta\theta$. For instance, a 0.83 arcsec seeing translates into a typical $\sim$0.67 arcsec FWHM at 500nm (Martinez *et al.*, 2010) at Cerro Paranal, where the ESO VLT is located.

If you are using HST images and cannot get stars with large enough S/N or because the FoV is too small, you can use *TinyTim*[1] to generate simulated PSFs.

### 2.2.1.1 *PSF shape*

In ground-based images obtained with seeing-limited conditions, the PSF is often assumed to be Gaussian in shape. This is a good approximation for the PSF core, but not for the PSF wings, which are significantly more extended (King, 1971). In relatively shallow images with a S/N that is not high enough for the PSF wings to be determined accurately, a simple Gaussian model can be used, while in deeper imaging surveys, a more accurate model is required. One such model is described using a Moffat function (Moffat, 1969; Trujillo *et al.*, 2001):

$$\mathrm{PSF}(r) = \frac{\beta - 1}{\pi\alpha^2}\left(1 + \left(\frac{r}{\alpha}\right)^2\right)^{-\beta},$$

with $\alpha = \mathrm{FWHM}/(2 \times \sqrt{2^{1/\beta} - 1})$, and $\beta = 4.765$.

[1] See `http://www.stsci.edu/hst/observatory/focus/TinyTim`.

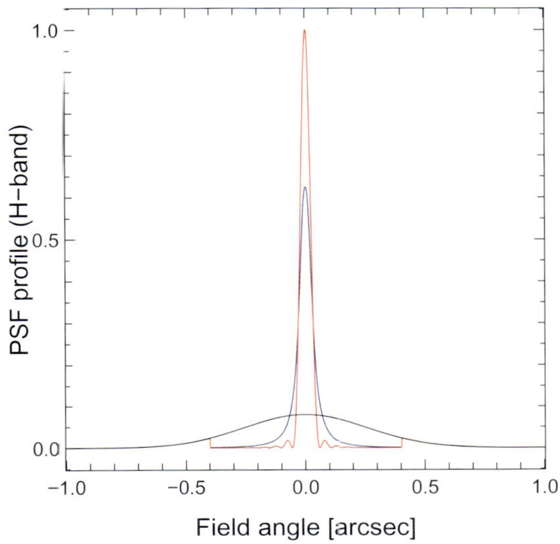


Fig. 2.1: Examples of different ground-based PSF profiles. The black line represents a simulated seeing-limited PSF in H band for an 8-m telescope with a 0.8 arcsec seeing (Neichel *et al.*, 2008). The red line represents the PSF obtained assuming correction by a simple Single-Conjugate Adaptive Optics (SCAO) system; note the diffraction-limited core with the first Airy rings appearing, superimposed to a residual uncorrected halo at radii larger than 0.4 arcsec. The blue line represents the PSF obtained with a more advanced AO system (Multi-Conjugate AO), whereby the correction is applied to a larger field of view but with the turbulence only partially corrected: note the extended wings compare to the SCAO case; the structure and amplitude of these wings can vary across the field-of-view, which can complicate accurate astrometry and photometry (see Sec. 2.2.4).

In images obtained using AO systems (see, e.g., Roddier, 1981), the PSF is partially (or totally) corrected for the effect of the atmospheric turbulence (see Fig. 2.1). As the correction improves, the PSF core converges towards the diffraction-limited shape which is described by an Airy function (with FWHM $\sim \lambda/D$). This core is superimposed to a halo with FWHM $\sim \lambda/r_0$, where $r_0$ is the Fried parameter of the turbulence (Fried, 1966). This parameter corresponds to the coherence length of the atmospheric turbulence, i.e., the length over which the turbulence remains significantly correlated.

Empirical measurements of the PSF should be preferred over simulations. However, in low S/N conditions and cosmological fields in which bright stars are rare, it can be difficult to get a reliable measurement (e.g., with space-borne telescope images of distant galaxies).

In such cases, the PSF, which depends on the exact characteristics of the instrument and detector, can be simulated using PSF simulation tools, like *TinyTim* in the case of the HST. Often, typical PSF can be downloaded from the website of space based missions. These can be useful to characterize images and test extraction methods.[2]

#### 2.2.1.2 *Resolution and sampling*

Spatial *resolution* must not be confused with spatial *sampling*. In modern numerical images, spatial *sampling* is imposed by the size of the detector pixel projected on sky (i.e., in arcsec). A very common mistake is to characterize the spatial *resolution* of an image as the detector pixel size. Indeed, the Nyquist–Shannon rule states that to be resolved, a spatial element of resolution must be sampled by more than two elements of sampling. The image is said to be oversampled when the PSF FWHM is sampled by more than two pixels. On the other hand, when the spatial resolution is limited by the pixel size, the resulting images are said to be under-sampled.

A classical example is the WFPC2 camera on-board the HST with 0.1 arcsec pixels, while the PSF size is 85% smaller at 4000Å (Holtzman, 1990). The new HST NIR WFC3 camera has 0.13 arcsec pixels for a ∼0.18 arcsec FWHM PSF. Under-sampling is often used to increase the size of the field of view. The sampling rate $p = \mathrm{FWHM}/\Delta_{\mathrm{pixels}}$, where $\Delta_{\mathrm{pixel}}$ is the pixel size of the detector, can be used to characterize the sampling rate of the PSF. Under-sampling is said to be critical in terms of photometric and astrometric precision when $p \leq 1.5$ (Howell *et al.*, 1996), while $p \geq 2.35$ guarantees a detection threshold that is equal to three times the background r.m.s. noise with a S/N = 5 on point-like sources (Harris, 1990).

[2]For instance, PSFs for the Spitzer IR satellite can be found at `http://irsa.ipac.caltech.edu/data/SPITZER/docs/irac/calibrationfiles/psfprf/`.

#### 2.2.1.3 *Improving the resolution: dithering vs. deconvolution*

The choice of the pixel size of an instrument is generally the result of a trade-off between sensitivity and resolution (Puech *et al.*, 2008). Under specific conditions, the spatial resolution can however be increased beyond that of the nominal resolution of the instrument.

*Image reconstruction* techniques rely on observational dithering, which consists in shifting the different exposures of the same field of view using small telescope offsets. Sub-pixel dithering can be used to increase the spatial resolution because part of the sub-pixel flux distribution can be recovered by interlacing the different exposures onto a finer grid (e.g., Lauer, 1999). The most widely used technique is drizzling (Fruchter and Hook, 2002; Fruchter, 2011; see also the *multidrizzle* package within IRAF,[3] or the AstroDrizzle package in Python), which was extensively used to re-map HST exposures onto a sub-pixel grid (e.g., ACS images in the GOODS field were drizzled down to 0.03 arcsec per pixel from a nominal 0.05 arcsec per detector pixel). Drizzling takes into account both the rotation between exposures and geometric distortions. Drizzling conserves astrometry and photometry but can introduce different amount of correlated noise, depending on the specific kernel that is used to distribute the flux onto the finer grid. Other algorithms using the Bayesian approach have been also developed (see, e.g., Marsh *et al.*, 2004 and references therein). In very large FoV imaging surveys, dithered exposures can be combined using dedicated softwares such as SWARP.[4]

*Image restauration* techniques are *a posteriori* methods that rely on being able to deconvolve the image and its associated PSF in order to improve the observed spatial resolution (Starck *et al.*, 2002; Puetter *et al.*, 2005) of the image. Deconvolution is more effective with point-like sources than with extended objects, and requires relatively high S/N ratios. These are not conditions that are generally present in distant galaxy images.

[3]See `http://www.stsci.edu/hst/HST_overview/documents/multidrizzle`.
[4]See `http://www.astromatic.net/software/swarp`.

### 2.2.2 *Depth*

Low S/N ratios (see definition in Sec. 2.3) are commonplace in observations of distant galaxies. In such a case, artifacts caused by statistical fluctuations of the background noise can easily be confused with faint surface brightness sources or galaxy sub-regions.

On the other hand, real but low surface objects can remain undetected because the exposures times are too short. Low surface brightness galaxies are obviously very difficult to detect at large distances: as of 2015, the most distant spectroscopically confirmed low surface brightness galaxy is only at $z = 0.53$ (Disseau *et al.*, 2016). At such a distance, the lowest surface brightness material starts to vanish below the background level of typical HST images (see Fig. 2.2). When analyzing images of distant galaxies, it is therefore crucial to properly characterize the sensitivity limits (i.e., depth) of these images.

The depth of an image is defined as the lowest flux level that can be reliably detected in an image. It is usually defined as a multiple of the background noise r.m.s. fluctuations in the image. Depth is parametrized as a limiting surface brightness in magnitudes per square arcsec for extended objects, or as magnitude (see Sec. 1.2) for point-source objects (see Ex. *What are the deepest images ever?*).

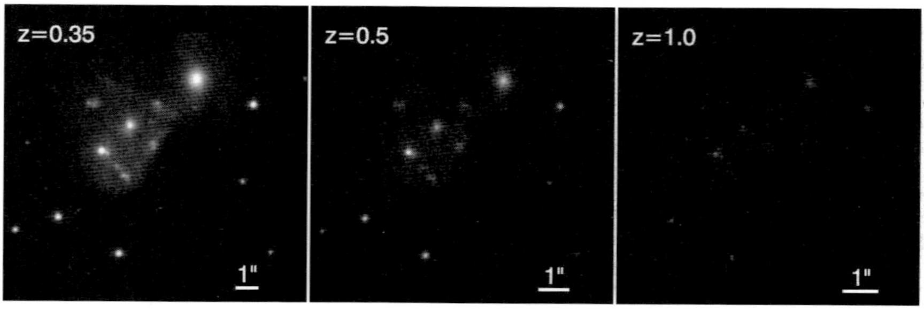


Fig. 2.2: Effect of the cosmological surface brightness dimming on the morphology of galaxies as a function of redshift (from Barden *et al.*, 2008, ©AAS, reproduced with permission). From left to right: HST/ACS F606W-F812W color images at $z = 0.35$, 0.5, and 1.0 simulated from SDSS images of local galaxies involved in a merging process. The physical scale is constant across the stamps.

Since the limiting magnitude for extended objects depends on the geometry of sources, depth is often only given for point sources.

Note that depth is a measure of the flux detection limit of a given image, which is different than the completeness of a photometric survey, which quantifies the probability that all sources brighter than a given limit are detected and extracted (see Sec. 2.2.2.1).

**Example 2.2: *What are the deepest images ever?***

As of 2013, the deepest "cosmological field" ever observed is the Hubble Ultra Deep Field (HUDF. The stacked HST image reach a depth of 29.45 (AB) mag for extended sources in a 0.2 $arcsec^2$ aperture both in $V$ band with S/N = 10 and 37.6 h of exposure time (HST/ACS camera, see Beckwith *et al.*, 2006) and in $H$ band with S/N = 5 and 19.9 h of exposure time (HST/WFC3 camera, see Koekemoer *et al.*, 2013).

2.2.2.1 *Detecting and analyzing galaxy structures*

Only galaxy structures detected at levels S/N $\geq$ 3 can be considered to be reliably detected,[5] whereas fainter structures cannot be considered to be statistically distinguished from noise fluctuations.

In principle, the detection threshold could be fixed in an absolute manner, i.e., using a constant surface brightness threshold in mag/$arcsec^2$. However, distant galaxies are often faint and well below the sky background (e.g., $z \sim 7$ source candidates have $J_{AB} \sim 26-28$, see Grazian *et al.*, 2012, whereas the sky continuum background[6] is $J_{AB} \sim 19$ mag/$arcsec^2$, see, e.g., Sullivan and Simcoe, 2012). NIR images are of particular interest to study distant galaxies because the NIR samples the stellar mass distribution of these objects, or

[5]This corresponds to a 99.7% confidence interval for Gaussian statistics. For Poisson distributions, this corresponds to slightly lower confidence intervals in the case of small counts.

[6]Note that the zodiacal light, which will be the main background contributor for space images obtained with the JWST, is ~3–4 mag fainter. Space images obtained with the HST in the same band have an extra contribution to the background of ~0.3 mag due to the earth-shine (which peaks in the optical).

some important spectral feature that has been redshifted, such as the Lyman Break (see Fig. 2.4). In the NIR, the sky background varies significantly over scales of a few arseconds and arcminutes (see also Sec. 3.3.3). It is therefore a better approach to use a *relative* detection threshold that is defined as a multiplicative factor of the *local* sky background standard deviation (generally at least $\sim$2–3$\sigma$). This does not strictly correspond to a constant isophotal level but it guarantees a given statistical significance level above local background noise fluctuations. In order to assemble a reliable source catalog, a good estimate of the background variations and noise properties is therefore essential (see Ex. *How to construct a source catalog from an image*?). When the noise is not constant across the image, constructing variance maps and using them as weights during the detecting and extraction process is more accurate.[7]

**Example 2.3: *How to construct a source catalog from an image*?**

The most widely used method for estimating and subtraction the sky background is to use SEXTRACTOR in combination with a weight image. SEXTRACTOR is a software that automatically detects, extracts, and measures the photometry and basic shape parameters of objects in images. SEXTRACTOR can be used in different ways depending on the nature of the background:

(1) If the background is relatively constant over the image (e.g., the sky background in optical images obtained using the HST/ACS), then a simple grid can be used to sample the background across the image. This can be done using the GLOBAL mode of SEXTRACTOR (see manual). The size of the grid, which depends on the image size and density of objects, should be carefully tested and one should try

(*Continued*)

[7]Weight maps can be constructed and/or combined using Weightwatcher, see `http://www.astromatic.net/software/weightwatcher`.

**Example 2.3: *Continued***

different grid sizes and compare the resulting background map with the original image. A grid that is too dense will be affected by the objects in the field (i.e., look for correlations between bright background regions and objects) and the noise. A grid that is too sparse will miss the background fluctuations on small spatial scales;

(2) If the background varies significantly over the image size (e.g., the sky/thermal background in the NIR or at larger wavelengths, even with space telescopes), then a better approach is to derive local estimates of the background around each detected source. This can be done using the LOCAL mode of SEXTRACTOR (see manual), which uses a rectangular annulus around each object to estimate the local background level. One should try different values of annulus thickness and check that there are no correlations between bright background regions and objects in the field. This mode can cause an overestimation of the background when objects have large extended faint wings (e.g., low surface brightness material surrounding galaxies).

Details are discussed in the SEXTRACTOR manual.[8] As noted there, when using this software, it is important to keep in mind that the SEXTRACTOR photometric uncertainties are underestimates since they do not include the dominant uncertainties associated with the local background estimate. The SEXTRACTOR photometric measurements should always been checked using another code such as, e.g., DAOPHOT. More detailed photometric measurements can be performed using semi-automatic tasks such as *daophot/polyphot* within IRAF.

(*Continued*)

[8]see also the wealth of information contained in the very useful SEXTRACTOR for Dummies guide available at `http://star-www.dur.ac.uk/~pdraper/extractor/Guide2source_extractor.pdf`.

**Example 2.3: *Continued***

This program allows the user to define polygon apertures of arbitrary shapes:

(1) Display the image of the object in DS9 using IRAF;
(2) Set up the photometric zero point using the Daophot/Sky routine (*skypar* and *datapar* parameters);
(3) You can then either use *polyphot* in interactive mode to define a region with the cursor into which the photometry will be measured, or use *polyphot* in a fully automatic mode, which requires an additional ASCII file listing the coordinates of the polygons;
(4) Finally, test the quality of the measured photometry using different regions for the sky and, e.g., the galaxy as a whole if you are doing aperture photometry.

More advanced IRAF programs like *ellipse* allow one to perform photometry of an extended source using concentric isophotes which can then be used to measure the photometric profile of a resolved galaxy.

Very distant galaxies are intrinsically faint and small objects (e.g., the expected $R_{\rm half}$ is $\sim$100–150 mas at $z \sim 6-7$, Grazian *et al.*, 2012), that can appear to be point sources in poorly resolved images. The detection of such an object is determined by the integrated S/N of the source (at least 3-5 per element of resolution[9]), and the relative size of the object compared to the image PSF. Such unresolved objects can be identified above a given detection threshold by verifying that their light profile are consistent with the the image PSF.

For the cases of very low S/N, simulations are needed to check whether a given structure is above the detection threshold, and to determine a reliable confidence interval for the detection. This is

[9]The S/N should always be specified by element of resolution, which is the observable quantity, rather than per element of sampling.

particularly necessary in the IR, where source confusion adds a significant source of noise (see Sec. 2.5.2.3).

The statistical significance of a source detection can be highly improved at low S/N by analyzing images of the same field taken using different band-passes. Since one is testing the null hypothesis that there is no object in a given pixel in all bands, the individual probability of detection in each band can be considered to be independent and these can be directly combined. This approach makes it significantly easier to detect faint galaxies than by simply stacking all the individual images into a single combined image (Szalay *et al.*, 1999; see example in Fig. 2.3). This is of particular interest when detecting faint and distant sources such as Lyman Break Galaxies (see, e.g., Bouwens *et al.*, 2015).

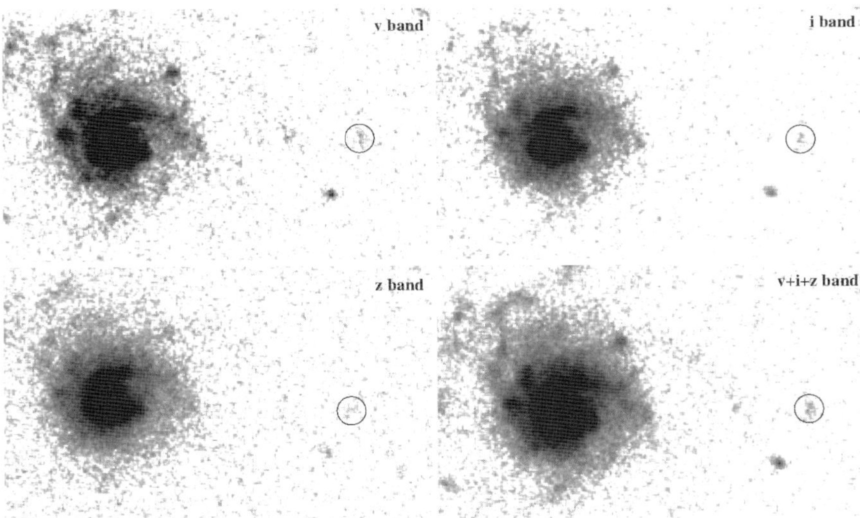


Fig. 2.3: Example of how the detection threshold of sources can be improved using several images in different bands. *From left to right and top to bottom*: HST/ACS image of J033214.97-275005.5 at $z = 0.66$ and its outskirt region in v, i, z, and $\mathrm{i+v+z}$ band (log stretch). The very faint encircled object corresponds to very low signals in the individual images (corresponding to only 2, 2, and $1.5\sigma$ above the corresponding backgrounds). If the images were to be interpreted individually, it would be impossible to consider the detection of this object as significant. Simply stacking them into a single combined $\mathrm{v+i+z}$ image provides an improved detection threshold $\gtrsim 3$. Considering the three images as independent measurements results in an improved detection threshold $\sim 4.4\sigma$ (as computed from a $\chi^2$ probability distribution with 3 degrees of freedom).

#### 2.2.2.2 *Image processing techniques*

Given a set of images, image processing techniques can be useful tools to improve the S/N ratio prior to analysis. The most widely used technique is spatial smoothing using a simple two-dimensional (2D) Gaussian. This is a particularly useful approach with oversampled images, where smoothing does not sacrifice spatial resolution. Alternatively, spatial resolution can be sacrificed to increase S/N by re-binning the image, helping with the detection of extended sources.

Finally, adaptive binning techniques using variable bin sizes across the image in order to reach a specified S/N per final bin have been developed (Cappellari and Copin, 2003; Diehl and Statler, 2006). This approach is preferred to noise filtering techniques (e.g., Starck and Murtagh, 2006), since it avoids interpolating the data.

### 2.2.3 ***Spectral Bandwidth***

#### 2.2.3.1 *Photometric filters*

Photometric filters can be separated into different classes depending on the width of their bandpass. Narrow-band filters can be used to increase the contrast between the continuum and an emission line (such as H$\alpha$ or [O II]) to detect star-forming galaxies. The Lyman $\alpha$ emission line, for example, has been widely used to detect very high-$z$ galaxies, called Ly$\alpha$ emitters (LAEs, see, e.g., bottom panel of Fig. 2.4). Broad-band filters integrate the galaxy flux over a larger range of wavelength (see top panel of Fig. 2.4). A widely used application of broad-band filters is to detect dropouts in the rest-frame UV continuum flux in order to identify Lyman Break Galaxies at high redshifts (LBGs, see Sec. 1.6; see also an example below in Fig. 2.18). While broad-band filters provide a good approximation of the overall continuum flux, they can be significant affected by emission lines (from a few percents to up to $\sim$60%, see Zackrisson *et al.*, 2008) and this effect becomes more common in high redshift, star forming galaxies.

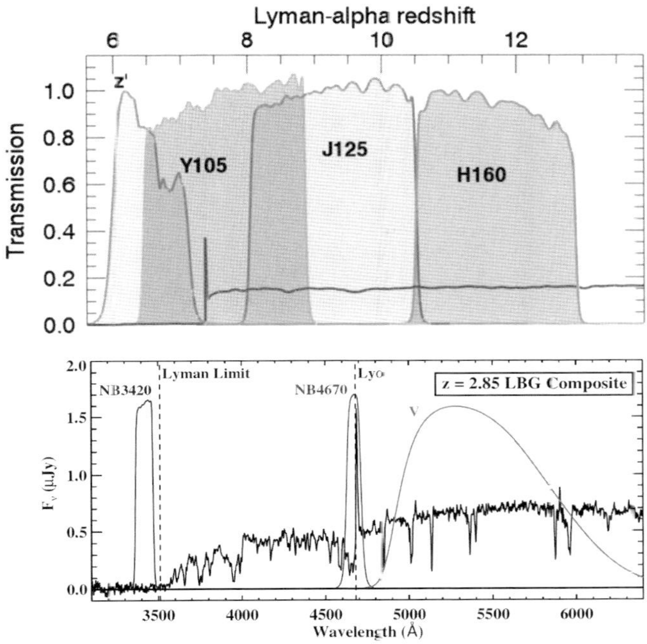


Fig. 2.4: Transmission curves of broad and narrow-band filters used for the detection of very high redshift sources. *Top panel:* Broad-band Subaru $z'$, HST/WFC3 $Y_{150}$, $J_{125}$, and $H_{160}$ filters used in the FourStar Galaxy Evolution survey as a function of wavelength (from Tilvi *et al.*, 2013, ©AAS, reproduced with permission). The black line shows a template SED of a $z = 7.4$ galaxy with Ly$\alpha$ in emission (note also the break on the continuum between the two sides of the line). The top axis indicates the redshift of the Ly$\alpha$ emission line as a function of wavelength. *Bottom panel*: Keck/LRIS broad-band ($V$) and narrow-band NB3420 and NB4670 filters used by Mostardi *et al.* (2013). The black line show an LBG template redshifted to $z = 2.85$. The NB4670 filter is centered on the Ly$\alpha$ emission line for objects at $z \sim 2.85$ and was used together with the $V$ band to select a sample of LAEs (©AAS, reproduced with permission).

Independent measurements using narrow-band filters or spectroscopy are eventually needed to get accurate continuum flux estimates (e.g., Waller, 1990).

There are many photometry filters (e.g., Bessell, 2005) and these are associated with numerous and distinct star systems that are used to calibrate photometric measurements.

The most important characteristics of a photometric filter are its spectral width, its central wavelength, and its transmission curve. During the past years, significant effort was made to collect these characteristics for a wide range of filters.[10] Transformations between the most used sets of filters are available in, e.g., Fukugita *et al.* (1996), Smith *et al.* (2002), Carpenter (2001), Bessell (2005), or Bilir *et al.* (2008).

#### 2.2.3.2 *Rest-frame photometry*

Measurements of galaxy flux and morphology are affected by the observed spectral bandwidth of the observations (see Secs. 2.6 and 2.4).

This is due to the SED, the distribution of flux as a function of wavelength, see Sec. 2.6) potentially having significant variations.

Blue and UV filters (i.e., wavelengths shorter than the 4000Å break) are dominated by main sequence OB stars (see top-left panels of Fig. 2.5). These filters are therefore sensitive to recent (i.e., <10–100 Myr) star formation episodes. This UV light distribution is strongly affected by dust, which absorbs light and re-emits it in the IR (see also Sec. 2.6). As a result, galaxy morphology should never be characterized using only (rest-frame) UV filters. This is a serious drawback to morphological studies of very distant galaxies since the rest-frame UV light is progressively shifted into the optical and NIR with redshift. However, the WFC3 camera on-board the HST can now provide well-resolved rest-frame optical images for galaxies up to $z \sim 3$.

Red (optical) filters (i.e., wavelengths longer than the 4000Å break) are dominated by main-sequence A-to-M and RGB stars, while NIR filters (i.e., $\geq 1$ $\mu$m) are dominated by post-main sequence stars like RGB and AGB stars (see bottom-left panel of Fig. 2.5). The fraction of NIR light due to main sequence stars (which represent most of the stellar mass) is approximately constant at all ages and is not very sensitive to dust. Therefore, and although NIR filters do not capture the light from the stellar populations that account for the largest fraction of the stellar mass, they are often considered to

[10]See `http://voservices.net/filter` or `http://ulisse.pd.astro.it/Astro/ADPS/enter2.html`.

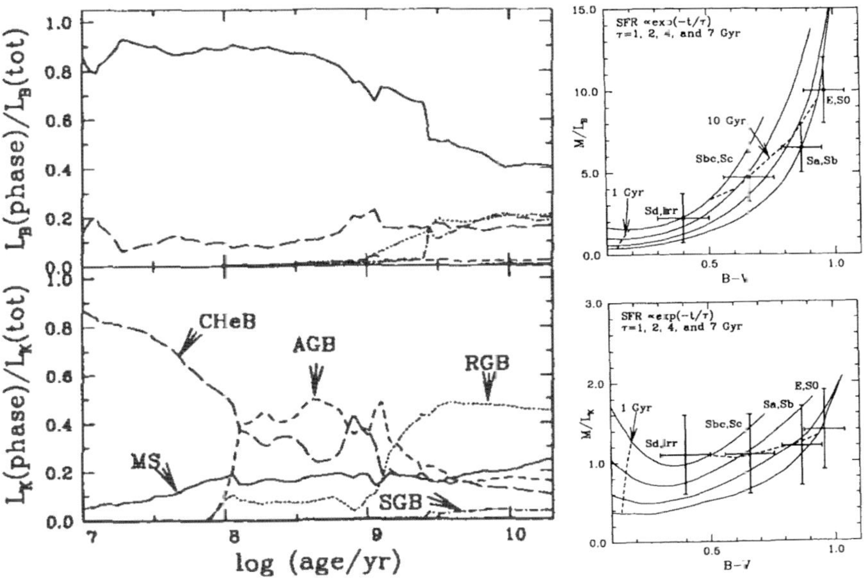


Fig. 2.5: Impact of stellar populations on the luminosities and mass-to-light ratios of galaxies (adapted from Figs. 6, 9, and 10 of Charlot (1996), reproduced with kind permission from Springer Science + Business Media) *Left panels*: Fractional contributions of stellar populations to the B (top) and K (bottom) luminosity assuming an instantaneous burst, a solar metallicity, and a Salpeter IMF. MS, SGB, RGB, CHeB, AGB stand for main sequence, sub-giant, red-giant, core-He burning phase, and asymptotic giant stars, respectively. *Right panels*: $M/L$ vs. $B-V$ in B (top) and K (bottom) bands for exponentially decreasing star formation. Time-scales of star formation are indicated on the upper left of each panels and correspond to solid lines from top to bottom (the infinite time-scale correspond to a constant star formation). Note the strong variation of the $M/L$ in the B band and the much smaller variation in the K band within a vertical scale that is five times smaller than the former one. The dashed lines show the model values at 1 and 10 Gyr, respectively.

be the best tracers of the stellar mass distribution (e.g., Kuchinski *et al.*, 2000; see right panels of Fig. 2.5 and Sec. 2.6).

When comparing different samples of galaxies taken at different redshifts, one should always compare two filters which correspond as much as possible to the same rest-frame wavelengths, otherwise the comparison might turn out to be meaningless. For instance, SDSS and HST/ACS images are a good match in terms of bandpass and spatial resolution to compare $z \sim 0$ and $z \sim 0.65$ objects (see Table 2.2 and Fig. 1.9).

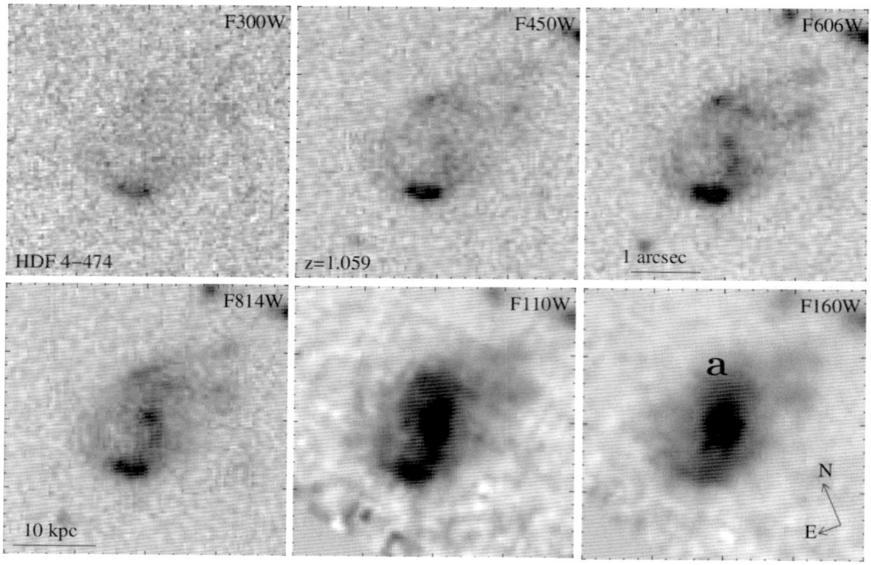


Fig. 2.6: Comparison of the morphology of a $z \sim 1$ galaxy as a function of wavelength from Bunker *et al.* (2000) reproduced with kind permission. *From left to right and top to bottom*: HST images at ~300, 450, 606, 814, 1100, and 1600nm. In the optical (rest-frame UV), the morphology is dominated by a star-forming knot, while the extended nature of the object is revealed in the NIR (rest-frame red optical). This figure also illustrates how the PSF increases with wavelength for a diffraction limited telescope.

Table 2.2: Comparison between central wavelengths of SDSS and GOODS filters.

| **Filters** | — | **u band** | **g band** | **r band** | **i band** | **z band** |
|---|---|---|---|---|---|---|
| **SDSS** | — | 3551Å | 4686Å | 6165Å | 7481Å | 8931Å |
| — | **b band** | **v band** | **i band** | **z band** | — | — |
| **HST/ACS** | 4313Å | 5915Å | 7697Å | 9103Å | — | — |
| **Rest-frame** | 2582Å | 3542Å | 4609Å | 5451Å | — | — |

A redshift of $z = 0.65$ was assumed for calculating the rest-frame wavelengths. Both sets of images have also comparable resolution (0.74 kpc at $z = 0.025$ with SDSS images vs. 0.81 kpc at $z = 0.65$ with ACS) and depth (within ~1 mag), see (Delgado-Serrano *et al.*, 2010) for details.

### 2.2.4 *Observing with imagers*

#### 2.2.4.1 *Space vs. ground-based imagers*

Imagers can be classified into two broad types: space-based and ground-based imagers, which provide very different depths, resolutions, and spectral bandwidths. Therefore, choosing a particular instrument strongly depends on the desired scientific objective.

Space-based telescopes have two major advantages compared to ground-based telescopes:

- Given a constant aperture, they reach much larger depths than ground-based telescopes, since they are not affected by sky emissions. The background light in space images is limited to the zodiacal light, which is roughly one to two orders of magnitude fainter that the sky background continuum in H band (and almost three order of magnitudes smaller if ones includes OH sky lines, see Leinert *et al.*, 1998; Noll *et al.*, 2012);
- They deliver seeing-free images, which means that their PSF is limited by the optical system, and not by atmospheric turbulence. Therefore, space-based images generally achieve a much higher spatial resolution that ground-based images (a factor ∼10 smaller).

These two reasons are why detecting and then measuring morphological properties of distant galaxies should always be done using space-based images, since they deliver incomparable depths and resolutions. Additionally, deep space imaging is extensively used out of the optical/NIR window, where atmospheric absorptions and/or emissions prevents efficient photometry from the ground.

Ground-based telescopes, with their larger apertures are however an attractive way to conduct large photometric surveys. A useful list of photometric surveys from ground-based telescopes can be found on the web.[11]

[11] See `http://home.strw.leidenuniv.nl/~jarle/Surveys/DeepFields/index.html`.

#### 2.2.4.2 *Dithering*

Dithering has become a common practice to produce deep images in the NIR as well as in the optical. Besides being helpful to improve the spatial resolution of images (see Sec. 2.2.1), dithering also helps mitigating the impact of detector defects such as bad pixels or dead columns, since a given region on the sky is imaged on different areas of the detector, as well as to fill possible gaps between chips.

Dithering is also used to reduce the effective noise of a set of observations (see below), which can be essential when observing very faint sources below the background level, particularly in the NIR (see, e.g., Vaduvescu and McCall, 2004 for practical considerations on the dithering spatial pattern and time sampling).

#### 2.2.4.3 *Observing with AO*

AO systems allow ground-based imaging (Davies and Kasper, 2012) to produce images with improved spatial resolution compared to seeing-limited images, especially in the NIR where they are most efficient.

However, even in the NIR, AO images can suffer from two important limitations: (1) photometry accuracy can be strongly limited by the variability of the PSF across the FoV; (2) morphological studies of extended and faint sources can be limited by the shape of the AO-corrected PSF, which consists in a central diffraction-limited core superimposed on an extended halo (see Fig. 2.1). At low S/N conditions such as those encountered when imaging distant galaxies, it becomes particularly difficult to estimate the exact shape of the PSF wings. The impact of these effects on the morphological decomposition of faint and extended objects (see Sec. 2.4) remains largely unclear and will likely be a major challenge of the future Extremely Large Telescopes.

#### 2.2.4.4 *Observing in the IR with ground-based telescopes*

At wavelengths larger than 1 $\mu$m, the S/N of images is dominated by the high level and by the variability of the sky background. Additionally, the level of atmospheric transmission is strongly affected by variations in water vapor (see Fig. 3.10). To produce accurate photometric measurements in the IR, it is therefore essential to monitor the airmass and the water vapor content of the atmosphere during the entire course of the observations.

In the NIR, it can be useful to use *quadrant jittering observations*, which consists in shifting the target at the center of each image quadrant in a row. Each other three quadrants contain empty sky, which can be used to subtract the background contribution in each image. When the target size is too large to do so, one has to dedicate 50% of the observing time to sample nearby empty sky regions using offset sky positions.

Above $\sim$2 $\mu$m, thermal emission from the telescope is the dominant source of background and limits the S/N. In this case, the exact same optical path has to be used when observing the target and the sky using *chopping and nodding observations*.[12]

While *chopping*, the telescope secondary is offset so that a nearby empty sky region is observed instead of the target and the resulting two frames, respectively called the $ON$ and $OFF$ positions, are subtracted.

Since the optical path is not strictly the same between the two positions, the subtraction leaves a residual, which can further be corrected using *nodding*. *Nodding* consists in moving the whole telescope so that the position of the target switches between the two nod positions. The two resulting pairs of subtracted images between the two nod positions can then be used to correct for the residual background signal.[13] Either way, sky images need to have the same S/N as the science images in order to avoid noise propagation during the sky subtraction step.

## 2.3 Data Reduction

### 2.3.1 *Basic principles*

Data reduction consists in removing systematic effects caused by imperfections in the observing instrument, which can take several

[12]See, e.g., `http://www.gemini.edu/sciops/instruments/midir-resources?q=node/10139#chopping`.
[13]See more practical considerations at `http://www.gemini.edu/sciops/instruments/mid-ir-resources/mid-ir-observing`.

forms: additive biases (e.g., detector pre-charge offsets, scattered light, fringing), multiplicative transmission effects (e.g., absorption due to dust on optical surfaces, pixel-to-pixel variations of the quantum efficiency [QE] in the detector, vignetting), local detector deflects (broken column on the CCD, hot/dark pixels, or transient events like cosmic rays.

For these effects to be measured and corrected, calibration frames such as bias, dark, and flat-field (FF) frames (see description below) are required. Since all of these frames are affected by Poisson and readout noises, it is necessary to average several individual raw calibration frames[14] to reduce the impact of random noise. The main difficulty during the data reduction process is then twofold: (1) separate additive from multiplicative effects to correct them in the right order and reduce systematic errors efficiently (e.g., Gilliland, 1992), and (2) average a large enough number of raw calibration frames when constructing the average frames (Newberry, 1991). In practice, all average frames are affected by residual noise, which propagates onto the science frames during the reduction process. Dithering the science frames is a method to average out this residual calibration noise.

When an instrument is stable enough, one can average frames from previous nights, or over an even larger period of time, to reduce the random noise further. Modern observatories have quality check procedures in place to monitor the stability of their instruments. A good first step in data reduction is to check whether such "static calibrations" can lead to an accurate enough reduction process.

This can be done by dividing the static calibration frames by daily average frames and by looking for spatial and temporal variations. Stability is generally easier to achieve for instruments installed at a Nasmyth focus than at a Cassegrain focus, where mechanical flexures can prevent stability beyond a few minutes. Note that every time the instrument setup is changed (e.g., new filter), one has to make sure that the corresponding calibration frames are taken or are available.

[14]The resulting average frame is sometimes referred to as a "master frame".

### 2.3.2 *Description of basic calibration frames*

The minimal set of calibration frames consists of bias/dark and FF frames. Bias frames measure the pre-charge offset current used on the detector to prevent false zeros. Dark frames measure, in addition to the bias, the dark current resulting from thermal emission by the detector. With liquid nitrogen cooled camera, the dark current is negligible in the optical. Otherwise, and especially in the NIR, dark frames have to be taken. Bias frames are obtained using zero second exposures with the shutter closed. Dark frames differ in that they must have the same exposure time as the scientific frames.[15]

Correcting pixel-to-pixel variations and larger-scale illumination variations requires FF frames. A difficulty in obtaining FF frames is generally related to having an illumination that is uniform over the whole detector surface.

Another difficulty is that pixel-to-pixel variations are usually dominated by QE inhomogeneity on the detector. Since QE can vary as a function of wavelength, one must be very careful about the differential response between the scientific target and the source used to produce FF frames. This results in the FF correction to generally be the limiting factor in broad-band photometry. There are two main flavors of FF frames:

- Dome/Screen FF: the telescope is pointed towards a region of the dome or a screen inside the dome, which is used to scatter the light from a lamp. Such FFs can be done during day-time but it can be hard to avoid shadows, as well as spatial and spectral variations. These FFs are the most efficient in term of S/N since the illuminating source can be very bright, and such FFs are therefore widely used to correct pixel-to-pixel variations;
- On-sky FF: a blank field is used during twilight (preferentially in the direction opposed to the sun), which implies that the light distribution can change rapidly. By construction, the spectral

[15]Although dark frames can often be taken with shorter integration times and rescaled to the proper exposure time since they are expected to scale linearly with the square root of the exposure time (which should be carefully tested before being attempted).

response is closer to the observational conditions, but twilight time can be too short for several filters to be calibrated (Tyson and Gal, 1993). Bright, individual stars are filtered out and the remaining blank regions are used to produce a FF frame. To reach a better accuracy, attached night sky FF (i.e., FF frames taken close to the scientific field just before and/or after the scientific exposures and produced after defocusing of the telescope) can be used. They are by construction ideal in term of spectral response, and show residual spatial gradients of only $\sim 1\%$ (Chromey and Hasselbacher, 1996).[16] Sky FF are mandatory when fringing or scattered light need to be corrected (see below).

Bias (or dark) frames depend mainly on the temperature of the detector. As long as no update are made to the detector and the temperature environment is well-controlled, using a *static* average bias, i.e., an average bias frame from a different night (or averaged over several nights), is often accurate enough. Conversely to bias frames, FF frames are more easily affected by short-term variations such as dust or telescope flexures, which implies that using static FF correction should generally be controlled carefully. Professional instruments are now stable enough so that FF frames taken at the beginning and the end of the night (during day-time) allows one to do an accurate enough reduction process. A given set of FF can usually be used over a few days or even months around the observations date if no fringing or scattered light corrections are required.

### 2.3.3 *Reducing and calibrating images*

The data reduction process can be summarized as follows (see Ex. *How to reduce an images?*):

$$S = \frac{R - D}{F - D}, \tag{2.1}$$

where $S$ is the reduced image, $R$ is an individual raw scientific frame, $F$ is the average Flat, and $D$ is the average Dark (or Bias when the

[16] An average night sky FF can also be constructed using the dithered scientific field directly (see, e.g., Tyson and Seitzer, 1988; Tyson, 1990). Such a "supersky" image is possible only in low density regions of the sky and requires a large number of individual images, which is often difficult to obtain in practice.

dark current can be neglected, see above). Note that the FF needs to be normalized by its average value for flux to be conserved in Eq. (2.1).

Each arithmetic operation in Eq. (2.1) results in an increase of random noise from the calibration average frames into the raw scientific image. A very large number of individual raw calibration frames should be used to limit the noise propagation, in particular in the case of the FF (Newberry, 1991), for which one should use a number of raw frames at least one order of magnitude larger that the number of individual scientific frames (Tyson, 1990). Any frame with deviant statistics (such as deviant mean and r.m.s. values when compared to the other frames) should be discarded. A median or $\sigma$-clipping combination should be preferred in order to better eliminate cosmic rays. Finally, the individually reduced scientific images must be normalized by the exposure time before they are combined, i.e., in units of $e^-/s$.

Once reduced, the scientific images must also to be corrected for optical instrumental distortion. Most reduction pipelines now integrate automatic procedures to do this, using frames that measure the optical response of the instrument to a matrix of pinhole light sources. After being geometrically corrected, the images are ready to be astrometrically calibrated, which can be done relatively to reference catalogs like the USNO, GSC, or UCAC catalogs. These catalogs provide a large enough number of stars with accurate positions, i.e., typically better than a few hundreds of milli-arcsec (a few tens of mas for the most accurate ones). These reference positions are generally used to fit a polynomial, which describes the shift, rotation, and distortion of the image. The accuracy of the astrometric solution is limited by that of the catalog used to construct it. Accuracies of one tenth of a pixel is generally reached using such a procedure. A useful software for setting up relative astrometric (and photometric) solutions is SCAMP.[17] Further details about astrometric calibration can be found in the SCAMP manual.

Photometric calibration can be done relatively to standard stars,[18] that already have well-calibrated magnitudes $m_{\mathrm{standards}}$.

[17] See `http://www.astromatic.net/software/scamp`.

[18] Which requires images of the standard stars to be reduced in the same manner as the scientific frames.

The integrated counts $N_{\text{standards}}$ of such stars can be measured in images to be calibrated using PSF-fitting techniques (see Sec. 2.5.2.3). The photometric zero point $ZP$ of the image (see definition in Sec. 1.2) can then be derived with $ZP = m_{\text{standards}} + 2.5 \log (N_{\text{standards}}) - k \cdot Z$, where $k$ is the atmospheric extinction coefficient (which depends on airmass, wavelength, and altitude), and $Z$ is the airmass.

While average extinction coefficients $k$ are generally provided by observatories, more accurate photometric calibration requires specific measurements of the atmospheric extinction (Sterken and Manfroid, 1992). The latter require a standard to be observed immediately before and/or after scientific exposures and must be taken in similar conditions (i.e., same airmass).

**Example 2.4: *How to reduce an image?***

All the operations described here can be conducted within IRAF using tasks provided in the *noao:imred:ccdred* package. The most relevant tasks are *ccdproc*, *imcombine*, *imstat*, *imarith*, *implot*, *imexam*, *zerocombine*, and *flatcombine*, amongst many others. We refer to their respective manuals for further detail.[19]

(1) Inspect all individual bias frames and discard invalid frames. To do this, look for any frame with strange mean and/or r.m.s. value when compared to other frames. Trim the images to remove the pre-scan and over-scan regions. Combine them using a median or a $\sigma$-clipping average to construct an average-bias for each night. If a static average bias is available from the observatory, compare with your own-made average bias and check for differences. Compare the average-bias from one night to another. If there are no significant variations, construct an average-bias with all frames, regardless of the date, or use the static average frame from the observatory;

(*Continued*)

[19]See iraf.noao.edu/iraf/docs/ccdred.ps.Z.

**Example 2.4: *Continued***

(2) Do the same for dark frames. Check whether dark correction is needed (i.e., whether the mean value of the average dark frame is significantly larger than the mean of the average bias frame);

(3) Do the same for FF frames. Depending on the previous step, correct for bias or dark contributions to produce an average FF frame and normalize it by its median value;

(4) Reduce all individual science frames one at a time: subtract the average bias (or dark if needed) to each raw frame and divide by the de-biased average flat (preferentially a night sky FF). Make sure you always use the same filter;

(5) Normalize the individual frames by the exposure time to have units of ADU/s. Convert ADU in electrons using the detector gain $G$ value;

(6) Combine the individual reduced frames to produce the final average field frame. First correct for misalignments (using IRAF tasks to measure the position of stars in the field and the *imalign* task to shift the images). Use interpolation between frames to correct bad pixels and columns. If the number of frames is low (i.e., $\leq 3$), image processing procedures such as *lacosmic* task within IRAF can be used to eliminate cosmic rays. When dithered exposures are available, you can use software such as SWARP to produce the final mosaic.

#### 2.3.3.1 *Fringing and scattered light*

The procedure described above can at best flatten images to a level of ~1% (Chromey and Hasselbacher, 1996) and typically results in a photometric accuracy of ~5–10%. If one requires a better accuracy for, e.g., detecting very faint surface-brightness features, a few tenth of percents in flatness can be needed. At these levels, residual scattered light, fringing, or flat-field effects can be the limiting factors (e.g., Andersen *et al.*, 1995; Grundahl and Sorensen, 1996; Moehler *et al.*, 2010).

When light is close to monochromatic (e.g., because a narrow-band filter is used or there are strong OH sky lines in the filter passband), fringing, in particular at red wavelengths, caused by interferences between multiple reflections of light within thin detectors, can be present. This limits the flatness of the reduced images (Howell, 2012). A fringing pattern is generally spatially stable but varies in amplitude as a function of night sky brightness. While effects due to flat-fielding must be corrected multiplicatively, fringing, as well as sources of scattered light in the instrument, should be corrected additionally (Tyson and Seitzer, 1988).

All of the required corrections can be derived using a *supersky* frame, which can be obtained from dithered sky FF frames (see Footnote 16). The resulting average image in the detector frame (e.g., pixel-to-pixel) only contains features stable in time, with fringes corresponding to the smallest scales, and FF residuals to the largest ones. The two can be separated by smoothing the *supersky* frame using kernels of different sizes. Using a large kernel will isolate the largest scale gradients (using, e.g., the *mkskycor* task within IRAF) by producing a *superflat* frame, which can be used as an *illumination correction* and reduce the multiplicative residuals. The difference between the supersky and the superflat frames produces a *background model* that contains the fringe pattern, which can be used to correct additive effects provided it is scaled to the proper exposure time.

The ultimate limit in flatness is due to scintillation effects, which limit the photometric accuracy to a few tenths of percent on a 10 meter class telescope (Kornilov, 2012). At $\lambda > 1.6$ $\mu$m, subtracting the thermal background signal will also place an additional limit to the flatness of the final image.

### 2.3.4 *S/N ratio*

In a reduced image with a given integration time $\Delta t$, the S/N per spatial $(x, y)$ and spectral $(\lambda)$ pixel is given by the "full CCD equation"

(Howell, 2006; Merline and Howell, 1995)[20]:

$$\mathrm{S/N} = \frac{O.\Delta t}{\sqrt{O.\Delta t + B.\Delta t + \mathrm{RON}_{\mathrm{eff}}^2.ndit + G^2.\sigma_q^2 + D.\Delta t + \sigma_f^2}}, \tag{2.2}$$

where $O$ and $B$ are the object and background counts (in $e^-$) per individual exposure, respectively, $\mathrm{RON}_{\mathrm{eff}}$ is the effective detector read-out noise (in $\mathrm{e}^-$/pix, see below), $G$ is the gain in units of $\mathrm{e}^-$/ADU (Analogue-to-Digital Unit), $\sigma_q$ ~0.5 $\mathrm{e}^-$/ADU is the $1-\sigma$ error introduced by the A/D converter (i.e., the quantification or digitization error, see Howell, 2006), $D$ is the dark current (in $\mathrm{e}^-$/s/pix), $\sigma_f$ is the residual error from the flat-fielding, and $ndit$ is the number of exposures with individual $dit$ integration time so that $\Delta t = \mathrm{dit} \times \mathrm{ndit}$. Note that this equation neglects the impact of the uncertainty in estimating the background signal $B$.

There is a fundamental difference between optical and NIR detectors. With NIR detectors, charges are read pixel-by-pixel, while in optical CCDs, charges are read after being transferred to the side of the detector. This results in the *effective* $\mathrm{RON}_{\mathrm{eff}}$ in CCDs to actually be driven by the *Charge Transfer Efficiency* (CTE), while in the NIR, the effective $\mathrm{RON}_{\mathrm{eff}}$ is simply the individual pixel $\mathrm{RON}_{\mathrm{pix}}$. Typical individual pixel $\mathrm{RON}_{\mathrm{pix}}$ values are a few electrons per pixel in the optical, with slightly larger values in the NIR (Rieke, 2007).

In CCD, the effective $\mathrm{RON}_{\mathrm{eff}}$ per pixel can be estimated using:

$$\mathrm{RON}_{\mathrm{eff}} = \sqrt{\mathrm{RON}_{\mathrm{pix}}^2 + \mathrm{RON}_{\mathrm{CTE}}^2} \sim \mathrm{RON}_{\mathrm{CTE}}, \tag{2.3}$$

$$\mathrm{RON}_{\mathrm{CTE}} = \sqrt{2 * (1 - \mathrm{CTE}^p) \times N_\gamma}, \tag{2.4}$$

where $p$ is typically one fourth of the total number of pixels on the CCD, CTE is the CCD CTE, and $N_\gamma$ is the number of photons in a given pixel. A typical value of CTE is 0.99999 and typical astronomical devices often reach 0.999999. With typical $4k \times 4k$ detectors, this results in a $\mathrm{RON}_{\mathrm{eff}} \sim 1.4N_\gamma$. This is usually the dominant source of noise at visible wavelengths (i.e., between ~400 nm and

[20]In the following equation, the $(x, y, \lambda)$ variables have been dropped for simplicity.

700 nm), where the background is generally dominated by photon noise $\sim \sqrt{N_\gamma}$.

When the data reduction is accurate enough, Eq. (2.2) reduces to:

$$\mathrm{S/N}(x,y,\lambda) = \frac{O(x,y,\lambda).\Delta t}{\sqrt{O(x,y,\lambda).\Delta t + B(x,y,\lambda).\Delta t + \mathrm{RON}^2_{\mathrm{eff}}.ndit}}. \tag{2.5}$$

With optical ground-based telescopes, the main contributors to the background light $B$ are the sky background during dark time and the moonlight during bright time.[21] Faint targets are therefore usually observed during dark time (new moon $\pm 5$ days). This is not required in the NIR because there is a sharp decrease in moon flux with wavelength. Beyond $\sim$ 720nm, the main contributors are emission sky lines, in particular airglow (i.e., OH sky lines). Above $\sim$1.6 $\mu$m, thermal emission from the telescope and/or instrument limit the achievable S/N. The total background level typically increases monotonically with wavelength. For space telescopes, the S/N is usually limited by the zodiacal light, at any wavelength, except in the far IR (i.e., above 20 $\mu$m), where, at lower Galactic latitudes the galactic, background light can become the significant contributor (Leinert *et al.*, 1998).

#### 2.3.4.1 *S/N regimes*

In optical wavelengths, observations of galaxies are generally limited by photon noise, regardless of the dominant source of noise (i.e., the background or the source). This means that the S/N scales as the square root of the collecting surface of the telescope, the pixel surface,[22] the spectral resolution, the global throughput, or the total integration time. In this regime, there is a relation between the S/N

[21] Sky and moon brightnesses are generally listed on observatory websites. See `http://www.eso.org/gen-fac/pubs/astclim/paranal/skybackground/` for the example of the ESO Paranal Site.

[22] This is not true for compact sources (i.e., sources with sizes significantly smaller than the image PSF) when observed in large apertures. In this regime, S/N *decreases* with pixel scale, since most of the object flux is already contained in the aperture, while the background contribution keeps increasing with aperture size.

and the flux measurement uncertainty in magnitudes $\Delta m$, or the relative uncertainty on the flux density $f_\nu$ (see Sec. 1.2):

$$\Delta m = 1.08574\,\frac{\Delta f_\nu}{f_\nu} = \frac{1.08574}{\mathrm{S/N}}.$$

Hence, a S/N of 5/10/50 corresponds to a flux measurement uncertainty of ±20/10/2% (and of $\Delta m$ of ±0.22/0.11/0.022 mag), assuming that the residuals from the data reduction process are negligible (see above).

The cosmological dimming effect makes distant galaxies very low surface brightness objects (see Sec. 1.2). Sampling their rest-frame optical continuum flux distribution to study their morphology (see Sec. 2.4) requires an additional difficulty, which is observing them in the NIR. This is because the contribution of the NIR background due to the OH sky lines is very high and can saturate the images, imposing relatively short *dit*. In this regime, S/N $\propto \sqrt{ndit \times dit}$ so that short *dit* can be easily compensated by increasing *ndit* without degrading the S/N.

**Example 2.5: *How to choose* dit *in the NIR?***

Assuming a $z \sim 2$ galaxy with $K_{AB} = 22.0$ and a diameter of 1.5 arcsec observed using HAWK-I at the VLT (with a 0.1 arcsec pixel size, airmass = 1.2, and a 0.8 arcsec seeing), K band exposures are saturated when *dit* ≥ 40s. The same object observed with the E-ELT assuming the same pixel scale would be saturated with *dit* ≥ 2s! A more realistic pixel size of 5 mas for the E-ELT will allow unsaturated exposures if *dit* ≤800s. Note that the detector linearity and the FF accuracy are further limitations of large *dit* exposures. Such calculations can be easily done using Exposure Time Calculators (ETCs) such as those available on the ESO website. Always keep counts between 1/3 to 1/2 of the saturation limit. This results in a good photometric response and avoids saturation of luminous transient effects.

## 2.4 Galaxy Morphology

Galaxy morphology reflects the numerous physical processes that affect a galaxy during its lifetime. In the local Universe, galaxies show a remarkably simple sequence, namely the Hubble morphological classification sequence (Hubble, 1926). Sandage (2005) has reviewed in detail how this sequence was established, including the first attempts by Jeans (1919) and Reynolds (1920). In this sequence (as shown in Fig. 1.9), galaxies show a decrease in the flux ratio of the bulge, where stars have random ellipsoidal orbits, and of the disk, where stellar orbits follow a well-organized rotation (see an example of a Bulge/Disk decomposition in Fig. 2.7).

Of course, galaxies are more complex objects than a simple superimposition of these two components. Historically, the morphological sequence was organized through various properties of the disk, arms, bars, and even dust (see, e.g., Buta, 2011; see also Fig. 2.7). The disk can itself be decomposed into a thin disk (with a rotation velocity-to-velocity dispersion ratio $V/\sigma \sim 10$) generally made of younger stars, and a thick disk ($V/\sigma \sim 2$–$3$). It has been shown that a significant fraction of the NIR light (up to 50%) of edge-on spirals is produced in a thick disk (Comerón *et al.*, 2012). The central part of galaxies is often the host of bars, which are the origin of the double fork split in the Hubble Sequence.

Perhaps it is the apparent simplicity of galaxy morphologies in the nearby Universe that have encouraged many astronomers to attempt simple, multiple parameter classification methods, using asymmetry, compactness or other characteristics of the light distribution. Such methods are easy to automate to investigate large numbers of objects. However one should keep in mind that galaxies have very complex structures containing hundreds of billion stars as well as gas and dust. The above simplifications become particularly inaccurate when interpreting distant galaxies. In this chapter, we will investigate several proposed techniques for classifying galaxy morphologies, while keeping in mind the considerable knowledge accumulated from Hubble (1936), Sandage (1961), Sandage (2005, and references therein), de Vaucouleurs (1959), van den Bergh (1960b),

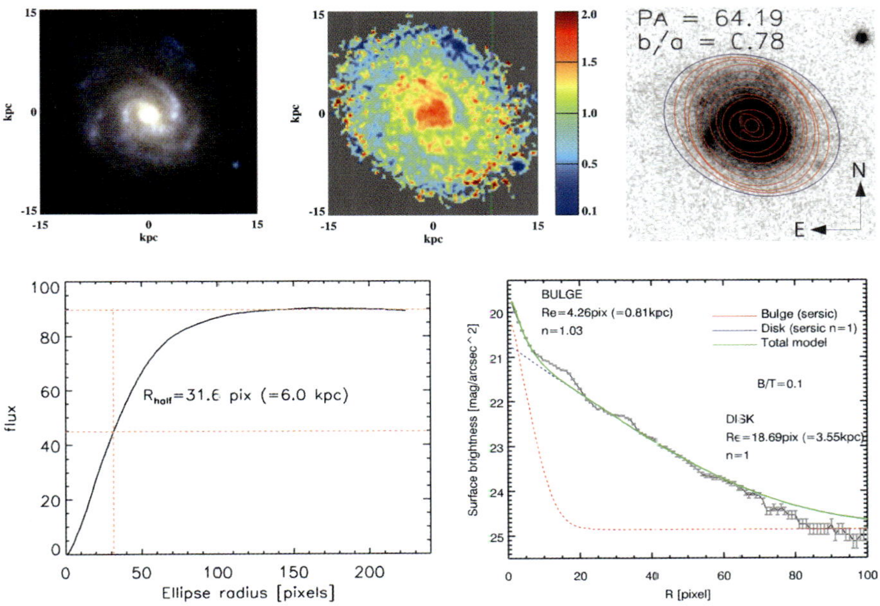


Fig. 2.7: Example of morphological analysis on a distant ($z = 0.575$) spiral galaxy. The HST/ACS v+i+z image (*top-left panel*) and the v–z color image (*middle-top panel*) reveal a spiral galaxy with blue spiral arms and a central red bulge, similar to local spiral galaxies. The IRAF *ellipse* task is used to fit isophotal ellipsoidal apertures (*top-right* panel) from which the light curve-of-growth (*bottom-left panel*) and surface brightness profile (*bottom-right panel*) can be derived. The outer isophote is used to measure the $b/a$ axis ratio from which the disk inclination *inc* and P.A. can be estimated (see Ex. *Measuring the surface brightness profile and basic properties of a distant galaxy*). The light curve-of-growth converges to a well-defined plateau at large radius, which shows that the background has been correctly subtracted and that the half-light radius $R_{\mathrm{half}}$ can be accurately measured. The surface brightness profile shows three components: an outer exponential disk, a central pseudo-bulge (with a Sersic index $n = 1.03$), for which the result of the GALFIT fitting is represented as blue and red dash lines, respectively, and a bar (not fitted by GALFIT), which is also revealed by the central ellipses in the top-right panel. The Bulge-to-Total ($B/T$) flux ratio is found to be $B/T = 0.1$, typical of late-type spiral galaxies.

van den Bergh (2002, and reference therein), and Kormendy (2013, and references therein).

### 2.4.1 *Morphological classifications*

Morphological classifications consist in assigning a given morphological type to a galaxy. Often, this morphological type is taken to be a Hubble type, which depends on well-defined morphological criteria that can be subdivided (for a spiral galaxy) according to the degree of central concentration, openness of the arms, resolution of the arms into complexes of star formation (all three criteria determine the position along the fork), and the presence or absence of a bar (determining the appropriate branch on the fork). Other, more exhaustive classification schemes exist (de Vaucouleurs, 1959; Morgan, 1958, 1959; Morgan and Mayall, 1957; van den Bergh, 1960a and 1960b, see also the review by Sandage, 2005). Given the faintness and compactness of distant galaxies, most of them focus on details that cannot be observed at high redshifts. This is why the Hubble sequence remains widely used for galaxy morphological classification.

Approximately 10% of the local galaxies that are more massive than a fourth of the Milky Way mass cannot be classified in the Hubble Sequence (Delgado-Serrano *et al.*, 2010). These galaxies have too many irregularities and/or lack recognizable sub-structures, and are called Peculiar galaxies (Pec). It is important to distinguish peculiar galaxies, which are massive galaxies with morphology likely affected by interactions, from irregular galaxies (Irr), which are dwarf galaxies for which morphologies have not been shaped by gravitation but are instead likely affected by stellar phenomena (van den Bergh, 2007).

To be meaningful, a morphological classification should be (Delgado-Serrano, 2012):

- Homogeneous: all galaxies in the sample must be classified using the same information;
- Reproducible: the criteria used to separate galaxies into distinct classes must assign a unique morphological class to each galaxy and avoid degeneracies during the classification process;
- Objective: the classification process must limit human subjectivity and establish clearly the classification procedure;
- Physical: the parameters used during the classification process should be as closely as possible correlated with galaxy physical properties (e.g., the bulge-to-disk ratio). This helps limit biases and ensures that the resulting classification is meaningful and not purely descriptive.

The last point is also a reason why the Hubble sequence is still widely used today: the Hubble type is indeed correlated not only with dynamics but also with a wide range of physical properties (Roberts and Haynes, 1994), although sometimes with a large amount of scatter.

### 2.4.2 *Characterizing morphology at high redshifts*

The cosmological dimming effect and the decrease in spatial resolution limit the morphological characterization of distant galaxies.

For example, determining the openness of the spiral-arm pattern becomes increasingly inaccurate with increasing redshift. At high redshift it has been proposed (Hammer, 2005; Zheng *et al.*, 2005) to define morphological classes as simply as possible, i.e., with only three broad classes, e.g. spheroidal galaxies (E and S0 with $B/T$ luminosity ratio $B/T > 0.5$), spirals (with bulge-to-disk luminosity ratio $B/T < 0.5$, see Fig. 2.7), and peculiar galaxies, which generally correspond to interacting or merging galaxies in the local Universe. van den Bergh (2002) distinguished early type spirals from galaxies with a type later than Sbc using $B/T \leq 0.1$, but only at $z < 1$. The advantage of this simplification is that it remains a "physical morphological classification", as suggested by Kormendy (2013), since bulge and disk are differently supported components.

With increasing redshifts, more irregular structures appear, such as clumps or overall morphologies that resemble comas, tadpoles, or chains (see a description by van den Bergh, 2002; Elmegreen *et al.*, 2005, 2007), while the detection of bars become increasingly difficult at $z \geq 0.8$ (Zheng *et al.*, 2005; Sheth *et al.*, 2008). It is essential to sample optical rest-frame wavelengths to accurately characterize galaxy morphology (see Fig. 2.6; see also Rawat *et al.*, 2009). The new WFC3 camera on-board the HST now allows us to study the rest-frame optical morphology of galaxies, including surface brightness profiles and colors up to $z \sim 3.5$ (Cameron *et al.*, 2011), though many galaxies are too faint or compact to be morphologically classified.

Measuring the surface brightness profile of distant galaxies provides us with essential information about their structure. Indeed, a simple analysis (see Ex. *Measuring the surface brightness profile and basic properties of a distant galaxy*) allows one to directly estimate several key structural parameters (as shown in Fig. 2.7), i.e., their P.A., inclination, and size (half-light radius). These parameters are important for morpho-kinematic classifications (see Sec. 4.5.4).

**Example 2.6: *Measuring the surface brightness profile and basic properties of a distant galaxy***

(1) Open an image of the object that samples the stellar mass content, i.e., taken at optical-NIR rest-frame wavelengths. Make sure you use an image that more than 4 times larger than the galaxy diameter to have a good sampling of the background;

(2) Clean up all neighboring structures around the galaxy using, e.g., the *imedit* task within IRAF, or SEXTRACTOR, and replace them with the average sky background;

(3) **Measuring the surface brightness profile:** Use the *ellipse* task within IRAF to fit concentric ellipses with increasing size at different isophotal levels that cover the whole galaxy surface. In the *geompar@* set the step between ellipse to be linear and give a first guess of the optical center. Carefully inspect the results in the outer and faintest regions, and take care of any variation in the ellipses axes as a function of radius in these regions (which can be associated to, e.g., a warp or an interaction). Measure the object flux enclosed within each ring aperture and plot the surface brightness of the object as a function of radius (see, e.g., the black curve in the bottom-right panel of Fig. 2.7);

(4) **Measuring the P.A.:** To measure the P.A. of the object, adopt the major axis of the ellipse in the outer regions that corresponds to the faintest reliable surface brightness threshold above the background (e.g., see the $1\sigma$ ellipse shown in blue in the top-right panel of Fig. 2.7). Measure the angle between the North direction and this axis.

(*Continued*)

**Example 2.6: *Continued***

Count it positive towards East: this gives you an estimate of the P.A. angle. The uncertainty on P.A. can be estimated as the maximal value between the difference of P.A. measured at the ellipses corresponding to $1\sigma$ and $2\sigma$ above the sky background (e.g., compare the blue and last red ellipses in the top-right panel of Fig. 2.7) and the uncertainty on the P.A. provided by the *ellipse* task at $1\sigma$. The P.A. should never been measured in the inner isophotes which can be dominated by a bar or ring;

(5) **Measuring the inclination:** An estimate of the inclination *inc* can be obtained from the b/a ratio at the outer $1\sigma$ ellipse provided by the *ellipse* task. Galaxy disk are never perfectly thin so that you need to account for their intrinsic thickness $q$: $\cos(\mathrm{inc})^2 = ((b/a)^2 - q^2)/(1 - q^2)$ with $q \sim 0.1$–0.25, possibly depending on morphological type (Hubble, 1926; Heidmann *et al.*, 1972; Unterborn and Ryden, 2008). A face-on disk will have inc $= 0°$, while an edge-on disk will have inc $= 90°$;

(6) **Measuring the half light radius $R_{\mathrm{half}}$:** Construct the light curve-of-growth of the object by integrating the surface brightness profile as a function of radius (see, e.g., the bottom-left panel of Fig. 2.7). Check that the outer part of the curve-of-growth is flat. If not, iterate on the sky background measurement (by, e.g., taking care of background spatial variations) and on the cleaning of neighboring structures until you get a plateau. Then estimate the galaxy total flux by integrating the flux curve-of-growth. Interpolate the radius $R_{\mathrm{half}}$ at which one gets half the total flux of the galaxy.

All these measurements are illustrated in Fig. 2.7.

Measuring the size of galaxies at different redshifts requires a metric that is independent of galaxy shape and type. This is not the case with the Kron or Petrosian radii that have been widely used in the local Universe. This can however be done homogeneously using the half-light radius $R_{\text{half}}$, which is the radius containing half of the total galaxy flux (see Ex. *Measuring the surface brightness profile and basic properties of a distant galaxy*). For a thin exponential disk, one has the useful relation $R_{\text{half}} = 1.67R_d$. Since $R_{25} \sim R_{\text{total}} = 3.2R_d$ in local disks, this leads to $1.92R_{\text{half}} \sim R_{\text{total}}$. The half-light radius is also conveniently related to the optimal size of aperture to perform integrated photometry (see Sec. 2.5.2).

### 2.4.3 *Empirical methods — Eyeball classifications*

Distant galaxies were first classified using visual inspection in a single band (Brinchmann *et al.*, 1998; van den Bergh, 2002). Visual classifications are highly dependent on the spectral band, resolution, and depth of images. Moreover, visual classifications are highly subjective, and comparison between different classifiers reveal substantial biases from one to the other, even at $z \sim 0$. This makes any visual classification hard to reproduce (Naim *et al.*, 1995). To limit this bias, visual classifications can be done relatively to typical members of each class, which are used to guide the eye. To minimize the uncertainties, it is crucial to combine three independent examinations by experimented astronomers (e.g., Zheng *et al.*, 2005).

So far the most ambitious visual classification at high redshifts is the one conducted by the CANDELS team (Kartaltepe *et al.*, 2015) with 7,634 galaxies at $0 < z < 4$ classified by 3 to 15 independent classifiers (to be compared to the 14,000 local galaxies visually classified by Nair and Abraham, 2010 in the SDSS). However, this classification scheme does not include all of the available information, such as B/T ratios to separate disk and spheroidal dominated galaxies (see Sec. 2.4.4), nor color maps. Moreover, it does not use nonexclusive morphological classes (see Sec. 2.4.6).

### 2.4.4 *Light decomposition using software*

These decompositions are also known as "parametric methods" because they consist of fitting the galaxy light distribution using analytical models. The best models correspond to physically-motivated decomposition such as the bulge-disk decomposition, which uses an exponential function for the disk as well as a Sersic profile for the bulge. Sersic profile (Sersic, 1968) is defined as $I(r) = I_0 \exp\left(-\kappa(r/r_e)^{1/n} - 1\right)$, where $\kappa$ is a constant that depends on the Sersic index $n$ (Peng *et al.*, 2010). The Sersic profiles with $n = 1$ corresponds to exponentially declining disks with $I(r) = I'_0 \exp -(r/R_d)$ (Freeman, 1970), where $R_d$ is the disk scale-length, while profiles with $n = 4$ correspond to a classic bulge with a de Vaucouleurs profile with $I(r) = I''_0(r/R_e)^{1/4}$, where $R_e$ is the effective radius. Useful relations between Sersic index and concentration or specific aperture magnitudes are given in Graham and Driver (2005). The bulge-disk decomposition can be interpreted physically as two dynamically separated components (rotation-supported disks and dispersion-supported bulges).

Depending on the S/N, up to three profiles can be fitted simultaneously in 2D images, with good results (Peng *et al.*, 2010). Several softwares have been developed to automate functional fittings, some of which are listed in Table 2.3 (see, e.g., Häussler *et al.*, 2007). Over the past years, several wrappers such as GALAPAGOS (Barden *et al.*, 2012), PyMORPH (Vikram *et al.*, 2010), SIGMA (Kelvin *et al.*, 2012), or iGalFit (Ryan, 2011) were developed in order to fully automatize the fitting procedure and to integrate it in a single pipeline that includes source extraction, PSF measurement, and profile fitting. Used blindly, this approach can provide good mathematical fits that are however completely unphysical solutions. A visual inspection of the results and an iterative approach are absolutely

Table 2.3: Popular programs used for parametric morphological decomposition.

| Software | Reference | Minimization scheme |
|---|---|---|
| GIM2D | Simard (1998); Simard *et al.* (2002) | Metropolis |
| GALFIT | Peng *et al.* (2002, 2010) | Marquardt–Levenberg |
| FITGAL | Wadadekar *et al.* (1999) | Davidson–Powell–Fletcher |
| BUDDA | de Souza *et al.* (2004) | Downhill simplex |
| GASPHOT | Pignatelli *et al.* (2006) | Marquardt–Levenberg |
| GALPHAT | Yoon *et al.* (2011) | Markov chain |

*From left to right*: Software name, reference paper, and Minimization algorithm used.

required (see Ex. *How to decompose a galaxy into bulge and disk using GALFIT?*). Simply relying on the minimization of the $\chi^2$ of the residual map is not enough since the fitted model is often an oversimplification of the galaxy structure. For instance, spiral arms or symmetric features like bars can result in large $\chi^2$ values when they are not accounted for by the model (see Fig. 2.8). To avoid systematic effects in the modeling, residuals must be examined to verify that they do not show symmetric features suggestive of well-known features in disk galaxies (see Ex. *How to decompose a galaxy into bulge and disk using GALFIT?*).

**Example 2.7: *How to decompose a galaxy into a bulge and a disk using GALFIT?***

(1) Prepare the data: Construct (if not available) a noise image, which can be done by considering Poisson noise (square root of the counts in each pixel) quadratically convolved with readout noise; in case of a drizzled image, one should use the weight image sometimes provided with the science frames ($\propto 1/\sigma^2$; see details in Appendix of Casertano *et al.*, 2000); then identify and measure profiles of stars that are as close as possible to the galaxy to get an empirical characterization of the PSF (see Ex. *How to estimate the spatial resolution of an image?*);

(2) Construct the object surface brightness profile to estimate the total magnitude, sky background, axis ratio, radius, and P.A. of the galaxy (see Ex. *Measuring the surface brightness profile and basic properties of a distant galaxy*);

(3) Identify, using the light profile, the number of components needed to model the galaxy. For this, look for change of slopes in the profile above the background noise (see Fig. 2.8). Pay attention to the central component, which should be replaced by a PSF profile if it is not resolved;

(4) If the central component is resolved, fit it by hand using a Sersic component to have it sticking as much as possible to

(*Continued*)

**Example 2.7: *Continued***

the very central part of the observed profile. Then add an exponential disk to model the outer parts. Once you have found initial conditions corresponding to a suitable model, let GALFIT converge (never let GALFIT converge without providing meaningful initial conditions to start with);

(5) Visually inspect the residual image for mismatches in P.A., ellipticity, or size and update the priors if needed. Be careful about the use of a simple reduced $\chi^2$ criterion: a value that is not close to 1 does not mean that the fit is not good if the residuals show structures typical of disks that are not part of the model (i.e., spiral arms; see Fig. 2.8). Closely inspect the central region: if you get large residuals with elongated elliptical isophotes, it probably indicates that there is a bar present. Depending on the resolution and depth of the image, you can try to add a third Sersic profile (with small axis ratio and small Sersic index) to model it explicitly. If you see extended ring-like residuals, it probably is a sign that you have not converged to the proper disk scale length value;

(6) Check whether the solution is physical: are all the fitted profiles centered on the same position within two elements of spatial resolution? If $B/T \leq 0.5$, does the central bulge have an effective radius smaller than that of the surrounding disk? If GALFIT does not converge towards a reliable solution, check whether you are not trying to fit too many parameters which are degenerated. First, try to fix parameters for which you can have good first-guess estimates independently (e.g., profile center, P.A., or ellipticity). The goal is to let the most degenerate parameters (i.e., intensity, or scale-length) converge first, and then try to set the fixed parameters free again. *Do no try to fix the Sersic index of the central component*: it can take a range of value from

(*Continued*)

**Example 2.7: *Continued***

disky-structures like pseudo-bulges to de Vaucouleurs-like profiles for classical bulges (Kormendy and Kennicutt, 2004). Be careful about dust when interpreting residuals: dust lanes in edge-on galaxies can easily mimic off-centered structures or asymmetric components;

(7) Inspect the radial profile of the resulting fit and check whether the underlying envelope of the observed light profile is well-fitted and not affected by small-scale variations such as spiral arms, clumps, etc. The profile is well-fitted when the fit closely follows the wings of the light profile in the outer regions of the galaxy down to the noise floor (see Fig. 2.8). Attribute a quality flag to the result, which will record whether the fit is robust, likely robust, or unreliable. For this, inspect visually the residual map and look for systematic structures (spiral arms, residual bulge, etc.).

The main difficulty with these programs is associated with the loss of low surface brightness material in observations of high redshift galaxies. This effect can in part, or completely, erase the galaxy substructures that are visible in the local Universe, such as spiral arms, or extended disks (see, e.g., Hammer *et al.*, 2009). Additional issues can arise from the apparent de-centering of sub-structures due to the more irregular shapes of distant galaxies. All of these programs generally result in low level of systematic errors provided that the object surface brightness is well above that of the background.

### 2.4.5 *Automatic methods with two or more parameters*

There are also called "nonparametric methods" that rely on structural parameters describing the statistical properties of the light distribution, without prior assumptions about the distribution of light in the objects.

Many distinct parameters have been developed over the past few years, the most commonly used being:

- The CAS system, which measures the light Concentration $C$, Asymmetry $A$, and Clumpiness/Smoothness $S$ (Schade *et al.*, 1995; Abraham *et al.*, 1996; Conselice *et al.*, 2000). $A$ is roughly correlated with morphological type (from Elliptical to Spirals and Mergers), $C$ is correlated with the bulge size, while $S$ is sensitive to galaxies undergoing star formation. Their combination provides, in principle, a statistical way of isolating different morphological type in local galaxies, including mergers (Conselice *et al.*, 2003, see Fig. 2.9);
- The G-$M_{20}$ system (Abraham *et al.*, 2003; Lotz *et al.*, 2004), is based on the Gini coefficient $G$, which measures the uniformity of the light amongst pixels, and on the second order moment of the brightest pixels $M_{20}$, which quantifies how the brightest pixels in the image are spatially distributed. The combination of these two parameters is also sensitive to different morphological types in the local Universe, including mergers (Lotz *et al.*, 2011, see Fig. 2.9).

#### 2.4.5.1 *Can we classify galaxy morphologies with automatic methods?*

The main advantage of automated, nonparametric methods is that they can be used very rapidly for a large number of galaxies. Nonparametric parameters however have the drawback of being dependent on the S/N (e.g., Conselice *et al.*, 2000; Lotz *et al.*, 2004; Lauger *et al.*, 2005; Lisker, 2008; Chen *et al.*, 2010). Large correction factors are also needed to account for differences in wavelength and resolution between the morphological templates, used to set the boundaries of the classification, and the observations, as well as for the redshift evolution of the reference templates (see, e.g., Burgarella *et al.*, 2001; Conselice *et al.*, 2003; Lauger *et al.*, 2005; Taylor-Mager *et al.*, 2007).

The systematic errors of these morphological classification can be quite large. For instance, Conselice *et al.* (2003, 2008) and Neichel *et al.* (2008) found that automatic methods overestimate the fraction

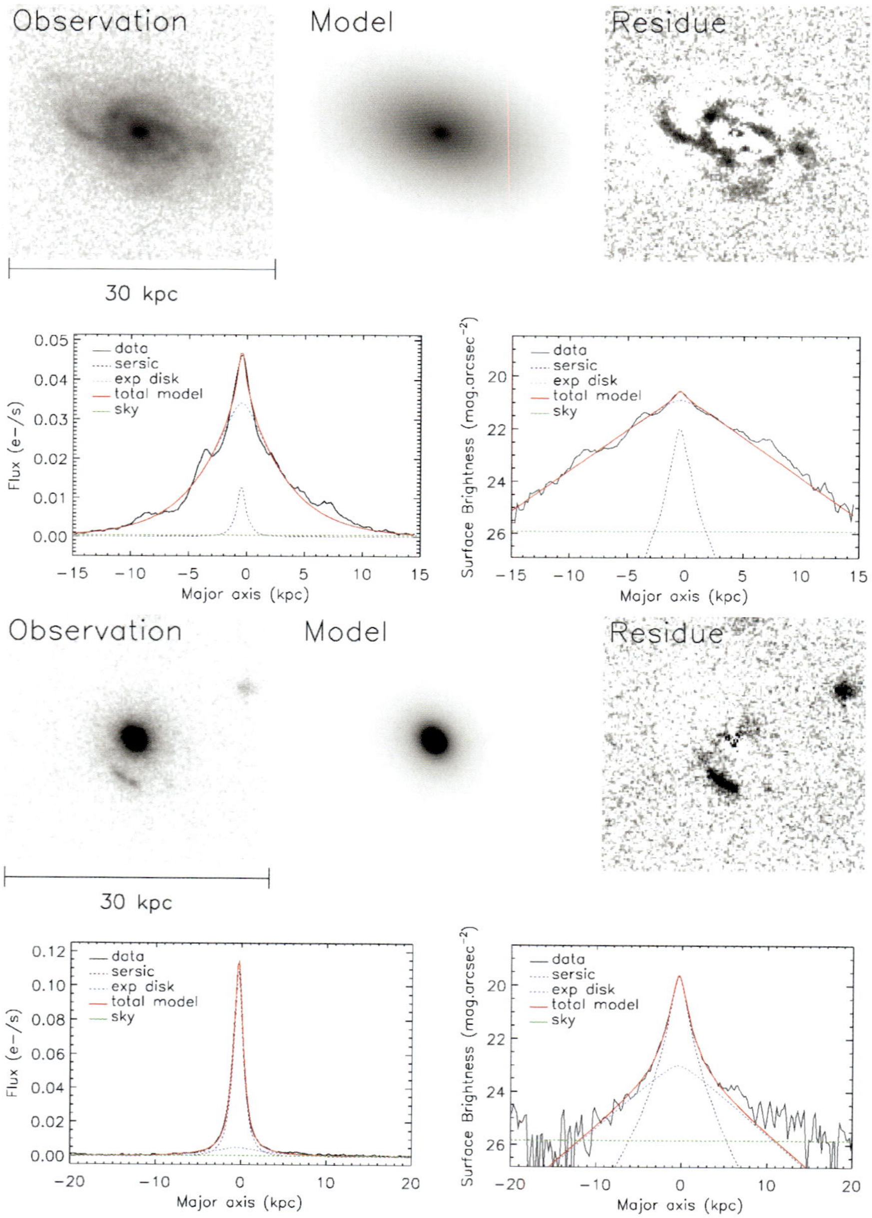
Observation
Model
Residue
30 kpc
data
sersic
exp disk
total model
sky
Flux (e−/s)
Surface Brightness (mag.arcsec$^{-2}$)
Major axis (kpc)
Observation
Model
Residue
30 kpc
data
sersic
exp disk
total model
sky
Flux (e−/s)
Surface Brightness (mag.arcsec$^{-2}$)
Major axis (kpc)

Fig. 2.8: (figure on facing page). Examples of use of a radial profile to check the results of GALFIT fittings in two galaxies at $z \sim 0.5$ (see Disseau *et al.*, 2016). *For each galaxy, from top left, clockwise*: HST image, GALFIT model, residual map, flux profile along the major axis in linear and magnitude scales. Note how the wings of the underlying envelope of the observed profile is well-fitted by the model and not affected by fluctuations due to spiral arms or clumps in the disk, which are clearly visible in the residual map. The first example on the top is a galaxy dominated by a disk (the spiral arms are clearly seen in the residual image because they are not considered by the fitted model, which can result in artificially high $\chi^2$ values). The galaxy on the bottom is dominated by a central spheroidal component and is fitted with a lower $\chi^2$, while the additional structure visible below the source makes it somewhat peculiar.

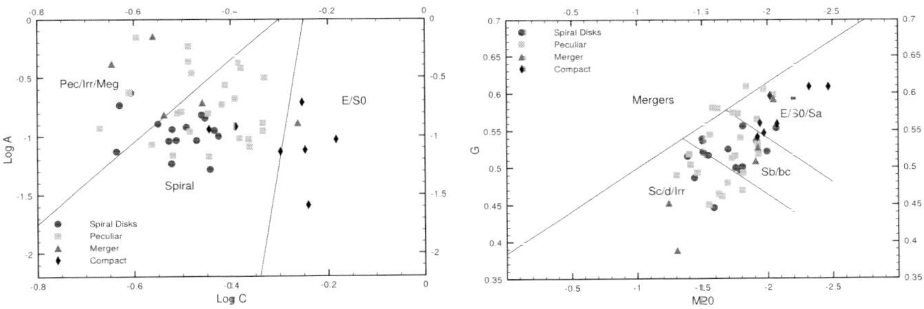


Fig. 2.9: Comparison between parametric and nonparametric morphological classification methods. Credit: Neichel *et al.* (2008), A&A 484, 159, reproduced with permission ©ESO. *From left to right*: A–C and G-$M_{20}$ planes. The different symbols refer to the semi-empirical morphological classification defined by Neichel *et al.* (2008), from which a good correlation between morphology an kinematics is found (see also Chapter 4). The automatic nonparametric methods are particularly inaccurate in separating mergers/peculiar galaxies from disks.

of spiral galaxies by a factor two at $z < 1$ (see Fig. 2.9); Franceschini *et al.* (2006) found that 30% of $z \leq 1.5$ galaxies classified as ellipticals look more like late-type spirals.

Such systematic errors start to dominate the statistical Poisson variance when the number of galaxies in the sample $N_{\rm gal} \geq 20$, making such methods relatively useless when applied to very large galaxy samples.

Principal Component Analysis has been used to attempt to reduce these errors by combining a large number of parameters and find their optimal combinations but only with relatively limited success, in particular at low S/N where large uncertainties remain (~30%, see Scarlata *et al.*, 2007). Train & Learn methods, like neural networks and support vector machines, have also been applied to try to further reduce these uncertainties which account for nonlinear boundaries between decision regions (Huertas-Company *et al.*, 2008). These methods can certainly reduce the uncertainty of the classification, but they remain subject to template variations when comparing galaxies at different look-back times.

Therefore, automatic nonparametric techniques cannot be recommended since they bring significant systematic uncertainties to their resulting morphological classifications.

In very large dataset, where a high degree of automatization is nevertheless required, one can verify the importance of biases and systematics by examining a sub-sample of objects for which a detailed morphological analysis (see below) has been performed. It is important to bear in mind that these issues are currently limiting the accuracy of all morphological studies inferred from the largest surveys.

#### 2.4.5.2 *Color bi-modality: an over-simplified approach*

Color has long been known to be a rough indicator of morphology (Roberts and Haynes, 1994), with elliptical galaxies being "old and red", and spiral galaxies being "young and blue". Several attempts have been made to try to use color to classify galaxies. The basic idea of classifying galaxies using color is rooted in the fact that color is a good evolutionary indicator of stars. There are however large scatters in color between different galaxy morphological types (e.g., Oh *et al.*, 2013). Galaxies are nonlinear systems that are much more complex than the simple combination of stars: color is highly dependent on geometric and extinction effects. For instance, dust can easily make a "young" galaxy "red"; a spiral galaxy seen edge-on will appear much more red than face-on; a very small star formation burst amongst an underlying old stellar population can easily turn an "old" galaxy blue (Schade *et al.*, 1999).

Despite this, a number of studies have attempted to use colors to separate galaxies into early and late types, very roughly corresponding to E/S0 and spiral galaxies, respectively. The method relies on color bi-modality: in an absolute magnitude vs. color plane, galaxies define a red-sequence and a blue cloud, where most of the early and late type galaxies lie, respectively (Strateva *et al.*, 2001; Baldry *et al.*, 2004; Bell *et al.*, 2004).

This color bi-modality has been widely used to separate galaxy populations into early and late types, by defining a statistical threshold separating the red sequence from the blue cloud. However, at any redshift, both the red sequence and the blue cloud are contaminated by a substantial fraction of late and early type galaxies, respectively. For instance, Cassata *et al.* (2008) found that the red sequence contains only 60–70% of true early-type galaxies at $z \leq 0.5$, 50% at $z \sim 2$, and ~25% at $z \geq 2$, while about 5% of relatively massive galaxies on the blue cloud are actually spheroidals at $z \sim 1$ (Kannappan *et al.*, 2009). Splitting galaxies into a red sequence and a blue cloud is therefore too simplistic and is not accurate enough for morphological studies.

### 2.4.6 *A pragmatic and conservative approach to classify distant galaxies*

Classifying the morphology of distant galaxies is usually with the goal to study galaxy evolution. In order to do this, one must assume that distant galaxies are indeed similar to nearby galaxies, i.e., that their morphologies can be characterized using a similar sequence as in the Local Universe. This conservative assumption is the only one allowing to properly calibrate morphological evolution, and has been demonstrated to be robust through comparison with kinematic studies (e.g., Neichel *et al.*, 2008; Hammer *et al.*, 2009).

**Example 2.8: *How to classify a galaxy morphology using a semi-empirical decision tree?***

(1) Extract stamp and color images (see Ex. *How to construct a color image?*) for all galaxies to be classified;

*(Continued)*

**Example 2.8: *Continued***

(2) Derive $R_{\text{half}}$ for all galaxies to be classified (see Ex. *Measuring the surface brightness profile and basic properties of a distant galaxy*);

(3) Determine the compactness criterion: there is a limit below which galaxies are too small to be reliably analyzed because of the limited spatial resolution of the images. A strict lowest limit to the half-light radius is given by the PSF size, i.e., $R_{\text{half}} \sim 0.5 \times \text{FWHM}$. However, because the measurement process of the other steps introduce an additional intrinsic scatter, the real limit is larger. This limit must therefore be derived empirically, by testing the whole decision tree on a test sample. In the case of $z \sim 0.6$ galaxies observed with HST/ACS images, Delgado-Serrano *et al.* (2010) found that this limit is $R_{\text{half}} \sim 1.33 \times \text{FWHM} = 1$ kpc, corresponding to 0.15 arcsec;

(4) Isolate compact galaxies in the sample using the compactness criterion;

(5) Decompose all of the remaining galaxies into disk and bulge (e.g., using GALFIT, see Ex. *How to fit a galaxy with GALFIT?*);

(6) Assess the bulge/disk decomposition reliability using the residual map and profile cut. Examine whether residuals show asymmetric and/or multiple components (see Fig. 2.10). Check for off-centered bulges with respect to the disk, i.e., whether the bulge and disk component are consistent;

(7) With each reliable fit, derive the $B/T$ flux ratio (e.g., as given by GALFIT) and isolate sub-classes depending on the $B/T$ value (see Fig. 2.10);

(8) With each sub-class, examine the color maps to check whether all peculiar sub-structures identified in the residual map and/or stamp image are associated with physically distinct components. In cases of reliable bulge/disk

(*Continued*)

**Example 2.8: *Continued***

decomposition, check whether a central bulge (if present) is redder than the disk, which is necessary to be able to classify a galaxy as a spiral. This is similar to the appearance of spiral galaxies in the local Universe and ensures that galaxies classified as spirals are indeed comparable in terms of color gradient at any redshift;

(12) Based on the result of the last step, assign a final morphological class to the galaxy;

(13) Classify the whole sample using at least three different classifiers and compare the results;

(14) Repeat the classification until all classifiers agree on the final morphological class, while keeping track of the initial rate of disagreement, which will give an order of estimate of the classification accuracy.

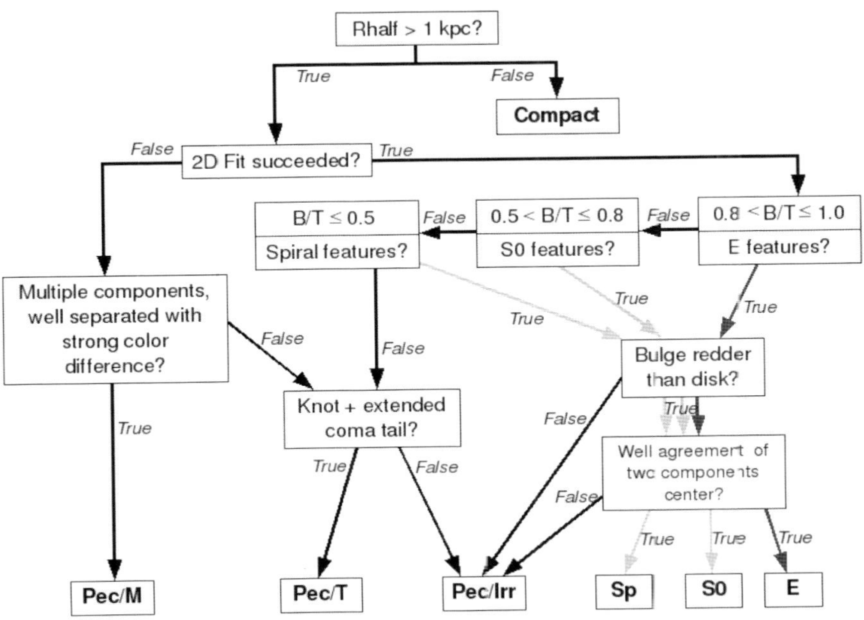


Fig. 2.10: Semi-Empirical morphological classification tree defined by Delgado-Serrano *et al.* (2010) to classify the morphology of $z \sim 0.6$ galaxies. Credit: Delgado-Serrano *et al.* (2010), A&A, 509, 78, reproduced with permission ©ESO.

Such a method was first developed by Zheng *et al.* (2004, 2005); Neichel *et al.* (2008) for spiral galaxies, and extended by Delgado-Serrano *et al.* (2010) to early-type galaxies. It is a semi-empirical method that relies on visual inspection of images using a well defined sequence of procedures in order to limit the subjectivity and improves the reproducibility of the classification. The sequence of procedures is clearly defined and ordered according to a decision tree that must be followed every time a galaxy morphology is being classified (see Ex. *How to classify a galaxy morphology using a semi-empirical decision tree*?). Specific decision trees must be created based on what observables are available for the classification. As the $B/T$ light ratio is one of the most correlated parameter to the Hubble sequence (e.g., de Lapparent *et al.*, 2011), this parameter should be at the core of any semi-empirical classification, as illustrated in Fig. 2.10.

A decision tree based approach requires the following preliminary steps:

(1) Ensure that all images are deep enough to avoid surface brightness effects and sufficiently resolved to limit the fraction of compact galaxies that are not properly resolved;
(2) Ensure that all quantities ($B/T$ and $R_{\rm half}$) are defined in an optical rest-frame (i.e., at wavelengths larger than ~500nm) as well as the decomposition of the galaxy light profile to estimate residuals, and scale lengths of the bulge and of the disk components;
(3) Prepare at least one color map (see Ex. *How to construct a color image of a galaxy*?).

It is only by combining all these different pieces of information and by comparing the classifications established by three experienced astronomers that one can achieve a robust galaxy morphological classification. Even simpler classifications compared to Fig. 2.10 can be obtained by merging all peculiar galaxies in a single category, i.e., Peculiar. Indeed, it can be relatively difficult to detect merging features such as tidal tails because they rapidly vanish within less than few hundred years after the merger, they depend on the exact merger orbit, and they correspond to extended and low-surface brightness features strongly affected by cosmological dimming.

The decision tree (see Fig. 2.10), especially in the simplified form described above, has been shown to lead to a morphological classification that is in good agreement with (1) manual classification made by world experts in galaxy morphology (van den Bergh, 2002) and (2) with a fully independent kinematic classification (Neichel *et al.*, 2008). While the resulting classification is limited by the relatively small size of the sample, it leads to a new and more modern galaxy classification scheme, i.e., a morpho-kinematic classification (Hammer *et al.*, 2009 and see Sec. 4.5.4).

#### 2.4.6.1 *Color maps*

Magnitude-scale color-images can be created by combining two images in different bands (Abraham *et al.*, 1999; Menanteau *et al.*, 2001; Zheng *et al.*, 2004). A color image is only useful when the spatial resolution of the original images is similar. Color maps are clearly limited to objects that are properly resolved, i.e., with sizes larger than 2–3 times the PSF size. For distant galaxies, such images should be constructed using space telescope images. Since the integration times of the respective band images are not necessarily the same, a S/N image must also be generated to identify reliable pixels in the color space (see Zheng *et al.*, 2004 and Ex. *How to construct a color map of a galaxy*?). The advantage of the Zheng *et al.* (2004) method is that it allows one to also estimate colors in fully extinguished regions. Color maps have a wide range of applications, from pixel-by-pixel SED analysis (e.g., Abraham *et al.*, 1999; Menanteau *et al.*, 2001; Zibetti and Groves, 2011), to morphological classification (Zheng *et al.*, 2004; Delgado-Serrano *et al.*, 2010; see Fig. 2.11).

**Example 2.9: *How to construct a color map?***

(1) Construct two stamp images with the same physical size in the two filters $X$ and $Y$. You will need images that

(*Continued*)

**Example 2.9: *Continued***

include the sky background. Estimate their spatial resolutions FWHM (see Ex.: *How to estimate the spatial resolution of an image?*). If for instance the spatial resolution in $X$ is larger than in $Y$, then convolve the $Y$ image using a Gaussian kernel of size $W = \sqrt{\mathrm{FWHM}_X^2 - \mathrm{FWHM}_Y^2}$. The IRAF task *PSFmatch* can be a useful tool for determining the kernel W;

(2) Determine the spatial shift and rotation between the two images using at least three stars located as close as possible to the target galaxy and align the two images. Both images should be aligned down to a precision that is smaller than one pixel in both directions;

(3) Construct the color map $\mu_{X-Y}$ defined as[23]:

$$\mu_{X-Y} = -\frac{1}{2}\log\left(\frac{\sigma_X^2}{F_X^2}+1\right)+\frac{1}{2}\log\left(\frac{\sigma_Y^2}{F_Y^2}+1\right)+\log\left(F_X/F_Y\right)$$

The associated noise within each pixel can be calculated using Eq. (10) of Zheng *et al.* (2004):

$$\sigma_{X-Y}^2 = \log\left(\frac{\sigma_X^2}{F_X^2}+1\right)+\log\left(\frac{\sigma_Y^2}{F_Y^2}+1\right)$$

Here, both the signal and noise include *both* the source and background. These relation are based on the property that $\sigma_X^2$ and $\sigma_Y^2$ are dominated by Poisson noise, which are close to being Log-normal distributions, which therefore allows one to estimate the associated S/N map as $\mu_{X-Y}/\sigma_{X-Y}$.

(4) Threshold the color map using a relative threshold in S/N to select reliable pixels, e.g., those with a S/N larger than four times the mean S/N in the background.

[23]Note that the corresponding Eq. (9) of (Zheng *et al.*, 2004) misses a square factor in the denominators, and that $\mu_{X-Y}$ is positive for Poisson statistics with at least two counts.

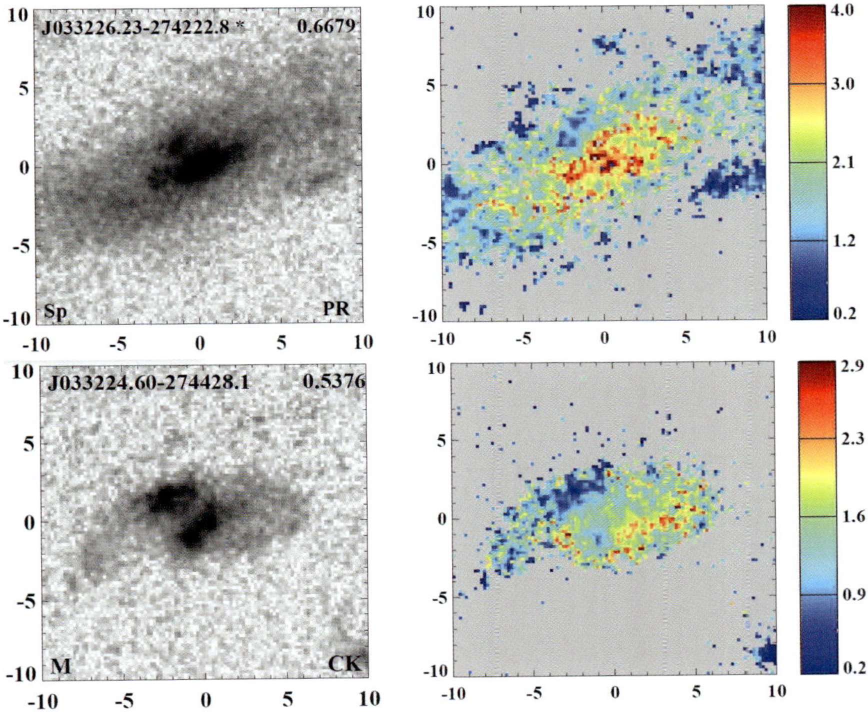


Fig. 2.11: Examples of $B$–$z$ color maps for two $z \sim 0.6$ galaxies from the IMAGES survey. Credit: Neichel *et al.* (2008), A&A, 484, 159, reproduced with permission ©ESO. *Top panel:* J033226.23-274222.8, which was classified as a spiral galaxy. The inner red ellipse corresponds to $\sim R_{\rm half}$ with $B - z = 2.2$, while the outer blue ellipse corresponds to $\sim 2.R_{\rm half}$ with $B - z = 1.8$. The $B - z$ color between $R_{\rm half}$ and $2.R_{\rm half}$ is found to be ~1.5. This difference of color as a function of aperture size, typical of $z \sim 0.6$ spirals (see Neichel *et al.*, 2008, Fig. 10), can significantly bias the photometric aperture corrections (see Ex. *Biases in colors caused by uncontrolled aperture corrections*); *Bottom panel*: J033224.60-274428.1, which was classified as a peculiar galaxy because of the strong color gradient between the two brightest components, which could be indicative of the two progenitors of an on-going merger.

## 2.5 Multi-wavelength photometry

This section explains how to extract useful information from photometric measurements of distant galaxies at different wavelengths and how to cross-correlate data in images taken at different wavelengths, depths, and with different resolutions (see Sec. 2.2).

Extensive cross-correlations of broad-band and narrow-band filters can be used to produce (SEDs, see Sec. 2.6) of galaxies. These, combined with Stellar Population Synthesis (SPS) models, have been successfully used to produce (indirect) estimates of the star formation history (SFH), metallicity, age of the stellar population, and of the total stellar mass of galaxies (see discussion about the validity of this method in Sec. 2.6).

### 2.5.1 *Cross-correlating images at different wavelengths*

The main difficulties with combining observations taken using different instruments and at different wavelengths are the varying depths and PSFs of the images. These issues are particularly important in the sub-millimeter, mid and far-infrared wavelengths. For instance, the Spitzer/MIPS camera has a 6, 18, and 40 arcsec FWHM, at 24, 70, and 160 $\mu$m respectively, which is much larger than the typical size of galaxies at intermediate and high redshifts (1 to 3 arcsec). It is therefore likely that more than one galaxy will be found in the optical image within one or two PSF size of the sources detected at 24 or 70 $\mu$m (see Fig. 2.12).

Astrometric accuracy also limits the comparison of two images. It is dependent on the number and the brightness of point-like sources in the images (bright enough to sample the PSF), and this can become a problem in some cosmological fields.

One should also keep in mind that the astrometric precision of fainter sources will be always smaller. Deriving accurate positions of the sources requires to take into account the pixel size, the size of the

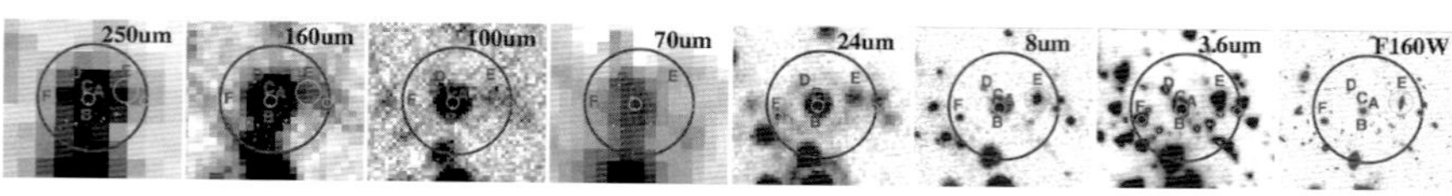


Fig. 2.12: Comparison between the same field observed at different wavelengths (from Yan *et al.*, 2014, ©AAS, reproduced with permission). Optical counterparts of IR sources are more and more difficult to identify as one compares longer wavelength images to optical images. There are six candidate counterparts within the 250 $\mu$m beam represented as a red circle, labeled from A to F, with $z = 1.225$, 0.93, 1.225, 1.34, 0.562, and 0.962, respectively.

PSF wings, and the astrometric errors of the data reduction process (in particular those due to distortion corrections).

**Example 2.10: *How to identify optical counterpart of IR sources?***

Consider the example shown in Fig. 2.12, with an *Herschel* beam at 250 $\mu$m encompassing six possible optical counterparts. If the later were limited to only one or two candidates, the probability of pure geometric coincidence could have been used to identify the most likely candidate. However it is of no help in cases with such a large number of candidates. Other methods have instead been used, such as modeling of the candidate SEDs in the optical. The SEDs are used to extrapolate the candidate fluxes in the IR and infer their respective probability to account for the flux detected in all bands. This often requires strong extrapolations and/or unjustified assumptions on the galaxy SEDs, such as constraining the FIR flux from the UV slope (see Sec. 2.6.2).

When several potential counterparts are detected, one should attempt to identify the best candidate (see Ex. *How to identify optical counterpart of IR sources?*). When the number of counterparts is not too large, one can derive the (Poisson) probability of pure geometric coincidence, i.e., $P(d, m) = 1 - e^{-n(m)\pi d^2}$, in which $d$ is the angular distance between the object at the poorest spatial resolution and the optical source in degrees, and $n(m)$ is the integrated surface density of galaxies at magnitude $< m$ (Flores *et al.*, 1999). This method is limited by nonGaussian position uncertainties due to extended emission, source confusion, and residuals (see Fig. 2.13). Other methods using *a priori* assumption about the nature of the galaxy SED (i.e., whether it is for instance a starburst galaxy) or redshifts measurements have been used, but these must be handled very carefully (see Ex. *How to identify optical counterpart of IR sources?*).

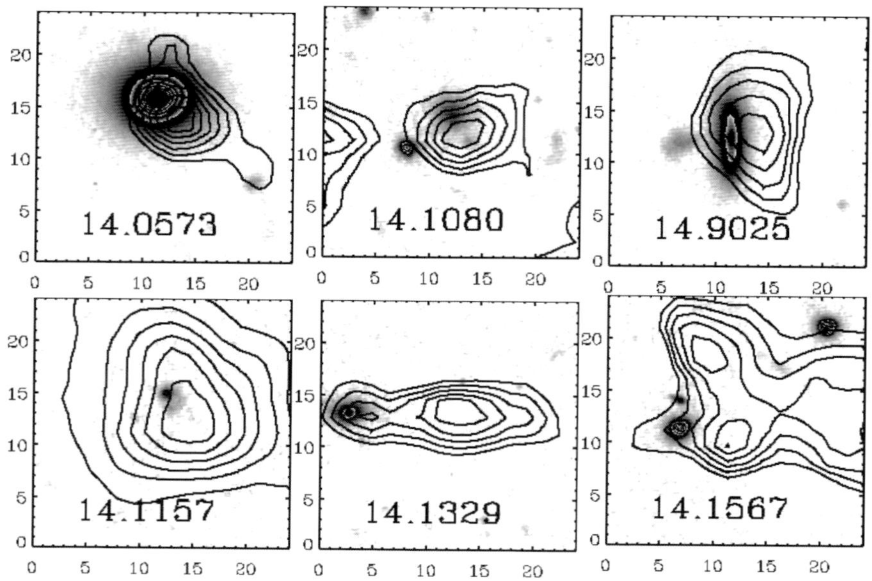


Fig. 2.13: Images ($24'' \times 24''$) of ISO/ISOCAM 7 $\mu$m sources, with 7 $\mu$m contours at $>3\sigma$ overlaid (resolution is $\sim 11''$ FWHM). Credit: Flores *et al.* (1999), A&A, 343, 389, reproduced with permission ©ESO. Residual glitches combined with the large ISOCAM pixel size can create artificially extended sources, which greatly complicates the identification of the optical counterpart.

### 2.5.2 *Aperture photometry*

When comparing fluxes at different wavelengths, one needs to measure fluxes in apertures of constant size. One might be tempted to use an aperture large enough to include the total flux of the galaxy at all wavelengths. In practice, it is impossible with this method to guarantee that the total flux is actually measured in all bands since only the flux above a given detection threshold can be extracted (see Sec. 2.2.2). Using a well-defined constant aperture size is therefore the preferred approach.

#### 2.5.2.1 *Aperture shape*

The most widely used aperture shape is circular, although an elliptical aperture aligned with the galaxy P.A. should be preferred, unless the galaxy shape is too irregular (e.g., mergers). More complex apertures can be defined for specific purposes.

#### 2.5.2.2 *Aperture size*

In principle, one should use an aperture size that maximizes the S/N in a given filter (Howell, 1989). Very distant galaxies are intrinsically small so they will often have a size that is smaller or comparable to the size of the image PSF and appear as point-like objects. Such galaxies are also generally very faint, so the resulting images are generally photon-noise limited. In seeing-limited conditions, the S/N of a point-source reaches a maximum at a radius of $R_{\rm optimal} = 0.7 \times$ FWHM.[24] Note that for distant galaxies, the effective optical aperture should be derived from the expected luminosity profile convolved by the PSF FWHM. For average seeing conditions (FWHM $\sim 0.8$ arcsec) and $R_{\rm half} = 0.75$ arcsec (typical of $z \sim 0.6$ galaxies), $R_{\rm optimal} \sim 1.1$ arcsec, for an exponential disk. Apertures of 2–3 arcsec diameter are widely used for measuring galaxy fluxes at $z \sim 1$–$2$.

Another issue is that the optimal aperture for a given galaxy is not necessarily the same in all images at different wavelengths because these images can have different noise properties, exposure times, and pixel scales, or simply because the object shape varies as a function of wavelength.

However, a galaxy color should always compare two flux measurements made using the *same* aperture in order to compare similar regions. A compromise between aperture size and S/N as a function of wavelength is therefore necessary when comparing fluxes measured in different bands. A good strategy is to define the optimal aperture mid-way between the two extreme wavelengths that are being probed: this minimizes S/N loss at other wavelengths. For instance, if one is interested in studying a galaxy SED from the UV to the IR, a good choice is to define the optimal aperture at $\sim 1\,\mu$m, which

[24]For nearby extended objects, the optimal aperture in background-limited conditions is found to be $\sim 1.07 R_{\rm half}$ and $\sim R_{\rm half}/6$ for an exponential and De Vaucouleurs luminosity profiles with a half light radius $R_{\rm half}$, respectively (Phillipps, 1984).

has the additional advantage of tracing stellar mass with a relatively good accuracy.

Special attention must be made when the aperture is particularly small when compared to the image PSF size. This is often the case in the UV or in the IR (e.g., GALEX, Spitzer, etc.). In this case, flux measurements must be corrected for PSF broadening using the PSF light curve-of-growth, which characterizes how the flux measured for a point-source object grows as a function of aperture size (see, e.g., Morrissey *et al.*, 2007).

#### 2.5.2.3 *Flux measurement*

Making flux measurement consists of extracting and counting ADUs in a given aperture and converting them into calibrated fluxes or magnitudes using the zero point of the instrument (see Sec. 1.2). Since the 80s, many programs implemented using different programming languages have been developed to perform these steps automatically (e.g., FOCAS, see Jarvis and Tyson, 1981). The most widely used ones are IRAF (see the *daophot* task) and SEXTRACTOR (Bertin and Arnouts, 1996). SEXTRACTOR, in particular, performs very well when measuring the photometry of a large number of galaxies in a fully automatic way, when their sizes are large compared to the PSF (see Ex. *How to construct a source catalog from an image*?, in Sec. 2.2.2). SEXTRACTOR estimates the background levels and copes with crowding and deblending efficiently, at least in optical/NIR wavelengths where confusion (see Fig. 2.14) is not an issue.

Another source of variance in background measurements is caused by unresolved sources. This is called confusion noise and must be accounted for when extracting structures and measuring fluxes. One enters the confusion-limited regime when the total background noise is dominated by this effect. This situation can be caused by the density of objects (crowding) and by the PSF size. This can be an issue in the UV or in the IR, where the PSF is significantly larger than the intrinsic galaxy size. For instance, the 500 $\mu$m channel of

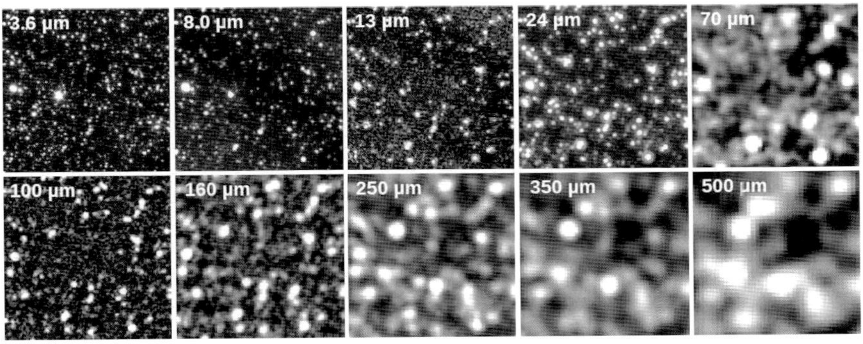


Fig. 2.14: Confusion at long wavelengths. The same region of the GOODS-N field (5×5 arcmin$^2$) is shown at different wavelengths. Credit: Elbaz *et al.* (2011), A&A, 533, 119, reproduced with permission ©ESO. Observations at 3.6, 8.0, 13, 24, and 70 $\mu$m are from Spitzer (top row), while those at 100, 160, 250, 350, and 500 $\mu$m are from Herschel (bottom row).

the Herschel satellite, with its 36.6 arcsec FWHM the PSF is clearly affected by confusion (see Fig. 2.14).

In such a case, one can use observations of the same field taken at other (better-resolved) wavelengths and use them as priors for the spatial distribution, shape, and fluxes of sources (see Ex. *How to measure photometry in the confusion limited regime*?). The main drawback of flux extraction methods when using priors is that sources that have no counterparts at the wavelength used to guide the extraction are not measured. This can however be mitigated by first reconstructing the field using priors at other wavelengths and by then using the residual image to look for sources that had no counterparts in the secondary image used to guide the extraction (see Fig. 2.15). The main drawback of this method is that, by construction, there is no knowledge of the SEDs for the sources that have no counterpart in high resolution images, which limit their study to a simple detection.

Another difficulty occurs when the size of galaxies is small compared to the size of the PSF (again, typical in the UV and the IR). In this case, methods designed specifically to perform point-source flux measurements (i.e., PSF-fitting techniques) should be used instead of simple aperture photometry. Such an approach, which performs the

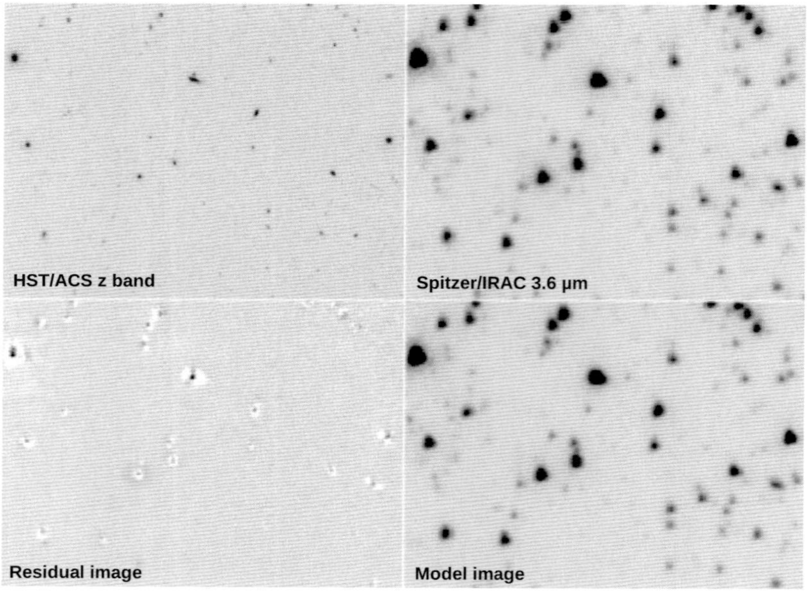


Fig. 2.15: Example of photometry with priors with the TFIT package (from Laidler *et al.* (2007), PASP, 119, 1325, published by the University of Chicago, ©The Astronomical Society of the Pacific). *Top panels, from left to right*: HST/ACS z-band image of a galaxy field used as high-resolution prior, and Spitzer IRAC 3.6 $\mu$m image of the same field from which one wants to extract accurate galaxy fluxes, model image obtained by photometric reconstruction, and residual image. Note how some galaxies overlap in the low resolution Spitzer image. One is in the confusion limited regime when such overlapping dominates the background of the image (see Fig. 2.14). The model image (*bottom-right panel*) is obtained by running SEXTRACTOR on the high resolution ACS image to detect as many sources as possible and convolving the result by the Spitzer PSF (see Laidler *et al.*, 2007 for details). The residual image (*bottom-left panel*) is obtained by subtracting the model image from the low resolution Spitzer image. Residuals are caused mainly by inaccuracies in the PSF shape and/or color gradients, which limit the accuracy of the derived photometry.

photometry by modeling a source with a PSF, has been implemented in programs like DAOPHOT (Stetson, 1987, see also the *daophot* task within IRAF) and DOPHOT (Schechter *et al.*, 1993). Comparisons between different programs suggest that SEXTRACTOR is now also competitive with PSF-fitting photometry (Annunziatella *et al.*, 2013).

**Example 2.11: *How to measure photometry in the confusion limited regime*?**

(1) Use stars in the field to measure the PSF profile in the confusion-limited primary image (see Ex. *How to estimate the spatial resolution of an image*?);

(2) Use another secondary image of the same field taken at another wavelength at which confusion is not an issue (e.g., optical from the HST). Run SEXTRACTOR on this image to produce a list of positions and shapes (size, P.A.; see *How to construct a source catalog from an image*? and see also Fig. 2.15) of the objects;

(3) For each object in the high resolution secondary image, convolve the object flux distribution extracted from this image so that the final resolution of the convolved image matches the PSF of the primary image (one can use a specific software such as, e.g., TFIT, see Laidler *et al.*, 2007 or PyGFiT, see Mancone *et al.*, 2013). If the PSF of the secondary image is much narrower than that of the primary image, this is often done by simply convolving the secondary image by the primary image PSF;

(4) For all sources detected in the low resolution primary image, select the high resolution sources that contribute to the low resolution source. Use a linear combination of the convolved image obtained at the previous step to fit the total low resolution source flux and decompose the low resolution flux into the different individual contributions;

(5) Subtract all fitted sources to the primary image to produce a residual image and check the flux measurement quality (see Fig. 2.15). It can be useful to proceed iteratively by running SEXTRACTOR on fainter and fainter objects until no further extraction is possible, and one might need to tune the extraction parameter accordingly.

#### 2.5.2.4 *Can we recover total, multi-wavelength magnitudes?*

Estimating the total stellar mass or luminosity of a galaxy requires an estimate of the total magnitude of that object in at least one band so that aperture measurements can be converted into total fluxes. One generally assumes that the *aperture correction*, i.e., the difference in magnitude, in a given band, between the magnitude in a given aperture and the total magnitude, is independent of wavelength, or that the effect is minor when compared to other factors of uncertainty and can be safely ignored. This is something that is often assumed when inferring physical quantities from SEDs (e.g., stellar mass), although it can actually be an additional source of systematic error (see below). The aperture correction should be estimated in a band that is deep enough to recover the faintest isophotal levels, and which samples rest-frame wavelengths larger than 500nm so that it is representative of the stellar mass distribution (see Sec. 2.2.3).

#### 2.5.2.5 *Flux and color uncertainties*

Uncertainties in flux measurements must include several factors and depend on the regime in which photometric measurements are conducted:

- For point-sources, uncertainties are generally dominated by photon noise, and these can be estimated by simulating artificial stars with increasing magnitudes, as shown in Fig. 2.16. This results in lower limit estimates of the flux uncertainty which can reach large values at faint magnitudes (i.e., several tenth of magnitudes, see Fig. 2.16);
- In the case of extended sources, additional sources of uncertainties are caused by aperture corrections and the estimate and subtraction of the background. Programs do not always account for all these sources of errors (i.e., SEXTRACTOR does not account for uncertainties in the estimate of the background, which is the dominant source of uncertainty). An additional uncertainty is due to the photometric calibration of the image (i.e., the photometric zero point), which can be important and even be the dominant

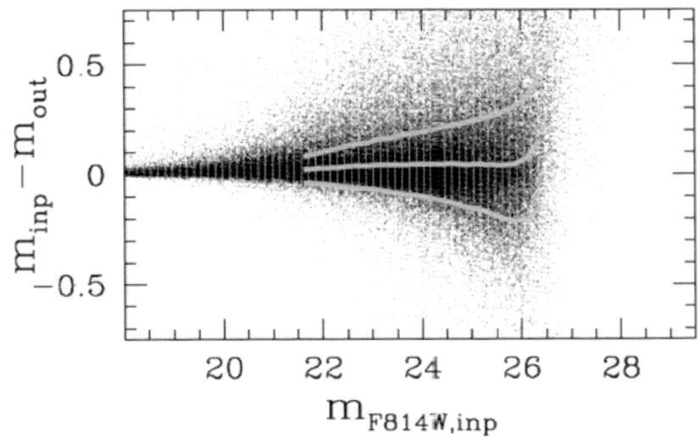


Fig. 2.16: Example of random error characterization for a photometric pipeline for HST/ACS $z$-band point sources (from Annibali *et al.*, 2008, ©AAS, reproduced with permission). The difference between the input magnitude of artificial stars and the measured value are shown as a function of the input magnitude. Note the increasing random uncertainty with fainter sources, which becomes prominent ($\sigma > 0.1$ mag) above $m \sim 22$. The grey lines show the $1 - \sigma$ scatter.

source of (systematic) error, particularly in ground-based NIR imaging;

- In the confusion-limited regime, confusion noise and source correlation between images at different wavelengths are important additional sources of error. These are usually estimated by comparing the resulting source catalog with a pre-existing and well-understood catalog at the same wavelength. Several iterations may be necessary to optimize the source extraction process and minimize the resulting uncertainties.

Additionally, there are other sources of systematic error that can be particularly problematic when measuring colors:

- With point-sources, PSF-fitting must be handled carefully when comparing photometry at several wavelengths, for instance in the IR, since the PSF size varies as a function of wavelength. This implies that in this regime, photometry is extracted with no control on aperture size, which can results in biased colors. To overcome this, one can use PSF-matching photometry to measure consistent photometry (see Ex. *How to measure photometry in the confusion regime*? and Fig. 2.15);

- With extended sources, the use of different apertures can result in significant biases when estimating colors, and hence the SEDs of galaxies. For example, the average $B - z$ in $z \sim 0.6$ spirals is found to vary significantly with radius, with a typical $\sim 0.4$ mag difference between $R_{\rm half}$ and $2 \times R_{\rm half}$ apertures (see Fig. 2.11). In turn, this can bias the photometric aperture corrections when they rely on different wavelengths (see Ex. *Biased colors resulting from uncontrolled aperture corrections*).

**Example 2.12: *Biases in colors caused by uncontrolled aperture corrections***

Let us assume that one observes a $z \sim 0.6$ rotating disk, in which one wants to measure photometry within two apertures of size $R_{\rm half}$ and $2 \times R_{\rm half}$. In such a galaxy, Neichel *et al.* (2008, see their Fig. 10) found that on average $B - z \sim 2.2$ and 1.8 within these two apertures, respectively (see Fig. 2.11). Let us also assume that one measures the photometry of this galaxy within $R_{\rm half}$ (which provides more S/N) and tries to estimate the total flux by determining the aperture correction between $R_{\rm half}$ and $2 \times R_{\rm half}$. If this aperture correction is estimated in $B$ band and also applied in $z$ band, this biases the extrapolated $z$-band magnitude by $\sim 0.4$ mag, corresponding to the $B - z$ difference between these two apertures. Such errors are very difficult to control since they depend on the exact shape of the SED and on how the different colors vary as a function of radius. If there is a single counterpart in all considered bands, one can use PSF-matching techniques to degrade all of the bands to the worst resolution before extract consistent photometry (see above). Otherwise, deriving global quantities from SED fitting (e.g., stellar mass or SFR, see next section) will be affected by systematic effects from the aperture correction. In our example, this could lead to an additional systemic of 0.16 dex in the derived total stellar mass.

It is important to test and quantify the precision and accuracy of the photometric pipeline used to measure the fluxes of distant galaxies in order to avoid underestimating uncertainties. This can be done by using numerical simulations of artificial sources as input test data.

## 2.6 Spectral energy distributions (SEDs)

### 2.6.1 *Impact of the stellar populations: which wavelength tells you what?*

A SED of a galaxy is a representation of the emitted light as a function of wavelength (or frequency). Galaxies can have very different SEDs depending on their stellar content and on the amount of dust present (see below), as illustrated in Fig. 2.17. While nearby elliptical galaxies are dominated by old stars and their SED peaks in the red visible/NIR, nearby spiral galaxies are basically a superimposition of an old population (mainly located in the central bulge) and a population of young and intermediate-age stars (often located in the disk). The UV and visible blue light is dominated by young stars in H II regions, with an increasing contribution from Sa to Sd and

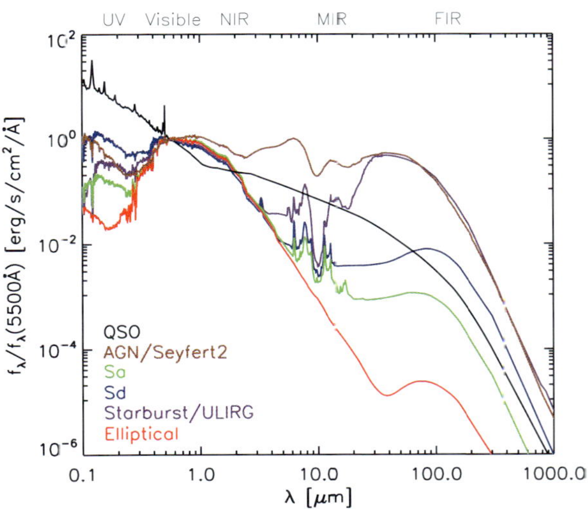


Fig. 2.17: Examples of galaxy SEDs (Polletta *et al.*, 2007) for different type of galaxies. The SEDs are plotted in units of flux density normalized at 5500Å as a function of wavelength, in logarithmic units. The templates were generated from the GRASIL code (Silva *et al.*, 1998) and/or observations (see Polletta *et al.*, 2007 for details).

starbursting galaxies (which translates into a progressive decreasing of the 4000Å break). Extreme starbursts such as Luminous Infrared Galaxies (LIRGs) emit most of their light in the mid infrared/far infrared (MIR/FIR) due to the reprocessing of the UV/blue visible light by dust (note also the numerous broad peaks dominating the MIR emission, which are due to Polycyclic Aromatic Hydrocarbons, see Leger *et al.*, 1989). This relatively simplistic picture can be complicated by the presence of an AGN, which can boost different parts of the SED due to nonthermal emission (see Sec. 3.6.2).

A low resolution estimate of the SED can be obtained using a series of narrow and/or broad-band filters (see Sec. 2.2.3). By comparing the observed SED with predictions from a grid of Stellar Population Synthesis (SPS) models (Tinsley, 1967, 1972; see also the review by Schaerer 2013), one can use SEDs to estimate accurate rest-frame luminosities (see Sec. 1.2) and in principle constrain the fundamental properties of the underlying stellar populations, i.e., stellar mass, SFR and history, metallicity, and ages (see Sec. 2.6.3.1). In practice, this technique faces considerable difficulties due to degeneracies between:

- Different ages; in the optical, the combination of a young and of an old stellar population results in identical signatures as of a single intermediate-aged population (see, e.g., Rodrigues *et al.*, 2014). This degeneracy can be broken by sampling the relative strengths of the Balmer and of the 400 nm breaks using the D4000 index (see Sec. 3.8.2), or determining the $EW$ of Balmer absorption lines;
- Age and metallicity; simultaneous variations of $\Delta\text{age}/\Delta Z \sim 2/3$ result in similar galaxy SEDs (Worthey, 1994). Breaking this degeneracy requires a spectral resolution of $R > 3000$ to resolve metal lines (see Sec. 3.8.2);
- Age and dust (Gordon *et al.*, 1997); distinguishing both effects requires sampling the SED both in the rest-frame UV and the IR (see below).

In principle, these limitations make the SED a rather limited tracer of age, metallicity, and dust. Useful constraints can be obtained in a relative sense and under specific conditions, e.g., if one compares the relative fraction of young vs. old stars assuming a given metallicity. However, metallicity remains very difficult to estimate using a simple SED when no constraints on age are available. Moreover, dust strongly depends on the specific object geometry, which makes it nearly impossible to estimate accurately.

More accurate constraints can be derived for the SFR but then the choice of the filters used to sample the SED is crucial. SFRs can in principle be estimated using the rest-frame UV (i.e., ∼1000–3000Å), which is sensitive to young stars. However, UV and optical photons are easily absorbed by dust grains and then re-emitted in the IR, so that sampling IR wavelengths is also required when estimating SFRs (see Sec. 2.6.5).

Figure 2.5 shows that the fraction of K band light due to MS stars is almost constant for ages larger than 0.1 Gyr (Charlot, 1996). Thus NIR light is far more reliable than UV or B luminosities when using an SED to estimate the stellar mass (see Sec. 2.6.4). With very high redshift galaxies, one must be particularly careful not to sample only the part of the SED that is below the 4000Å break (i.e., above $z \sim 1.5$), which is the case with the ACS camera on-board the HST, since this domain is largely dominated by the light emitted by the youngest stellar populations (see above and Fig. 2.21). The new NIR HST/WFC3 camera partially solves this problem by pushing it to $z \sim 3$. All of these considerations make the rest-frame UV, the optical-NIR, and the IR three essential regions that need to be sampled by the SED.

### 2.6.2 *Impact of dust and gas*

**Galactic extinction** The galactic extinction following a given line of sight can be estimated using the maps of Schlegel *et al.* (1998),[25] which were further corrected by Peek and Graves (2010).[26] For example, and according to the Schlegel *et al.* (1998) map, in the widely studied GOODS-South field, galactic extinction is only $E(B-V) = 0.008$.[27] This implies that in these fields, this correction

[25]See `http://astro.berkeley.edu/~marc/dust/data/data.html`.
[26]See `http://user.astro.columbia.edu/~jpeek/dust.html`.

can almost always be ignored (e.g., Nonino *et al.*, 2009). It remains nevertheless crucial to correct photometric measurements for Galactic extinction when the field of view is at moderate Galactic latitudes ($b^{II} < 45°$).

**Internal extinction** Estimating the internal extinction is far more complex. The only well-tested photometric method is to use multi-wavelength data and compare the ratio between the IR and UV fluxes (see details in Sec. 3.6.1).

Attempts at estimating this ratio directly using a tentative correlation between the UV SED slope $\beta$ (defined as $f_\lambda \propto \lambda^\beta$) and the "InfRared eXcess" IRX, defined as the ratio between the FIR and the UV flux, have shown that this so-called IRX-$\beta$ method can not be used, as it has been shown to fail as soon as the dust content becomes significant (i.e., in (U)LIRGs, see, e.g., Takeuchi *et al.*, 2012).

SED fitting is also often used to estimate the amount of extinction in the stellar phase. The most widely used empirical dust models are those of Cardelli *et al.* (1989); O'Donnell (1994); Calzetti *et al.* (2000); Fitzpatrick (1999); Fitzpatrick and Massa (2007), or the theoretical one of Charlot and Fall (2000). However, this requires *a priori* knowledge of the dust distribution and geometry, which is impossible for most distant galaxies (see Sec. 3.6.1)

**The specific case of extinction in regular galactic disks.** In the most inclined disk-galaxies, the line-of-sight integrates a larger part of the disk, resulting in thicker optical depth and a larger dust absorption. There is therefore a systematic change of internal absorption with inclination of disks (Tully *et al.*, 1998; Masters *et al.*, 2003; Wild *et al.*, 2011). To the first order, absorption in a given passband is often parameterized as $\Delta m = \gamma \log(a/b)$, where $a/b$ is the observed major to minor axis ratio, and $\gamma$ is a constant that depends on the wavelength (see Unterborn and Ryden, 2008 and Masters *et al.*, 2003 for more accurate parametrizations). The axis ratio is related to disk

[27] The color excess $E(B-V)$ is simply the difference between the observed $B-V$ color and the intrinsic (i.e., without extinction) one.

inclination and intrinsic disk thickness (see Ex. *Measuring the surface brightness profile and basic properties of a distant galaxy*). Another correction, defined as a function of mass and wavelength was also developed by Tully *et al.* (1998) and is based on the velocity width $W$ with $\Delta m = a_\lambda + b_\lambda(\log(W) - 2.5)$.

We emphasize that these corrections all assume a particular geometry for the dust distribution, and are therefore only applicable to galaxies that have the shapes and the kinematics of regular disk galaxies.

#### 2.6.2.1 *The impact of neutral HI absorption*

An abrupt flux decrease on the continuum appears at the "Lyman limit" of 912Å because this wavelength corresponds to an accumulation of a large number of Lyman series transitions for the hydrogen electron.[28] Because the Lyman series corresponds to transitions from the ground-state of the hydrogen electron, all photons with wavelengths shorter than the Lyman limit can ionize neutral hydrogen and are then absorbed even by clouds with tiny column densities. This implies that galaxies do not show thermal emission below this limit since all these photons are systematically absorbed by the gas lying along the line of sight of distant galaxies (see Chapter 3).[29] When two adjacent filters are used to measure the continuum flux just blue-ward and red-ward the Lyman limit, this results in a relatively strong *photometric break* (see Fig. 2.18). This "Lyman break" can be enhanced by intervening gas along the line of sight, which can also significantly absorb photons in the range 912–1216Å, i.e., between the Lyman limit and the Lyman $\alpha$ transition. Galaxies selected in this manner are called Lyman Break Galaxies[30] (see Sec. 1.6; Giavalisco 2002).

---

[28] Note that the other transition series of the hydrogen electron result in additional breaks on the continuum, such as for instance the break at 3646Å which corresponds to the limit of the Balmer series (see Chapter 3).

[29] Note, however, that nonthermal emission from, e.g., AGNs, can result in continuum flux below the Lyman limit.

[30] also called "dropout" galaxies.

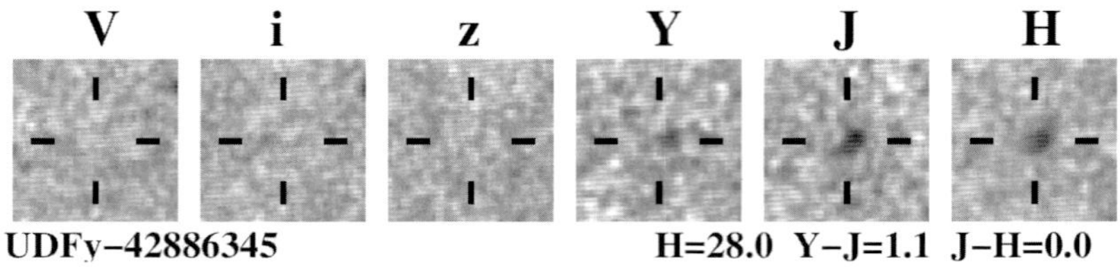


Fig. 2.18: $V_{606}$, $i_{775}$, $z_{850}$, $Y_{105}$, $J_{125}$, and $H_{160}$ image cutouts ($2.4'' \times 2.4''$) of an LBG candidate galaxy (from Bouwens *et al.*, 2010, ©AAS, reproduced with permission). Significant flux (i.e., $\geq 4\sigma$) is detected in the J and H bands, but not in the bluer bands ($<2\sigma$), which suggests that this LBG would be at $z \sim 8$. Uncertainties on the measured photometry is ~0.1–0.2 mag in each band.

Every HI structure along the line of sight can produce an absorption feature in the spectrum of background sources. This can result in a very dense structure of absorption lines called the Lyman alpha forest (see Sec. 3.7.2). Such absorption features were first modeled by Madau (1995). More recent studies have shown their model to over-predict the amount of absorption in the 912–1216Årange (see Meiksin, 2006; Inoue and Kamaya, 2008; see also Harrison *et al.*, 2011). A complete suppression of any emission blue-ward the Ly$\alpha$ line is expected if the IGM is filled with neutral HI, resulting in the so-called Gunn–Peterson trough (Gunn and Peterson, 1965). This effect is observed in the line-of-sights of quasars above $z \sim 6$ but not below, which suggests that the Universe reionization might be completed by $z \sim 6$ (Becker *et al.*, 2001).

#### 2.6.2.2 *The impact of nebular emissions*

Nebular emission can also strongly impact the shape of the SED in two ways:

- Nebular continuum emission, which arises from 2-photon and H or He free–free and free–bound emissions (Osterbrock and Ferland, 2006), increases with decreasing metallicity and increasing ionization parameter. It is predicted to dominate broad-band filters that sample wavelengths just short-ward of the Balmer break, and in the NIR for stellar ages younger than ~3 Myr (see Fig. 2.19;

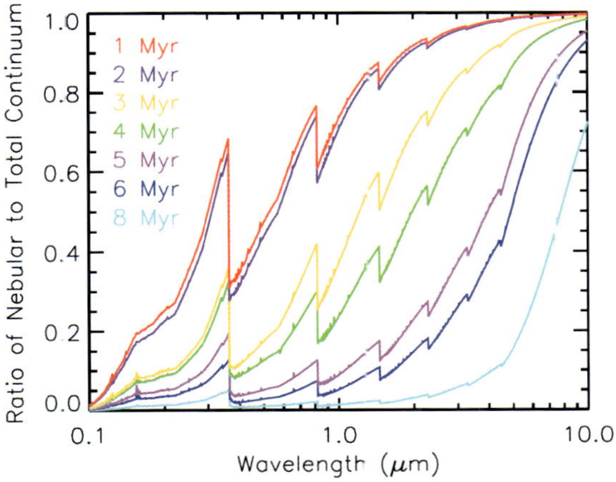


Fig. 2.19: Ratio between the nebular continuum and the total continuum flux as a function of wavelength and age for a Starburst99 SSP (from Reines *et al.*, 2010, ©AAS, reproduced with permission).

see also Leitherer *et al.*, 1999). It is therefore essential to take the nebular continuum emission contribution into account in relatively unevolved objects such as very distant young galaxies (e.g., Schaerer and de Barros, 2009) or local Blue Compact Dwarf galaxies (e.g., Krueger *et al.*, 1995). Neglecting nebular emission can lead to biases in the SFR calibrations by factors 0.05–0.15 dex (Raiter *et al.*, 2010);

- Emission lines: they can represent up to ∼50% of broad-band filter photometry in strong starbursting $z \sim 2$ galaxies (see Sec. 2.2.3). If neglected, this can lead to a bias in stellar mass estimates that is upward of a factor of 2–3 in ∼65% of $z \sim 3-6$ LBGs (e.g., Schaerer and de Barros, 2009; de Barros *et al.*, 2014, see also Zackrisson *et al.*, 2008).

When nebular emission is expected to be significant, it is essential to fit SED with SPS codes (see below) that include this contribution (e.g., Leitherer *et al.*, 1999; Fioc and Rocca-Volmerange, 1999; Schaerer and de Barros, 2009).

### 2.6.3 *Estimating galaxy properties from an SED*

#### 2.6.3.1 *Use of Stellar Population Models*

There are many Stellar Population Models libraries as well as associated SED fitting codes.[31] Their primary outputs are Single Stellar Populations (SSP) that describe the color of a stellar population with given age and metallicity. Since stars are continuously formed within a galaxy, one needs to assume a Star Formation History (SFH), from which a Composed Stellar Population (CSP) can be calculated, assuming a particular IMF (see, e.g., Conroy, 2013). The most widely used SFH parametrization is an exponentially declining burst with e-folding time $\tau$, which assumes the SFR of galaxies is decreasing, following their last burst. While this appears to be appropriate for most local galaxies, it has been suggested that $z \geq 2$ galaxies might instead be observed while they are undergoing a burst of star formation, i.e., before they reach their maximum SFR. This is why an "inverted $\tau$-model" was suggested to be more appropriate at high redshifts (e.g., Maraston *et al.*, 2010; Pforr *et al.*, 2012). A complete review of SPS models can be found in Conroy (2013) (see also Sec. 3.8), while the fitting process is reviewed by Walcher *et al.* (2011) to which we refer for details.

The limited knowledge of stellar evolution (i.e., the shape of the IMF, molecular cloud disruption timescale, and the treatment of advanced stages of stellar evolution such as blue straggler, the horizontal branch, or TP-AGB stars) and galaxy ISM properties (i.e., dust attenuation and clumpiness of the ISM) result in degeneracies in all these models and then in systematic uncertainties on the derived properties. Several extensive studies (e.g., Conroy *et al.*, 2009; Conroy and Gunn, 2010; Conroy *et al.*, 2010; Kriek *et al.*, 2010; Pirzkal *et al.*, 2012; Pforr *et al.*, 2012; Conroy, 2013) have shown that metallicity is by far the less constrained quantity and that ages and dust content remain difficult to estimate accurately using SEDs. This again is an indication that trying to infer

[31] See a exhaustive list at `http://www.sedfitting.org`.

such quantities from SED fitting alone should be avoided (see also Sec. 2.6).

### 2.6.3.2 *Issues related to SED fitting*

**Number of stellar populations.** It is difficult to fit several stellar populations together because of strong degeneracies between these stellar populations (see Sec. 2.6) especially when no spectral information is available (see Sec. 3.8.3).

It has been a common approach to only fit one stellar population to the whole galaxy SED in order to infer global average properties. This can lead to strong biases in all of the estimated quantities, especially when the light is dominated by the youngest stellar populations, while the oldest one will be almost completely out shined (see below).

**Aperture corrections.** One must be very careful to derive an SED that samples the same galaxy area as a function of wavelength, i.e., to use a constant aperture. This might turn out to be difficult in practice, in particular at IR wavelengths where the PSF is large compared to the size of distant galaxies (see Sec. 2.5.2). In addition, color gradients across galaxies complicate deriving the aperture corrections that are required to estimate global properties such as the total stellar mass or the SFR. This can result in important biases (see Ex. *Biased colors resulting from uncontrolled aperture corrections* in Sec. 2.5.2).

**Numerical methods and uncertainties.** The simplest method of fitting an SPS model to a SED is to use a simple $\chi^2$ least-square method (see Ex. *How to fit a galaxy SED?* and Fig. 2.20) while using photometric uncertainties that account for all sources of errors (see Ex. *How to construct a source catalog from an image?*). Several well-known difficulties are associated with this method (e.g., Papovich *et al.*, 2001): (1) systematics can arise from mismatches between models and reality (e.g., wrong dust model, IMF, or SFH, finite grid of parameters for the models, missing stellar populations, inaccurate

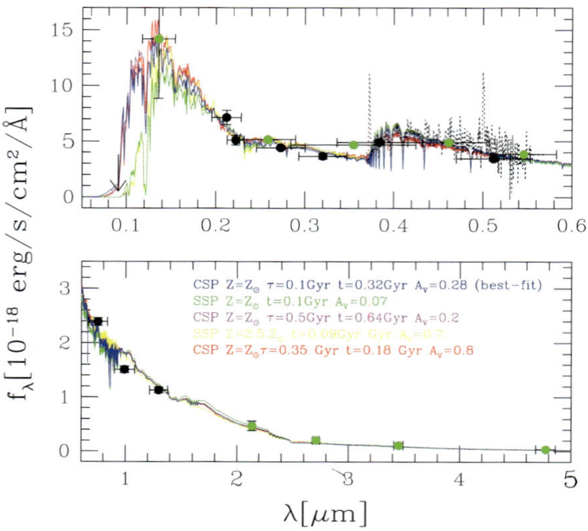


Fig. 2.20: Example of SED fitting using a simple least-square method. Credit: Puech *et al.* (2009), A&A, 493, 899, reproduced with permission ©ESO. The two panels show the rest-frame UV-optical (*upper panel*) and optical-IR (*lower panel*) SED of J033241.88-274853.9 at $z = 0.67$ using 3 arcsec, fixed-aperture photometry. Green points represent space-based photometry, while black-points represent ground-based photometry (see Puech *et al.*, 2009 for details). The GALEX/FUV $5 - \sigma$ detection limit is represented as a vertical arrow. A VIMOS low resolution spectrum is overlaid using a black dashed line. Because noise affects the $\chi^2$ estimation, all the SPS models shown are in principle an acceptable fit to the data. With the library of SED templates adopted here, the range of possible stellar mass within a 90% confidence interval is found to be $\log(M_{\rm stellar}/{\rm M}_\odot) = 9.45^{+0.17}_{-0.14}$.

stellar population templates, etc.), and (2) nonGaussian errors can bias the $\chi^2$ optimization.

Except for the case of very faint distant galaxies (for which random uncertainties can easily reach a few tenth of magnitudes), photometric uncertainties are generally small, meaning that SED fitting is actually not limited by photon noise but by systematic uncertainties between the different filters. A standard practice has been to add ∼0.05 mag in quadrature to all photometric uncertainties to account for systematic errors. These systematic errors should however be carefully estimated and accounted for first (see Ex. *How to fit a galaxy SED?*). In addition, when fitting SEDs with a large number of filters, different photometric points may not be independent.

Indeed, if a given region of the SED contribute to several photometric measurements (i.e., because it is included in several filters), then one should avoid to bias the SED fitting process with these regions. The latter can be handled using different weights for different bands.

**Example 2.13: *How to fit a galaxy SED?***

(1) Construct the galaxy empirical SED by cross-correlating photometric catalogs. Be careful to adopt a constant aperture size as a function of wavelength, which should be close to $R_{\rm half}$ at rest-frame optical-NIR in order to include sufficient light that constrains the stellar mass;
(2) Check whether PSF corrections are needed, especially in the observed UV and IR where PSF are large compared to galaxy sizes;
(3) Estimate the vertical and horizontal error-bars; properly account for systematic uncertainties (i.e., uncertainties on the photometric zero points or PSF corrections); Pay attention to avoid overlapping filters or weight them properly to avoid biasing the fitting in the region where filters overlaps;
(4) Construct a library of stellar population synthesis models using a range of metallicity $Z$, age $t$, and star formation history (e.g., the exponentially declining $\tau$ models);
(5) Adopt a dust model (e.g., the Calzetti extinction curve for starburst galaxies) which is characterized by two parameters, e.g., $A_V$ and $R_V$;
(6) For every set of parameters ($Z$, $\tau$) determine the age and dust values of the best-fit template from the library, and record the resulting normalization factors, the $\chi^2$ values, and the ($t$, $A_V$, $R_V$) best-fit values;
(7) Estimate the $\chi^2$ boundaries using confidence intervals or bootstrap re-sampling. All templates within these values are acceptable fits to the empirical SED (see Fig. 2.20). Try other dust models and SPS libraries to estimate systematic effects.

### 2.6.4 *Estimating stellar mass*

#### 2.6.4.1 *Using full SED fitting*

Estimating stellar mass from SED fitting requires sampling the rest-frame optical-NIR wavelengths, which provide the part of the SED most sensitive to stellar mass (see Sec. 2.6.1). Whether SPS models can accurately reproduce the NIR regions of the SED of a galaxy has been questioned (e.g., Taylor *et al.*, 2011). It appears that this issue can definitively be related to photometric inaccuracies in the NIR photometry when observed at ground-based telescopes without properly monitoring the airmass and water vapor content.[32] The inclusion of NIR bands definitively produce more precise and accurate estimates of the stellar mass, in particular at $z \geq 2$, with systematic errors that are reduced by a factor ∼4–10 (Pforr *et al.*, 2012, see Fig. 2.21).

The $\chi^2$ estimates are themselves affected by noise and therefore subject to uncertainties, which implies that the template with the minimal $\chi^2$ value does not necessarily provide the most accurate estimate of the stellar mass. There are two main methods to determine uncertainties on the resulting fitted result. Since one knows the exact number of parameters to be fitted, and then the degree of freedom during the fitting, a first method is to determine 68% and/or 90% confidence intervals using constant $\chi^2$ boundaries derived using the whole $\chi^2$ distribution (Avni, 1976). For instance in Fig. 2.20, considering all the templates that are within a 90% confidence interval result in a ∼0.3 dex uncertainty range. The second method is empirical and consists of using Monte Carlo simulations by randomizing the photometric measurements within the vertical error-bars and re-fitting the SED. Bayesian methods (see details in Walcher *et al.*, 2011) have been claimed to provide more accurate estimates by

[32]Indeed, atmospheric extinction in broad infrared filters is mostly due to saturation of molecular line absorption (see Milone and Young, 2011). As the airmass increases, more and more of the absorption within the passband is saturated because extinction varies strongly with wavelength. This has led the IAU to establish the Infrared Working Group Passbands in 1991, whose goal is to optimize the passband in NIR for minimizing these effects.

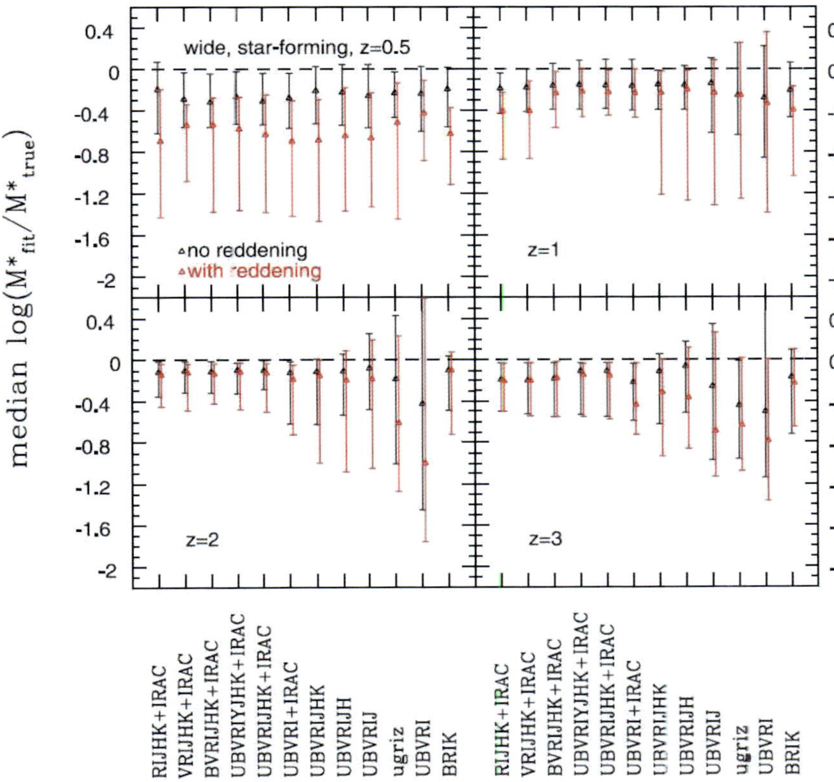


Fig. 2.21: Effect of the filter set on stellar mass estimates derived using SED fitting (from Pforr *et al.*, 2012, MNRAS, 422, 3285, reproduced with permission of Oxford University Press on behalf of the Royal Astronomical Society). The four panels show (clockwise) the median stellar mass recovery of star forming galaxies as a function of filter set (broad-band filters only) for redshifts 0.5, 1, 2, and 3. Red dots represent reddened solutions, while black dots represent un-reddened solutions and error bars represent 68% confidence levels. Note how the stellar mass estimates become particularly inaccurate at high redshifts where the filter set does not probe wavelengths above the 4000Å break at $z = 2$ ($z = 3$), which is redshifted at 1.2 (1.6) $\mu$m, i.e., in J (H) band.

~0.1–0.15 dex (see, e.g., Taylor *et al.*, 2011) when estimating properties from SED fitting (e.g., Kauffmann *et al.*, 2003; Walcher *et al.*, 2011; see also Ex. *How to estimate the stellar mass of a galaxy using SED fitting?*). An advantage of these methods is that they provide a way of estimating the confidence intervals properly, even in cases for which the input probability distribution functions (p.d f.) are not

Gaussian at all (Pirzkal *et al.*, 2013). However, they require an *a priori* p.d.f. for the SPS templates, which in turn implies that parameters estimated using the Bayesian approach depend on this choice. When there is no specific knowledge available about the priors, which is generally the case in SED fitting, the only reasonable choice is to use uniform distributions.[33] It should otherwise be strictly avoided when the *a priori* p.d.f. is not well-controlled.

**Example 2.14: *How to estimate the stellar mass of a galaxy using SED fitting?***

This example shows how to avoid various traps in estimating stellar mass using full SED fitting. It does not provide a way to estimate the systematic uncertainties associated with this method.

(1) Ensure that proper photometric measurements in the rest-frame NIR, i.e., above 0.8 $\mu$m, are available;
(2) If the empirical SED does not account for the total fluxes, correct for aperture effects using a total or isophotal magnitude (see Sec. 2.5.2.4);
(3) Fit a set of stellar population libraries to the empirical SED as described in Ex. *How to fit an SED?*
(4) For each template in the library, project the SED onto a given rest-frame optical/NIR filter, and use the ratio between the projected and model luminosities in this band to re-scale the tabulated template $M_{\rm stellar}$ (for instance in the Bruzual and Charlot, 2003 SED library, the templates are all normalized to $M_{\rm stellar}/{\rm M}_\odot = 1$);
(5) Derive, for each template, $e^{-\chi^2/2}$, which is proportional to the probability of measuring the observed fluxes assuming that this template is the true one. Using the Bayesian

(*Continued*)

[33]Note that strictly speaking, the p.d.f. is uniform only *within* the chosen SPS library. It is therefore recommended to test whether the chosen library samples a large enough parameter space in metallicity and age.

**Example 2.14: *Continued***

theorem with uniform priors, the probability that a particular template is correct given the observed fluxes is also proportional to $e^{-\chi^2/2}$ (Walcher *et al.*, 2011);

(6) Marginalize over the posterior probability distribution by estimating $<M_{\text{stellar}}> = \int M_{\text{stellar}} e^{-\chi_T^2/2} dT$. This will estimate the most likely value for $M_{\text{stellar}}$ amongst the set of libraries used;

(7) Estimate the statistical uncertainty on $<M_{\text{stellar}}>$ using $\sigma_{<M_{\text{stellar}}>} = \sqrt{<M^2_{\text{stellar}}> - <M_{\text{stellar}}>^2}$;

(8) Evaluate the impact of out shined old stellar populations (see Fig. 2.22).

In most situations, estimates of the stellar mass are probably affected by relatively large systematic uncertainties due to, for example, mismatch in the SFH (Maraston *et al.*, 2010; Pirzkal *et al.*, 2012), or the poorly understood effect of TP-AGB stars on galaxy SEDs, which remains a matter of debate (e.g., Maraston, 2005; Kriek *et al.*, 2010; Melbourne *et al.*, 2012; Zibetti *et al.*, 2013). The numerous and uncontrolled biases that can affect SED fitting (see Sec. 2.6.3.1) can result in errors in the stellar mass estimates that are as high as 0.6 dex (i.e., a factor 4) for $z \sim 2$ bright red galaxies (Conroy *et al.*, 2009) or even for star forming galaxies at $z \sim 0.5$ (Pforr *et al.*, 2012). The effect is expected to be the strongest with galaxies dominated by their youngest populations, whose light dominates the photometric budget by a large factor, preventing the oldest populations from being detected. It is therefore likely that stellar mass estimates from a single component SED fitting in increasingly distant blue galaxies become very inaccurate if not irrelevant. Useful methods have been proposed to estimate the maximal contribution of underlying old populations by fitting a two component model (Papovich *et al.*, 2001, see Fig. 2.22). This approach leads to more robust upper limits to the stellar mass value.

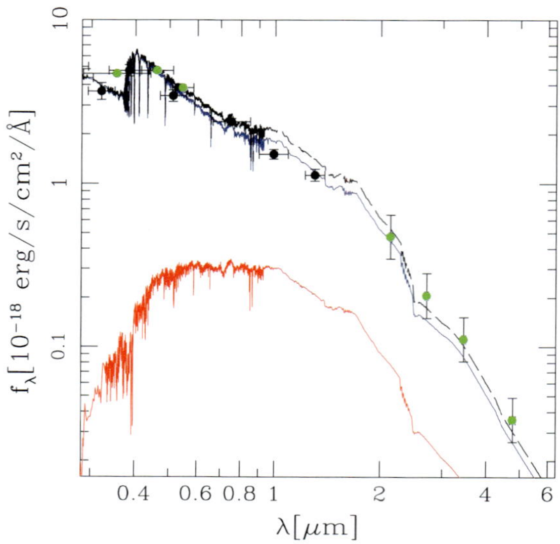


Fig. 2.22: Example of how one can estimate the maximal contribution of the oldest populations in SED fitting, following Papovich *et al.* (2001). Credit: Puech *et al.* (2009), A&A, 493, 899, reproduced with permission ©ESO. The figure shows the rest-frame SED of J033241.88-274853.9 above the Balmer break with dot colors identical to Fig. 2.20; the blue curve represents the best-fit model as shown in Fig. 2.20. If this galaxy was dominated by young stars only, the blue curve would represent a good SED model reliable for estimating the stellar mass. However, Balmer absorption lines in the best-fit model suggest the presence of an underlying older stellar population (A/F stars), which might be hidden below the much larger flux contribution emitted by the youngest populations. To try to estimate the maximal contribution of the oldest populations, one can fit a library of SSPs with identical parameters as that of the best-fit model but using a stellar age equal to the age of the Universe at the redshift of the object that is added to the best-fit model representing the contribution of the youngest populations. The resulting best-fit old population model is shown in red, while the combination of the best-fit young and old population templates is shown in black. The stellar mass estimated using the blue curve/youngest population is found to be $\log(M_{\rm stellar}/{\rm M}_\odot) = 9.45^{+0.17}_{-0.14}$ (Diet Salpeter IMF, see Fig. 2.20), while taking into account the possible contribution of the oldest populations increases the stellar mass up to 9.73 (see Puech *et al.*, 2009).

#### 2.6.4.2 *Using the single color method*

A well-known expectation of stellar population synthesis models is the existing correlation between the stellar mass-to-light ratio and the color of objects (see Fig. 2.5). This has been used by Bell and de Jong (2001) and Bell *et al.* (2003) to construct simplified prescriptions to statistically estimate the integrated stellar mass-to-light ratio

Table 2.4: Optimal color-luminosity $M/L$ ratio calibrations.

| $\lambda$ | $a_\lambda$ | $b_\lambda$ |
|---|---|---|
| $B$ | −0.942 | 1.737 |
| $R$ | −0.520 | 1.094 |
| $J$ | −0.261 | 0.433 |
| $K$ | −0.206 | 0.135 |

All these calibrations use the $B - V$ color as recommended by McGaugh and Schombert (2014). The calibrations are given as $\text{Log}_{10}(M/L_\lambda) = a_\lambda + b_\lambda \times (B - V)$. Other calibrations can be found in Table 7 of (Bell *et al.*, 2003). Note how $b_\lambda$ converges to 0 with increasing $\lambda$, which results in a weaker dependence on color at longer wavelengths.

in different bands using one or several observed colors (e.g., Zibetti *et al.*, 2009; see also Gallazzi and Bell, 2009).

Optical stellar mass-to-light ratios are well-correlated with optical-optical colors because of the well-known degeneracy between the metallicity and the age of the underlying stellar populations, which results in relatively tight correlations between these two quantities (see, e.g., Bell and de Jong, 2001). In practice, a NIR rest-frame magnitude combined with a UV–V color leads to more accurate estimates of the mass-to-light ratio, hence of the stellar mass (Kauffmann and Charlot, 1998; Bell and de Jong, 2001). Furthermore, McGaugh and Schombert (2014) found that the Bell *et al.* (2003) color-based prescriptions result in almost self-consistent stellar masses as a function of the wavelength of reference in the mass-to-light ratio. They found that the best correlations between color and mass-to-light ratio are with the B-V color making such combinations particularly attractive (see $b_\lambda$ values in Table 2.4).

An important drawback of these methods is that the calibrations may be sensitive to a possible evolution in $M/L$ as a function of redshift. Such an evolution is expected from the decrease in average maximal stellar age that galaxies can reach as lookbacktime (redshift) increases: for an $M/L$ calibration derived at $z = 0$ with galaxies supposed to form at a given reference epoch (e.g., Bell *et al.*, 2003 adopted a reference age of 12 Gyr for all local galaxies in their $z \sim 0$ sample), there is simply less time for stars to age compared to those in local galaxies. This translates, e.g., in a $\sim -0.16$ dex difference in

$M/L_K$ between $z = 0$ and $z \sim 0.6$ (see Appendix A.3 of Puech *et al.*, 2008).

**Example 2.15: *How to get simple estimates of the stellar mass?***

(1) Gather photometric measurements in two observed filters that bound a rest-frame optical-NIR broad-band filter (i.e., at least >500 nm) as closely as possible; Favor space over ground-based photometry and longer wavelengths;
(2) Measure the rest-frame optical-NIR luminosity in the chosen band as described in Sec. 1.4;
(3) Check the table provided by Bell *et al.* (2003) (see also Table 2.4) and find the filters that provide the best sampling of a rest-frame UV–V color, and use them to estimate the stellar mass-to-light ratio $M/L$ in the chosen band; Use the rest-frame optical-NIR luminosity measured in the previous step to get an estimate of the stellar mass;
(4) Convert to your favorite IMF by rescaling by the proper correction factor (see, e.g., Bell *et al.*, 2003). Note that the Salpeter + Bell *et al.* (2003) prescription result in stellar mass estimates that are in very good agreement with the local baryonic Tully–Fisher relation (Stark *et al.*, 2009).

Comparisons between the different methods (color-based or SED fitting) reveal some discrepancies up to a factor of two for local galaxies (Kannappan and Gawiser, 2007). While SED fitting can be prone to uncontrolled systematics, because of their statistical nature these simplified prescriptions provide estimates that are easier to control *a posteriori*.

### 2.6.5 *Estimating galaxy SFRs*

#### 2.6.5.1 *SFR tracers*

The most direct photometric tracer of star formation is the rest-frame UV, which samples the light emitted by young stars with masses above $\sim$5 $M_\odot$, i.e., with ages below $\sim$100 Myr (average sensitivity

~10 Myr). The range ~1250–2500Å, i.e , longward the Lyman alpha break and blue enough to avoid contamination by older stars, is found in SPS models to scale linearly with SFR. When the dust content is significant, the UV light emitted by young stars is absorbed and re-emitted in the IR between ~8–1000 $\mu$m (see Fig. 2.17). The Total IR (TIR) luminosity $L_{\rm TIR}$[34] is therefore found to correlate with the galaxy SFR and trace bursts shorter than 100 Myr (with an average sensitivity ~ 5 Myr). These calibrations can be contaminated by a so-called cirrus diffuse emission associated to a more extended dust component heated by normal optical stellar photons. which can even dominate the IR luminosity in galaxies with low SFR. UV SFR and IR SFR calibrations are generally accurate within ~0.3 and 0.2 dex, respectively (e.g., Kennicutt, 1998; Le Floc'h *et al.*, 2005; Rieke *et al.*, 2008).

Massive stars ($\geq$8 M$_\odot$) end up as supernovae in which electrons are accelerated by magnetic fields. This results in synchrotron emission at radio centimetric wavelengths (e.g., in the 1.4 Ghz channel) that is well correlated with FIR emission within ~0.2 dex and is therefore an indirect probe of star formation (Condon, 1992), at least up to $z \sim 1.3$ (Smolčić *et al.*, 2008; Ivison *et al.*, 2010). Radio wavelengths have the advantage of being insensitive to dust but they are affected by AGN emission.

In the following we refer to Kennicutt (1998) and Kennicutt and Evans (2012) and references therein for the exact photometric calibrations to estimate the SFR.

#### 2.6.5.2 *Estimating the SFR in distant galaxies*

The cosmic density of SFR is found to be three times higher when accounting for the IR luminosity and not only for the UV luminosity (Flores *et al.*, 1999; Le Floc'h *et al.*, 2005). This means that UV-based SFR estimates (SFR$_{\rm UV}$) strongly underestimates the total SFR in distant galaxies. On the other hand, SFR$_{\rm IR}$ estimates are found to fail in low metallicity (below 1/3 Z$_\odot$) and/or dust-free galaxies

[34]The 8–1000 $\mu$m range has been widely used in the past for integrating the IR luminosity (but see Kennicutt and Evans, 2012).

or galaxies with low specific SFR[35] because of the cirrus emission present in these objects (Kennicutt and Evans, 2012).

Getting an accurate estimate of the total SFR in a distant galaxy therefore requires estimating both the $\mathrm{SFR_{UV}}$ and the $\mathrm{SFR_{IR}}$ and combining them properly. The exact combination depends mainly on how well the UV photometry accounts for the total light emitted by young stars, which can introduce multiplicative factors for correcting the $L_{\mathrm{UV}}$ (see, e.g., Bell *et al.*, 2005) or the $L_{\mathrm{IR}}$ (see, e.g., Kennicutt and Evans, 2012). The uncertainty associated with these multiplicative factors is generally smaller than the systematic uncertainties associated with the estimation of the $\mathrm{SFR_{UV}}$ and the $\mathrm{SFR_{IR}}$, which are typically a factor $\sim$ two. We refer to Kennicutt (1998); Kennicutt and Evans (2012) and references therein for the exact existing calibrated combinations of the $\mathrm{SFR_{UV}}$ and the $\mathrm{SFR_{IR}}$.

**Example 2.16: *How to estimate the SFR in a distant galaxy?***

(1) Try to find a $z = 0$ calibration suitable for estimating the $\mathrm{SFR_{IR}}$ at a given $z$. Reliable rest-frame calibrations are between $\sim$8 (limited by PAH emissions) and 250 $\mu$m (limited by confusion). Check for depth, for aperture, and for confusion issues caused by redshift and object size. For instance, for $z \sim 0.5$–$0.7$ galaxies, the Spitzer 24 $\mu$m MIPS band roughly samples the rest-frame at $\sim$15 $\mu$m, which can be converted into $L_{\mathrm{TIR}}$ using the Chary and Elbaz (2001) calibration, i.e., $L_{\mathrm{IR}} = 11.1 \times L_{15\,\mu\mathrm{m}}^{0.998}$, and then into the $\mathrm{SFR_{IR}}$ using the Kennicutt (1998) calibration, i.e., $\mathrm{SFR_{IR}}[\mathrm{M_\odot} \cdot \mathrm{yr}^{-1}] = 1.71 \times 10^{-10} L_{\mathrm{IR}}[L_\odot]$ (Salpeter IMF). If you cannot find a known $z = 0$ calibration matching the observed bands, construct an IR SED using as many photometric bands that you have at your disposal. Fit this SED (see Ex. *How to fit an SED?*) using IR templates

*(Continued)*

[35]which is defined as $\mathrm{SFR}/M_{\mathrm{stellar}}$.

**Example 2.16: *Continued***

from, e.g., Chary and Elbaz, 2001; Dale and Helou, 2002; Lagache *et al.*, 2003; Rieke *et al.*, 2008) to estimate $L_{TIR}$;

(2) Construct a UV-optical SED by gathering photometric measurements from ground-based photometric surveys (e.g., EIS or GOODS) and/or space-borne images (e.g., HST/ACS and/or HST/WFC3 filters). Fit the SED and estimate a rest-frame $L_{UV}$ from the rest-frame best-fit template and then $\mathrm{SFR}_{\mathrm{UV}}$ using, e.g., the Kennicutt (1998) calibration $\mathrm{SFR}_{\mathrm{UV}}[\mathrm{M}_{\odot} \cdot \mathrm{yr}^{-1}] = 1.4 \times 10^{-28} L_{\nu}[\mathrm{erg.s}^{-1}.\mathrm{Hz}^{-1}]$, in which the $L_{\nu}$ is the monochromatic luminosity in the range 1500–2800Å. In this calibration, the SED is assumed to be flat over this wavelength range, otherwise a correction for the exact spectral shape is required (see, e.g., Bell *et al.*, 2005). Depending on the exact range over which $L_{\mathrm{UV}}$ is estimated, you might need an additional correction to account for the extra light emitted from young stars For instance in Bell *et al.*, 2005, a factor 2.2 is required to correct for the UV light missed blueward 1216Å and redward 3000Å;

(3) Combine the $\mathrm{SFR}_{\mathrm{UV}}$ and the $\mathrm{SFR}_{\mathrm{IR}}$ to estimate the total dust-corrected SFR;

(4) Estimate the SFR systematic uncertainty by quadratically combining the uncertainties between the measured fluxes and bolometric UV and IR luminosities, and the UV and IR to SFR calibrations. Estimate the random uncertainties by propagating the flux measurement uncertainties through the different calibrations.

Estimating the $\mathrm{SFR}_{\mathrm{UV}}$ in distant galaxies is possible for any redshift using photometry in blue bands, which samples more and more decreasing wavelengths in the rest-frame UV with increasing $z$. To estimate the $\mathrm{SFR}_{\mathrm{IR}}$, a number of correlations between the TIR and different MIR luminosities were used to develop secondary SFR calibrations using available MIR bands from satellites (i.e., ISOCAM, IRAS, Spitzer, AKARI, or Herschel; see Ex. *How to estimate the SFR in a distant galaxy*?). At $z \lesssim 1.5$, the Spitzer 24 $\mu$m band is relatively well-correlated with $L_{TIR}$ and has been therefore widely used to provide accurate $SFR_{IR}$ estimates (Shapley, 2011; Elbaz *et al.*, 2011). Above $z \sim 1.5$, it was shown that the 24 $\mu$m Spitzer band can overestimate $L_{TIR}$ by a factor 2 to 7 (Nordon *et al.*, 2010; Elbaz *et al.*, 2010). *Herschel* now offers spectral bands closer to the peak of IR emission ($\sim$100 $\mu$m), which provides better calibrations of the SFR (Galametz *et al.*, 2013). However, confusion noise starts limiting both the spatial resolution and the photometric accuracy past 250 $\mu$m (i.e., in the Herschel/SPIRE channels), particularly at 350 and 500 $\mu$m (see Fig. 2.14; Nguyen *et al.*, 2010; Berta *et al.*, 2010; Elbaz *et al.*, 2010, 2011). Comparisons between spectroscopic SFR measurements and photometric estimates are consistent within a factor two, when confusion is not an issue (e.g., Flores *et al.*, 2004; Domínguez Sánchez *et al.*, 2012, see Fig. 2.23).

Limitations of IR photometry has caused the total SFR at $z \geq 3$ to have often been estimated using the UV luminosity corrected for extinction using the IRX-$\beta$ technique. One should therefore be very

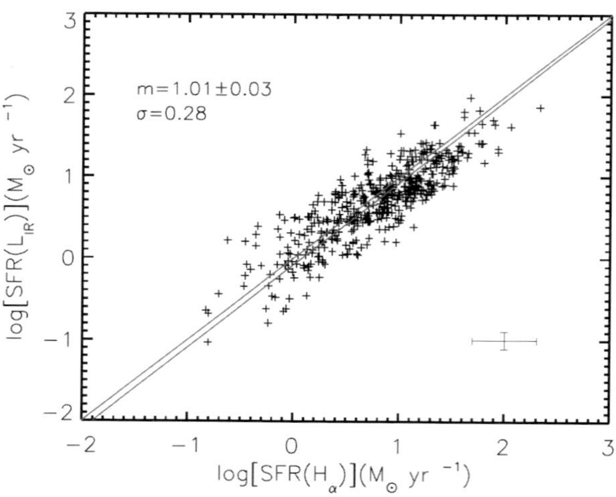


Fig. 2.23: Comparison between Herschel/IR SFR estimates and H$\alpha$ spectroscopic measurements (from Domínguez Sánchez *et al.*, 2012, *MNRAS*, **426**, p. 330, reproduced with permission of Oxford University Press on behalf of the Royal Astronomical Society; see also Flores *et al.*, 2004). The scatter is 0.28 dex, which shows that the SFR can be estimated within a factor two using IR estimates.

cautious in interpreting such SFRs since they can potentially be affected by large systematic errors that are related to such extrapolations (see 2.6.2).

## Bibliography


Abraham R. G., Ellis R. S., Fabian A. C. *et al.* (1999). The star formation history of the hubble sequence: Spatially resolved colour distributions of intermediate-redshift galaxies in the hubble deep field, *MNRAS* **303**, pp. 641–658.

Abraham R. G., Tanvir N. R., Santiago B. X. *et al.* (1996). Galaxy morphology to $i = 25$ mag in the hubble deep field, *MNRAS* **279**, pp. L47–L52.

Abraham R. G., van den Bergh S., Nair P. (2003). A new approach to galaxy morphology. i. Analysis of the sloan digital sky survey early data release, *ApJ* **588**, pp. 218–229.

Andersen M. I., Freyhammer L., Storm J. (1995). *Calibrating and Understanding HST and ESO Instruments, European Southern Observatory Conference and Workshop Proceedings*, Vol. 53 (ed.), Benvenuti, P., p. 87.

Annibali F., Aloisi A., Mack J. *et al.* (2008). Starbursts in the local universe: New hubble space telescope advanced camera for surveys observations of the irregular galaxy ngc 4449, *AJ* **135**, pp. 1900–1916.

Annunziatella M., Mercurio A., Brescia M. *et al.* (2013). Inside catalogs: A comparison of source extraction software, *PASP* **125**, pp. 68–82.

Avni Y. (1976). Energy spectra of x-ray clusters of galaxies, *ApJ* **210**, pp. 642–646.

Baldry I. K., Glazebrook K., Brinkmann J. *et al.* (2004). Quantifying the bimodal color-magnitude distribution of galaxies, *ApJ* **600**, pp. 681–694.

Barden M., Häußler B., Peng C. Y. *et al.* (2012). GALAPAGOS: From pixels to parameters, *MNRAS* **422**, pp. 449–468.

Barden M., Jahnke K., Häußler B. (2008). Ferengi: Redshifting galaxies from sdss to gems, stages, and cosmos, *ApJS* **175**, pp. 105–115.

Becker R. H., Fan X., White R. L. *et al.* (2001). Evidence for reionization at $z \sim 6$: Detection of a Gunn–Peterson trough in a $z = 6.28$ quasar, *AJ* **122**, pp. 2850–2857.

Beckwith S. V. W., Stiavelli M., Koekemoer A. M. *et al.* (2006). The Hubble ultra deep field, *AJ* **132**, pp. 1729–1755.

Bell E. F., de Jong R. S. (2001). Stellar mass-to-light ratios and the Tully–Fisher relation, *ApJ* **550**, pp. 212–229.

Bell E. F., McIntosh D. H., Katz N. *et al.* (2003). The optical and near-infrared properties of galaxies. i. Luminosity and stellar mass functions, *ApJS* **149**, pp. 289–312.

Bell E. F., Papovich C., Wolf C. *et al.* (2005). Toward an understanding of the rapid decline of the cosmic star formation rate, *ApJ* **625**, pp. 23–36.

Bell E. F., Wolf C., Meisenheimer K. *et al.* (2004). Nearly 5000 distant early-type galaxies in combo-17: A red sequence and its evolution since $z \sim 1$, *ApJ* **608**, pp. 752–767.

Berta S., Magnelli B., Lutz D. *et al.* (2010). Dissecting the cosmic infra-red background with herschel/pep, *A&A* **518**, L30.

Bertin E., Arnouts S. (1996). Sextractor: Software for source extraction. *A&AS* **117**, pp. 393–404.

Bessell M. S. (2005). Standard photometric systems, *ARA&A* **43**, pp. 293–336.

Bilir S., Ak S., Karaali S. *et al.* (2008). Transformations between 2MASS, SDSS and BVRI photometric systems: Bridging the near-infrared and optical, *MNRAS* **384**, pp. 1178–1188.

Bouwens R. J., Illingworth G. D., Oesch P. A. *et al.* (2010). Discovery of $z \sim 8$ galaxies in the hubble ultra deep field from ultra-deep wfc3/ir observations, *ApJ* **709**, pp. L133–L137.

Bouwens R. J., Illingworth G. D., Oesch P. A. *et al.* (2015). UV luminosity functions at redshifts z ~ 4 to z ~ 10: 10,000 galaxies from HST legacy fields, *ApJ* **803**, p. 34.

Brinchmann J., Abraham R., Schade D. *et al.* (1998). Hubble space telescope imaging of the cfrs and ldss redshift surveys. i. Morphological properties, *ApJ* **499**, p. 112.

Bruzual G., Charlot S. (2003). Stellar population synthesis at the resolution of 2003, *MNRAS* **344**, pp. 1000–1028.

Bunker A., Spinrad H., Stern D. *et al.* (2000). Stellar populations and galaxy morphology at high redshift, arXiv: astroph/0004348.

Burgarella D., Buat V., Donas J. *et al.* (2001). The ultraviolet visibility and quantitative morphology of galactic disks at low and high redshift, *A&A* **369**, pp. 421–431.

Buta R. J. (2011). Galaxy Morphology, arXiv:1102.0550v1.

Calzetti D., Armus L., Bohlin R. C. *et al.* (2000). The dust content and opacity of actively star-forming galaxies, *ApJ* **533**, pp. 682–695.

Cameron E., Carollo C. M., Oesch P. A. *et al.* (2011). Active and passive galaxies at $z \sim 2$: Rest-frame optical morphologies with wfc3, *ApJ* **743**, 146.

Cappellari M., Copin Y. (2003). Adaptive spatial binning of integral-field spectroscopic data using Voronoi tessellations, *MNRAS* **342**, pp. 345–354.

Cardelli J. A., Clayton G. C., Mathis J. S. (1989). The relationship between infrared, optical, and ultraviolet extinction, *ApJ* **345**, pp. 245–256.

Carpenter J. M. (2001). Color transformations for the 2mass second incremental data release, *AJ* **121**, pp. 2851–2871.

Casertano S., de Mello D., Dickinson M. *et al.* (2000). Wfpc2 observations of the hubble deep field south, *AJ* **120**, pp. 2747–2824.

Cassata P., Cimatti A., Kurk J. *et al.* (2008). Gmass ultradeep spectroscopy of galaxies at $z \sim 2$. iii. The emergence of the color bimodality at $z \sim 2$, *A&A* **483**, pp. L39–L42.

Charlot S. (1996). *The Universe at High-z, Large-Scale Structure and the Cosmic Microwave Background, Lecture Notes in Physics*, Vol. 470 (eds.), Martinez-Gonzalez, E. and Sanz, J. L., Berlin: Springer-Verlag p. 53.

Charlot S., Fall S. M. (2000). A simple model for the absorption of starlight by dust in galaxies, *ApJ* **539**, pp. 718–731.

Chary R., Elbaz D. (2001). Interpreting the cosmic infrared background: Constraints on the evolution of the dust-enshrouded star formation rate, *ApJ* **556**, pp. 562–581.

Chen C.-W., Côté P., West A. A. *et al.* (2010). Homogeneous ugriz photometry for acs virgo cluster survey galaxies: A non-parametric analysis from sdss imaging, *ApJS* **191**, pp. 1–31.

Chromey F. R., Hasselbacher D. A. (1996). the flat sky: calibration and background uniformity in wide field astronomical images, *PASP* **108**, p. 944.

Comerón S., Elmegreen B. G., Salo H. *et al* (2012). Breaks in thin and thick disks of edge-on galaxies imaged in the spitzer survey stellar structure in galaxies ($s^4g$), *ApJ* **759**, p. 98.

Condon J. J. (1992). Radio emission from normal galaxies, *ARA&A* **30**, pp. 575–611.

Conroy C. (2013). Modeling the panchromatic spectral energy distributions of galaxies, *ARA&A* **51**, pp. 393–455.

Conroy C., Gunn J. E. (2010). The propagation of uncertainties in stellar population synthesis modeling. iii. Model calibration, comparison, and evaluation, *ApJ* **712**, pp. 833–857.

Conroy C., Gunn J. E., White M. (2009). The propagation of uncertainties in stellar population synthesis modeling. i. The relevance of uncertain aspects of stellar evolution and the initial mass function to the derived physical properties of galaxies, *ApJ* **699**, pp. 486–506.

Conroy C., White M., Gunn J. E. (2010). The propagation of uncertainties in stellar population synthesis modeling. ii. The challenge of comparing galaxy evolution models to observations, *ApJ* **708**, pp. 58–70.

Conselice C. J., Bershady M. A., Dickinson M. *et al.* (2003). A direct measurement of major galaxy mergers at $z \leq 3$, *AJ* **126**, pp. 1183–1207.

Conselice C. J., Bershady M. A., Jangren A. (2000). The asymmetry of galaxies: Physical morphology for nearby and high-redshift galaxies, *ApJ* **529**, pp. 886–910.

Conselice C. J., Rajgor S., Myers R. (2008). The structures of distant galaxies — I. Galaxy structures and the merger rate to $z \sim 3$ in the Hubble Ultra-Deep Field, *MNRAS* **386**, pp. 909–927.

Dale D. A., Helou G. (2002). The infrared spectral energy distribution of normal star-forming galaxies: Calibration at far-infrared and submillimeter wavelengths, *ApJ* **576**, pp. 159–168.

Davies R., Kasper M. (2012). Adaptive optics for astronomy, *ARA&A* **50**, pp. 305–351.

de Barros S., Schaerer D., Stark D. P. (2014). Properties of $z \sim 3-6$ Lyman break galaxies. II. Impact of nebular emission at high redshift, *A&A* **563**, A81.

de Lapparent V., Baillard A., Bertin E. (2011). The EFIGI catalogue of 4458 nearby galaxies with morphology. II. Statistical properties along the Hubble sequence, *A&A* **532**, A75.

de Souza R. E., Gadotti D. A., dos Anjos S. (2004). BUDDA: A new two-dimensional bulge/disk decomposition code for detailed structural analysis of galaxies, *ApJS* **153**, pp. 411–427.

de Vaucouleurs G. (1959). Classification and morphology of external galaxies, *Handbuch der Physik* **53**, p. 275.

Delgado-Serrano R. (2012). *The Evolution of the Hubble Sequence: Morpho-Kinematics of Distant Galaxies*, Ph.D. thesis.

Delgado-Serrano R., Hammer F., Yang Y. B. *et al.* (2010). How was the hubble sequence 6 gyr ago? *A&A* **509**, A78.

Diehl S., Statler T. S. (2006). Adaptive binning of X-ray data with weighted voronoi tessellations, *MNRAS* **368**, pp. 497–510.

Disseau K., Hammer F., Puech M. *et al.* (2016). The 5 Gyr evolution os sub-M* galaxies, *MNRAS*, submitted.

Domínguez Sánchez H., Mignoli M., Pozzi F. *et al.* (2012). Comparison of star formation rates from h$\alpha$ and infrared luminosity as seen by herschel, *MNRAS* **426**, pp. 330–341.

Elbaz D., Dickinson M., Hwang H. S. *et al.* (2011). Goods-herschel: An infrared main sequence for star-forming galaxies, *A&A* **533**, A119.

Elbaz D., Hwang H. S., Magnelli B. *et al.* (2010). Herschel unveils a puzzling uniformity of distant dusty galaxies, *A&A* **518**, L29.

Elmegreen D. M., Elmegreen B. G., Ravindranath S. *et al.* (2007). Resolved galaxies in the hubble ultra deep field: Star formation in disks at high redshift, *ApJ* **658**, pp. 763–777.

Elmegreen D. M., Elmegreen B. G., Rubin D. S. *et al.* (2005). Galaxy morphologies in the hubble ultra deep field: Dominance of linear structures at the detection limit, *ApJ* **631**, pp. 85–100.

Fioc M., Rocca-Volmerange B. (1999). PEGASE.2, a metallicity-consistent spectral evolution model of galaxies: The documentation and the code, arXiv: astroph/9912179.

Fitzpatrick E. L. (1999). Correcting for the effects of interstellar extinction, *PASP* **111**, pp. 63–75.

Fitzpatrick E. L., Massa D. (2007). An analysis of the shapes of interstellar extinction curves. v. The ir-through-uv curve morphology, *ApJ* **663**, pp. 320–341.

Flores H., Hammer F., Désert F. X. *et al.* (1999). Deep galaxy survey at 6.75mu M with the ISO satellite, *A&A* **343**, pp. 389–398.

Flores H., Hammer F., Elbaz D. *et al.* (2004). Star formation rates of distant luminous infrared galaxies derived from h$\alpha$ and ir luminosities, *A&A* **415**, pp. 885–888.

Franceschini A., Rodighiero G., Cassata P. *et al.* (2006). Cosmic evolution of the galaxy's mass and luminosity functions by morphological type from multi-wavelength data in the cdf-south, *A&A* **453**, pp. 397–421.

Freeman K. C. (1970). On the disks of spiral and so galaxies, *ApJ* **160**, p. 811.

Fried D. L. (1966). Limiting resolution looking down through the atmosphere, *JOSA* (1917–1983) **56**, p. 1380.

Fruchter A. S. (2011). A new method for band-limited imaging with undersampled detectors, *PASP* **123**, pp. 497–502.

Fruchter A. S., Hook R. N. (2002). Drizzle: A method for the linear reconstruction of undersampled images, *PASP* **114**, pp 144–152.

Fukugita M., Ichikawa T., Gunn J. E. *et al.* (1996). The sloan digital sky survey photometric system, *AJ* **111**, p. 1748.

Galametz M., Kennicutt R. C., Calzetti D. *et al.* (2013). Calibration of the total infrared luminosity of nearby galaxies from spitzer and herschel bands, *MNRAS* **431**, pp. 1956–1986.

Gallazzi A., Bell E. F. (2009). Stellar mass-to-light ratios from galaxy spectra: How accurate can they be? *ApJS* **185**, pp. 253–272.

Giavalisco M. (2002). Lyman-break galaxies, *ARA&A* **40**, pp. 579–641.

Gilliland R. L. (1992). *Astronomical CCD Observing and Reduction Techniques, Astronomical Society of the Pacific Conference Series*, Vol. 23 (ed.), Howell, S. B., p. 68.

Gordon K. D., Calzetti D., Witt A. N. (1997). Dust in starburst galaxies, *ApJ* **487**, p. 625.

Graham A. W., Driver S. P. (2005). A concise reference to (projected) sérsic $r^{1/n}$ quantities, including concentration, profile slopes, petrosian indices, and kron magnitudes, *PASA* **22**, pp. 118–127.

Grazian A., Castellano M., Fontana A. *et al.* (2012). The size-luminosity relation at $z = 7$ in candels and its implication on reionization, *A&A* **547**, A51.

Grundahl F., Sorensen A. N. (1996). Detection of scattered light in telescopes. *A&AS* **116**, pp. 367–371.

Gunn J. E., Peterson B. A. (1965). On the density of neutral hydrogen in intergalactic space. *ApJ* **142**, pp. 1633–1641.

Hammer F. (2005). Growth of spirals: Secular or driven by mergers? arXiv:astroph/0509907.

Hammer F., Flores H., Puech M. *et al.* (2009). The hubble sequence: Just a vestige of merger events? *A&A* **507**, pp. 1313–1326.

Harris W. E. (1990). A comment on image detection and the definition of limiting magnitude, *PASP* **102**, pp. 949–953.

Harrison C. M., Meiksin A., Stock D. (2011). IGMtransmission: A Java GUI to model the effects of the Intergalactic Medium on the colours of high redshift galaxies, arXiv:astroph/1105.6208.

Häussler B., McIntosh D. H., Barden M. *et al.* (2007). GEMS: Galaxy fitting catalogs and testing parametric galaxy fitting codes: GALFIT and GIM2D, *ApJS* **172**, pp. 615–633.

Heidmann J., Heidmann N., de Vaucouleurs G. (1972). Inclination and absorption effects on the apparent diameters, optical luminosities and neutral hydrogen radiation of galaxies I. Optical and 21-cm line data, *MNRAS* **75**, p. 85.

Holtzman J. A. (1990). Stellar photometry with the wide field/planetary camera of the hubble space telescope, *PASP* **102**, pp. 806–834.

Howell S. B. (1989). Two-dimensional aperture photometry — signal-to-noise ratio of point-source observations and optimal data-extraction techniques, *PASP* **101**, pp. 616–622.

Howell S. B. (2006). Handbook of CCD astronomy, In: S.B. Howell. *Cambridge Observing Handbooks for Research Astronomers*, 2nd edn., Vol. 5, Cambridge, UK: Cambridge University Press. ISBN 0521852153.

Howell S. B. (2012). Fringe Science: Defringing CCD images with neon lamp flat fields, *PASP* **124**, pp. 263–267.

Howell S. B., Koehn B., Bowell E. *et al.* (1996). Detection and measurement of poorly sampled point sources imaged with 2-d array, *AJ* **112**, p. 1302.

Hubble E. P. (1926). Extragalactic nebulae. *ApJ* **64**, pp. 321–369.

Hubble E. P. (1936). *Realm of the Nebulae.* New Haven: Yale University Press. ISBN 9780300025002.

Huertas-Company M., Rouan D., Tasca L. *et al.* (2008). A robust morphological classification of high-redshift galaxies using support vector machines on seeing limited images. i. Method description, *A&A* **478**, pp. 971–980.

Inoue A. K., Kamaya H. (2008). Intergalactic dust and its photoelectric heating, arXiv:astroph/0810.5614.

Ivison R. J., Magnelli B., Ibar E. *et al.* (2010). The far-infrared/radio correlation as probed by Herschel, *A&A* **518**, p. L31.

Jarvis J. F., Tyson J. A. (1981). Focas — faint object classification and analysis system, *AJ* **86**, pp. 476–495.

Jeans J. H. (1919). *Problems of Cosmogony and Stellar Dynamics*, Cambridge, University Press, 1919.

Kannappan S. J., Gawiser E. (2007). Systematic uncertainties in stellar mass estimation for distinct galaxy populations, *ApJ* **657**, pp. L5–L8.

Kannappan S. J., Guie J. M., Baker A. J. (2009). E/s0 galaxies on the blue color-stellar mass sequence at $z = 0$: Fading mergers or future spirals? *AJ* **138**, pp. 579–597.

Kartaltepe J. S., Mozena M., Kocevski D. *et al.* (2015). CANDELS visual classifications: Scheme, data release, and first results, *ApJS* **221**, p. 11.

Kauffmann G., Charlot S. (1998). The k-band luminosity function at $z = 1$: A powerful constraint on galaxy formation theory, *MNRAS* **297**, p. L23.

Kauffmann G., Heckman T. M., White S. D. M. *et al.* (2003). Stellar masses and star formation histories for $10^5$ galaxies from the sloan digital sky survey, *MNRAS* **341**, pp. 33–53.

Kelvin L. S., Driver S. P., Robotham A. S. G. *et al.* (2012). Galaxy and mass assembly (GAMA): Structural investigation of galaxies via model analysis, *MNRAS* **421**, pp. 1007–1039.

Kennicutt R. C., Evans N. J. (2012). Star formation in the milky way and nearby galaxies, *ARA&A* **50**, pp. 531–608.

Kennicutt, Jr. R. C. (1998). Star formation in galaxies along the Hubble sequence, *ARA&A* **36**, pp. 189–232.

King I. R. (1971). The profile of a star image, *PASP* **83**, p. 199.

Koekemoer A. M., Ellis R. S., McLure R. J. *et al.* (2013). The 2012 hubble ultra deep field (udf12): Observational overview, *ApJS* **209**, 3.

Kormendy J. (2013). *Secular Evolution of Galaxies*, by Jesús Falcón-Barroso, and Johan H. Knapen, Cambridge, UK: Cambridge University Press.

Kormendy J., Kennicutt, Jr. R. C. (2004). Secular evolution and the formation of pseudobulges in disk galaxies, *ARA&A* **42**, pp. 603–683.

Kornilov V. (2012). Stellar scintillation on large and extremely large telescopes, *MNRAS* **426**, pp. 647–655.

Kriek M., Labbé I., Conroy C. *et al.* (2010). The spectral energy distribution of post-starburst galaxies in the newfirm medium-band survey: A low contribution from tp-agb stars, *ApJ* **722**, pp. L64–L69.

Krueger H., Fritze-v. Alvensleben U., Loose H.-H. (1995). Optical and near infrared spectral energy distributions of blue compact galaxies from evolutionary synthesis, *A&A* **303**, p. 41.

Kuchinski L. E., Freedman W. L., Madore B. F. *et al.* (2000). Comparing galaxy morphology at ultraviolet and optical wavelengths, *ApJS* **131**, pp. 441–463.

Lagache G., Dole H., Puget J.-L. (2003). Modelling infrared galaxy evolution using a phenomenological approach, *MNRAS* **338**, pp. 555–571.

Laidler V. G., Papovich C., Grogin N. A. *et al.* (2007). Tfit: A photometry package using prior information for mixed-resolution data sets, *PASP* **119**, pp. 1325–1344.

Lauer T. R. (1999). Combining undersampled dithered images, *PASP* **111**, pp. 227–237.

Lauger S., Burgarella D., Buat V. (2005). Spectro-morphology of galaxies: A multi-wavelength (UV-R) classification method, *A&A* **434**, pp. 77–87.

Le Floc'h E., Papovich C., Dole H. *et al.* (2005). Infrared luminosity functions from the chandra deep field-south: The spitzer view on the history of dusty star formation at $0 \leq z \leq 1$, *ApJ* **632**, pp. 169–190.

Leger A., D'Hendecourt L., Defourneau D. (1989). Physics of ir emission by interstellar pah molecules, *A&A* **216**, pp. 148–164.

Leinert C., Bowyer S., Haikala L. K. *et al.* (1998). The 1997 reference of diffuse night sky brightness, *A&AS* **127**, pp. 1–99.

Leitherer C., Schaerer D., Goldader J. D. *et al.* (1999). Starburst99: Synthesis models for galaxies with active star formation, *ApJS* **123**, pp. 3–40.

Lisker T. (2008). Is the gini coefficient a stable measure of galaxy structure? *ApJS* **179**, pp. 319–325.

Lotz J. M., Jonsson P., Cox T. J. *et al.* (2011). The major and minor galaxy merger rates at $z < 1.5$, *ApJ* **742**, p. 103.

Lotz J. M., Primack J., Madau P. (2004). A new nonparametric approach to galaxy morphological classification, *AJ* **128**, pp. 163–182.

Mackay C. D. (1986). Charge-coupled devices in astronomy, *ARA&A* **24**, pp. 255–283.

Madau P. (1995). Radiative transfer in a clumpy universe: The colors of high-redshift galaxies, *ApJ* **441**, pp. 18–27.

Mancone C. L., Gonzalez A. H., Moustakas L. A. *et al.* (2013). PyGFit: A tool for extracting PSF matched photometry, *PASP* **125**, pp. 1514–1524.

Maraston C. (2005). Evolutionary population synthesis: Models, analysis of the ingredients and application to high-z galaxies, *MNRAS* **362**, pp. 799–825.

Maraston C., Pforr J., Renzini A. *et al.* (2010). Star formation rates and masses of $z \sim 2$ galaxies from multicolour photometry, *MNRAS* **407**, pp. 830–845.

Marsh R., Young T. R., Johnson T. *et al.* (2004). Enhancement of small telescope images using super-resolution techniques, *PASP* **116**, pp. 477–481.

Martinez P., Kolb J., Tokovinin A. *et al.* (2010). Atmospheric image blur with finite outer scale or partial adaptive correction, *A&A* **516**, A90.

Masters K. L., Giovanelli R., Haynes M. P. (2003). Internal extinction in spiral galaxies in the near-infrared, *AJ* **126**, pp. 158–174.

McGaugh S. S., Schombert J. M. (2014). Color–Mass-to-light-ratio relations for disk galaxies, *AJ* **148**, 77.

Meiksin A. (2006). Colour corrections for high-redshift objects due to intergalactic attenuation, *MNRAS* **365**, pp. 807–812.

Melbourne J., Williams B. F., Dalcanton J. J. *et al.* (2012). The contribution of tp-agb and rheb stars to the near-ir luminosity of local galaxies: Implications for stellar mass measurements of high-redshift galaxies, *ApJ* **748**, 47.

Menanteau F., Jimenez R., Matteucci F. (2001). The origin of blue cores in hubble deep field e/s0 galaxies, *ApJ* **562**, pp. L23–L27.

Merline W. J., Howell S. B. (1995). A realistic model for point-sources imaged on array detectors: The model and initial results, *ExA* **6**, pp. 163–210.

Milone G., Young A. T. (2011). *Telescopes from Afar*, (eds.) Gajadhar, S. and Walawender, J. and Genet, R. and Veillet, C. and Adamson, A. and Martinez, J. and Melnik, J. and Jenness, T. and Manset, N., id.27.

Moehler S., Freudling W., Møller P. *et al.* (2010). Correction of field rotator-induced flat-field systematics — a case study using archived VLT–FORS data, *PASP* **122**, pp. 93–102.

Moffat A. F. J. (1969). A theoretical investigation of focal stellar images in the photographic emulsion and application to photographic photometry, *A&A* **3**, p. 455.

Morgan W. W. (1958). A preliminary classification of the forms of galaxies according to their stellar population, *PASP* **70**, p. 364.

Morgan W. W. (1959). Preliminary classification of the forms of galaxies according to their stellar population. ii, *PASP* **71**, p. 394.

Morgan W. W., Mayall N. U. (1957). A spectral classification of galaxies, *PASP* **69**, p. 291.

Morrissey P., Conrow T., Barlow T. A. *et al.* (2007). The calibration and data products of GALEX, *ApJS* **173**, pp. 682–697.

Mostardi R. E., Shapley A. E., Nestor D. B. *et al.* (2013). Narrowband Lyman-continuum imaging of galaxies at $z \sim 2.85$, *ApJ* **779**, p. 65.

Naim A., Lahav O., Buta R. J. *et al.* (1995). A comparative study of morphological classifications of apm galaxies, *MNRAS* **274**, pp. 1107–1125.

Nair P. B., Abraham R. G. (2010). A catalog of detailed visual morphological classifications for 14,034 galaxies in the sloan digital sky survey, *ApJS* **186**, pp. 427–456.

Neichel B., Fusco T., Conan J.-M. (2008). Tomographic reconstruction for wide-field adaptive optics systems: Fourier domain analysis and fundamental limitations, *JOSA* **26**, p. 219.

Newberry M. V. (1991). Signal-to-noise considerations for sky-subtracted CCD data, *PASP* **103**, pp. 122–130.

Nguyen H. T., Schulz B., Levenson L. *et al.* (2010). Hermes: The spire confusion limit, *A&A* **518**, L5.

Noll S., Kausch W., Barden M. *et al.* (2012). An atmospheric radiation model for cerro paranal. i. the optical spectral range, *A&A* **543**, A92.

Nonino M., Dickinson M., Rosati P. *et al.* (2009). Deep u band and r imaging of goods-south: Observations, data reduction and first results, *ApJS* **183**, pp. 244–260.

Nordon R., Lutz D., Shao L. *et al.* (2010). The star-formation rates of $1.5 < z < 2.5$ massive galaxies, *A&A* **518**, L24.

O'Donnell J. E. (1994). $R_{nu}$-dependent optical and near-ultraviolet extinction, *ApJ* **422**, pp. 158–163.

Oh K., Choi H., Kim H.-G. *et al.* (2013). Demographics of sloan digital sky survey galaxies along the Hubble sequence, *AJ* **146**, 151.

Osterbrock D. E., Ferland G. J. (2006). *Astrophysics of Gaseous Nebulae and Active Galactic Nuclei*, (eds.) Osterbrock D.E. and Ferland G.J., Sausalito, CA: University Science Books, 2006.

Papovich C., Dickinson M., Ferguson H. C. (2001). The stellar populations and evolution of Lyman break galaxies, *ApJ* **559**, pp. 620–653.

Peek J. E. G., Graves G. J. (2010). A correction to the standard galactic reddening map: Passive galaxies as standard crayons, *ApJ* **719**, pp. 415–424.

Peng C. Y., Ho L. C., Impey C. D. *et al.* (2002). Detailed structural decomposition of galaxy images, *AJ* **124**, pp. 266–293.

Peng C. Y., Ho L. C., Impey C. D. *et al.* (2010). Detailed decomposition of galaxy images. ii. Beyond axisymmetric models, *AJ* **139**, pp. 2097–2129.

Pforr J., Maraston C., Tonini C. (2012). Recovering galaxy stellar population properties from broad-band spectral energy distribution fitting, *MNRAS* **422**, pp. 3285–3326.

Phillipps S. (1984). Optimum signal to noise ratios for observations of faint galaxies, *A&A* **136**, p. 371.

Pignatelli E., Fasano G., Cassata P. (2006). Gasphot: A tool for galaxy automatic surface photometry, *A&A* **446**, pp. 373–388.

Pirzkal N., Rothberg B., Nilsson K. K. *et al.* (2012). A link to the past: Using Markov Chain Monte Carlo fitting to constrain fundamental parameters of high-redshift galaxies, *ApJ* **748**, 122.

Pirzkal N., Rothberg B., Ryan R. *et al.* (2013). Real or interloper? The redshift likelihoods of $z > 8$ galaxies in the HUDF12, *ApJ* **775**, 11.

Polletta M., Tajer M., Maraschi L. *et al.* (2007). Spectral energy distributions of hard x-ray selected active galactic nuclei in the xmm-newton medium deep survey, *ApJ* **663**, pp. 81–102.

Puech M., Flores H., Hammer F. *et al.* (2008). Images. iii. The evolution of the near-infrared Tully–Fisher relation over the last 6 gyr, *A&A* **484**, pp. 173–187.

Puech M., Hammer F., Flores H. *et al.* (2009). A forming disk at $z \sim 0.6$: Collapse of a gaseous disk or major merger remnant? *A&A* **493**, pp. 899–906.

Puetter R. C., Gosnell T. R., Yahil A. (2005). Digital image reconstruction: Deblurring and denoising, *ARA&A* **43**, pp. 139–194.

Raiter A., Schaerer D., Fosbury R. A. E. (2010). Predicted uv properties of very metal-poor starburst galaxies, *A&A* **523**, A64.

Rawat A., Wadadekar Y., De Mello D. (2009). Rest-frame uv versus optical morphologies of galaxies using sérsic profile fitting: The importance of morphological $k$-correction, *ApJ* **695**, pp. 1315–1326.

Reines A. E., Nidever D. L., Whelan D. G. *et al.* (2010). The importance of nebular continuum and line emission in observations of young massive star clusters, *ApJ* **708**, pp. 26–37.

Reynolds J. H. (1920). Nebula, photometric measures of the nuclei of some typical spiral nebulae, *MNRAS* **80**, p. 746.

Rieke G. H. (2007). Infrared detector arrays for astronomy, *ARA&A* **45**, pp. 77–115.

Rieke G. H., Blaylock M., Decin L. *et al.* (2008). Absolute physical calibration in the infrared, *AJ* **135**, pp. 2245–2263.

Roberts M. S., Haynes M. P. (1994). Physical parameters along the hubble sequence, *ARA&A* **32**, pp. 115–152.

Roddier F. (1981). The effects of atmospheric turbulence in optical astronomy, *Progress in optics*. Vol. 19. Amsterdam, North-Holland Publishing Co., pp. 281–376.

Rodrigues M., Patricio V., Rothberg B. *et al.* (2014). *Astronomical Society of India Conference Series*, *Astronomical Society of India Conference Series*, Vol. 11, pp. 143–149.

Ryan, Jr R. E. (2011). iGalFit: An interactive tool for GalFit, arXiv:astroph/1110.1090.

Sandage A. (1961). *The Hubble Atlas of Galaxies*, Washington: Carnegie Institution.

Sandage A. (2005). The classification of galaxies: Early history and ongoing developments, *ARA&A* **43**, pp. 581–624.

Scarlata C., Carollo C. M., Lilly S. *et al.* (2007). COSMOS morphological classification with the zurich estimator of structural types (ZEST) and the evolution since $z = 1$ of the luminosity function of early, disk, and irregular galaxies, *ApJS* **172**, pp. 406–433.

Schade D., Lilly S. J., Crampton D. *et al.* (1999). Hubble space telescope imaging of the cfrs and ldss redshift surveys. iii. Field elliptical galaxies at $0.2 < z < 1.0$, *ApJ* **525**, pp. 31–46.

Schade D., Lilly S. J., Crampton D. *et al.* (1995). Canada–france redshift survey: Hubble space telescope imaging of high-redshift field galaxies, *ApJ* **451**, p. L1.

Schaerer D. (2013). *Astrophysics and Space Science Library*, *Astrophysics and Space Science Library*, Vol. 396, (eds.) Wiklind, T. and Mobasher, B. and Bromm, V., p. 345.

Schaerer D., de Barros S. (2009). The impact of nebular emission on the ages of $z \approx 6$ galaxies, *A&A* **502**, pp. 423–426.

Schechter P. L., Mateo M., Saha A. (1993). Dophot, a ccd photometry program: Description and tests, *PASP* **105**, pp. 1342–1353.

Schlegel D. J., Finkbeiner D. P., Davis M. (1998). Maps of dust infrared emission for use in estimation of reddening and cosmic microwave background radiation foregrounds, *ApJ* **500**, p. 525.

Sersic J. L. (1968). *Atlas de Galaxias Australes*, Cordoba, Argentina: Observatorio Astronomico.

Shapley A. E. (2011). Physical Properties of Galaxies from $z = 2$–4, *ARA&A* **49**, pp. 525–580.

Sheth K., Elmegreen D. M., Elmegreen B. G. *et al.* (2008). Evolution of the bar fraction in cosmos: Quantifying the assembly of the Hubble sequence, *ApJ* **675**, pp. 1141–1155.

Silva L., Granato G. L., Bressan A. *et al.* (1998). Modeling the effects of dust on galactic spectral energy distributions from the ultraviolet to the millimeter band, *ApJ* **509**, pp. 103–117.

Simard L. (1998). *Astronomical Data Analysis Software and Systems VII, Astronomical Society of the Pacific Conference Series*, Vol. 145, (eds.) Albrecht, R. and Hook, R. N. and Bushouse, H. A., p. 108.

Simard L., Willmer C. N. A., Vogt N. P. *et al.* (2002). The DEEP groth strip survey. II. Hubble space telescope structural parameters of galaxies in the groth strip, *ApJS* **142**, pp. 1–33.

Smith J. A., Tucker D. L., Kent S. *et al.* (2002). The u'g'r'i'z' standard-star system, *AJ* **123**, pp. 2121–2144.

Smolčić V., Schinnerer E., Scodeggio M. *et al.* (2008). A new method to separate star-forming from AGN galaxies at intermediate redshift: The submillijansky radio population in the VLA-COSMOS survey, *ApJS* **177**, pp. 14–38.

Starck J.-L., Murtagh F. (2006). *Astronomical Image and Data Analysis, Astronomy and Astrophysics Library*. Berlin Heidelberg: Springer-Verlag. ISBN 978-3-540-33024-0.

Starck J. L., Pantin E., Murtagh F. (2002). Deconvolution in astronomy: A review, *PASP* **114**, pp. 1051–1069.

Stark D. V., McGaugh S. S., Swaters R. A. (2009). A first attempt to calibrate the baryonic Tully–Fisher relation with gas-dominated galaxies, *AJ* **138**, pp. 392–401.

Sterken C., Manfroid J. (eds.) (1992). *Astronomical Photometry, A Guide, Astrophysics and Space Science Library*, Vol. 175, pp. 288. Berlin, Heidelberg: Springer-Verlag, ISBN 0-7923-1653-3.

Stetson P. B. (1987). DAOPHOT — A computer program for crowded-field stellar photometry, *PASP* **99**, pp. 191–222.

Strateva I., Ivezić Ž., Knapp G. R. *et al.* (2001). Color separation of galaxy types in the sloan digital sky survey imaging data, *AJ* **122**, pp. 1861–1874.

Sullivan P. W., Simcoe R. A. (2012). A calibrated measurement of the near-ir continuum sky brightness using magellan/fire, *PASP* **124**, pp. 1336–1346.

Szalay A. S., Connolly A. J., Szokoly G. P. (1999). Simultaneous multicolor detection of faint galaxies in the Hubble deep field, *AJ* **117**, pp. 68–74.

Takeuchi T. T., Yuan F.-T., Ikeyama A. *et al.* (2012). Reexamination of the infrared excess-ultraviolet slope relation of local galaxies, *ApJ* **755**, p. 144.

Taylor E. N., Hopkins A. M., Baldry I. K. *et al.* (2011). Galaxy and mass assembly (gama): Stellar mass estimates, *MNRAS* **418**, pp. 1587–1620.

Taylor-Mager V. A., Conselice C. J., Windhorst R. A. *et al.* (2007). Dependence of galaxy structure on rest-frame wavelength and galaxy type, *ApJ* **659**, pp. 162–187.

Tilvi V., Papovich C., Tran K.-V. H. *et al.* (2013). Discovery of lyman break galaxies at $z \sim 7$ from the zfourge survey, *ApJ* **768**, 56.

Tinsley B. M. (1972). Stellar evolution in elliptical galaxies, *ApJ* **178**, pp. 319–336.

Tinsley B. M. H. (1967). *Evolution of Galaxies and its Significance for Cosmology.*, Ph.D. thesis, The University of Texas at Austin.

Tokovinin A. (2002). From differential image motion to seeing, *PASP* **114**, pp. 1156–1166.

Trujillo I., Aguerri J. A. L., Cepa J. *et al.* (2001). The effects of seeing on sérsic profiles — ii. The moffat psf, *MNRAS* **328**, pp. 977–985.

Tully R. B., Pierce M. J., Huang J.-S. *et al.* (1998). Global extinction in spiral galaxies, *AJ* **115**, pp. 2264–2272.

Tyson J. A. (1990). *CCDs in astronomy*, *Astronomical Society of the Pacific Conference Series*, Vol. 8 (ed.), Jacoby, G. H., pp. 1–10.

Tyson J. A., Seitzer P. (1988). A deep CCD survey of 12 high-latitude fields, *ApJ* **335**, pp. 552–583.

Tyson N. D., Gal R. R. (1993). An exposure guide for taking twilight flatfields with large format CCDs, *AJ* **105**, pp. 1206–1212.

Unterborn C. T., Ryden B. S. (2008). Inclination-dependent extinction effects in disk galaxies in the sloan digital sky survey, *ApJ* **687**, pp. 976–985.

Vaduvescu O., McCall M. L. (2004). Strategies for imaging faint extended sources in the near-infrared, *PASP* **116**, pp. 640–651.

van den Bergh S. (1960a). A preliminary liminosity classification for galaxies of type Sb. *ApJ* **131**, p. 558.

van den Bergh S. (1960b). A preliminary luminosity clssification of late-type galaxies. *ApJ* **131**, p. 215.

van den Bergh S. (2002). Ten billion years of galaxy evolution, *PASP* **114**, pp. 797–802.

van den Bergh S. (2007). Galaxy Morphology: Out of order, *Nature* **445**, p. 265.

Vikram V., Wadadekar Y., Kembhavi A. K. *et al.* (2010). Pymorph: Automated galaxy structural parameter estimation using python, *MNRAS* **409**, pp. 1379–1392.

Wadadekar Y., Robbason B., Kembhavi A. (1999). Two-dimensional galaxy image decomposition, *AJ* **117**, pp. 1219–1228.

Walcher J., Groves B., Budavári T. *et al.* (2011). Fitting the integrated spectral energy distributions of galaxies, *Ap&SS* **331**, pp. 1–52.

Waller W. H. (1990). Emission-line and continuum fluxes from narrow- and broad-band imagery, *PASP* **102**, pp. 1217–1223.

Wild V., Charlot S., Brinchmann J. *et al.* (2011). Empirical determination of the shape of dust attenuation curves in star-forming galaxies, *MNRAS* **417**, pp. 1760–1786.

Worthey G. (1994). Comprehensive stellar population models and the disentanglement of age and metallicity effects, *ApJS* **95**, pp. 107–149.

Yan H., Stefanon M., Ma Z. *et al.* (2014). Optical-faint, far-infrared-bright Herschel sources in the CANDELS fields: Ultra-luminous infrared galaxies at $z > 1$ and the effect of source blending, arXiv:1312.0902.

Yoon I., Weinberg M. D., Katz N. (2011). New insights into galaxy structure from galphat- i. Motivation, methodology and benchmarks for sérsic models, *MNRAS* **414**, pp. 1625–1655.

Zackrisson E., Bergvall N., Leitet E. (2008). The impact of nebular emission on the broadband fluxes of high-redshift galaxies, *ApJ* **673**, pp. L9–L12.

Zheng X. Z., Hammer F., Flores H. *et al.* (2004). Hst/wfpc2 morphologies and color maps of distant luminous infrared galaxies, *A&A* **421**, pp. 847–862.

Zheng X. Z., Hammer F., Flores H. *et al.* (2005). Hst/wfpc2 morphologies and bar structures of field galaxies at $0.4 < z < 1$, *A&A* **435**, pp. 507–519.

Zibetti S., Charlot S., Rix H.-W. (2009). Resolved stellar mass maps of galaxies — i. Method and implications for global mass estimates, *MNRAS* **400**, pp. 1181–1198.

Zibetti S., Gallazzi A., Charlot S. *et al.* (2013). Near-infrared spectroscopy of post-starburst galaxies: A limited impact of tp-agb stars on galaxy spectral energy distributions, *MNRAS* **428**, pp. 1479–1497.

Zibetti S., Groves B. (2011). Resolved optical-infrared spectral energy distributions of galaxies: Universal relations and their break-down on local scales, *MNRAS* **417**, pp. 812–834.

Chapter 3

# Integrated Spectroscopy

## 3.1 Introduction

Spectroscopy at moderate to high resolutions allows the analysis of electron transitions occurring both in the warm gas of star photospheres, as well as in the ionized or neutral interstellar gas. It is a critical tool in extragalactic studies to investigate the properties of the gas and stellar populations. Since the first redshift surveys of distant galaxies, e.g., CFRS (Lilly *et al.*, 1995), hundreds of thousands of distant galaxy spectra have been collected (see Chapter 1) which have allowed the characterization of their main properties (Hammer *et al.*, 1997b): temperature, density and chemical composition of the gas (Secs. 3.6 and 3.7) and stellar populations (Sec. 3.8). Table 3.1 summarizes the main properties of the ISM that can be estimated from the UV and optical spectra. The spectroscopic diagnostics to study distant galaxies are based on those implemented for nearby galaxies. The reader is referred to books such a "Astrophysics of gaseous Nebulae and active nuclei" by Osterbrock (1989), "Physics and Chemistry of the Interstellar Medium" by Kwok (2007), for the physical justification of the local gas diagnostics.

There are several difficulties in interpreting integrated spectra of distant galaxies. Galaxies are comprised of several billions of stars, gas, and dust, and exhibit complex star formation histories. In the past, galaxies presented irregular morphologies, were actively forming stars and have large gas fraction. The integrated, global spectrum collected from these distant galaxies is often dominated by light arising from a few giant star-forming regions, or from a limited number of giant or supergiant stars. As we have seen in Chapter 2, e.g., when

Table 3.1: Diagnostics to estimate the properties of the ISM in UV and optical spectra.

| Quantity | Diagnostic | |
|---|---|---|
| | UV | Optical |
| Ionization Source | Ratios of emission lines<br>C IV]/C III], C IV]/Ly$\alpha$ | Diagnostic diagrams<br>[N II]/H$\alpha$ vs. [O III]/H$\beta$ |
| Electron density | C III], Si III] and N IV doublet | [O II] or [S II]OptiOOp doublet |
| Electron temperature | Ratios of 3 auroral to nebular lines e.g., [O III] | |
| Dust extinction | "UV slope"† | Balmer decrement<br>also IR/UV ratio |
| Density H I | Curve of growth (COG) of Damped Ly$\alpha$ | |
| Abundance | Unsaturated lines of low-ionization ions | |
| Metallicities | From abundances of metallic ions and H I density | Direct $T_e$ method<br>Strong lines ratio ([N II], [O III], [O II]) |
| Gas motions | Interstellar absorption lines<br>Mg II doublet, Fe II, Si IV | Forbidden emission line profiles<br>Na I D |

†The UV slope is a controversial diagnostic of dust extinction, see Chapter 2.

deriving stellar mass from the SED, spectral analyses are limited to the dominant galactic emissions: a continuum driven by giant and supergiant stars (see Sec. 3.8) and emission lines dominated by the ionized gas from forming (super) star clusters or AGN (see Sec. 3.6).

### 3.1.1 *Desiccating a galaxy spectrum*

Figure 3.1 shows a broad-band spectrum of a typical star-forming galaxy in the distant Universe. This UV-to-optical spectrum was constructed by combining observations of star-forming galaxies at different redshifts obtained by several visible spectrographs. The main absorption and emission line wavelengths are listed in the Appendix A.

#### 3.1.1.1 *UV spectrum*

In the UV, the SED is dominated by starlight from hot and massive O- and B-type stars, as well as by dust extinction (See Fig. 2.20). The prominent spectral feature is the Hydrogen Lyman $\alpha$ line at 1215.67Å, corresponding to Hydrogen electronic transitions from $n = 2$ to $n = 1$. The Ly$\alpha$ photons are emitted by the ionized gas in star forming galaxies. Since Ly$\alpha$ photons are easily scattered by the foreground ISM, the Ly$\alpha$ line can also be seen in absorption even in star-forming galaxies (see Sec. 3.7.3). The majority of UV lines are absorption lines from low and highly ionized metals in the cold interstellar medium (see Sec. 3.7) and from the photosphere of hot stars (see Sec. 3.8). An abrupt decrease in the flux appears at wavelengths below 912Å, which is called the Lyman limit. This is because all energetic photons with $E > 13.6$ eV can ionize neutral hydrogen and are consequently absorbed. Meanwhile, at redshifts larger than 3 for which the Ly$\alpha$ appears in the optical window, most galaxies are bathed by neutral gas clouds along the line of sight, providing a plethora of Ly$\alpha$ absorption lines at $\lambda_0 < 1215$Å. This collection of Ly$\alpha$ lines is called the *Lyman forest* and it is more intense at larger redshifts since an increasing fraction of the IGM becomes neutral.

The UV spectrum also contains a plethora of absorption and emission lines from metals. This includes permitted collisionally excited

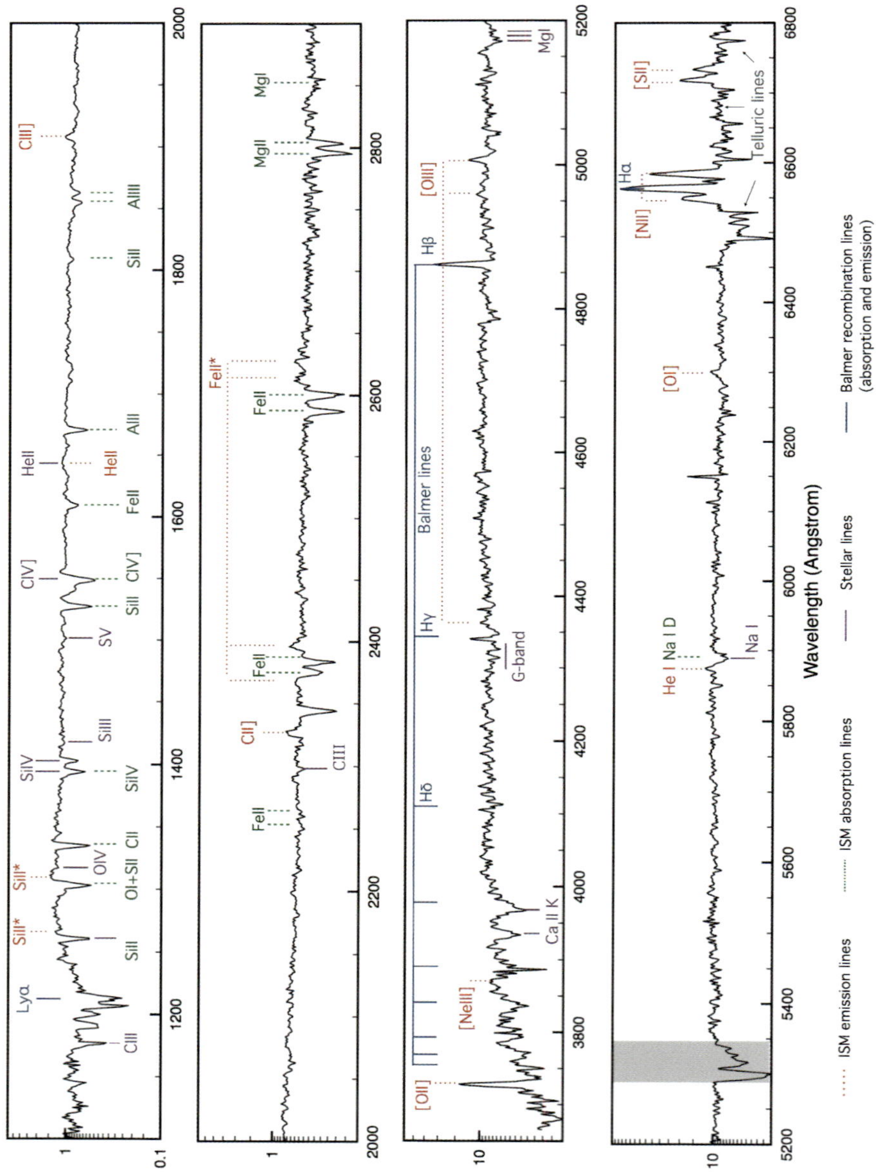
Lyα
SiII*
SiII*
SiIV
HeII
CIV]
CIII]
CIII
SiII
OI+SII
OIV
CII
SiIV
SiIII
SV
SiII
CIV]
FeII
HeII
AlII
SiII
AlIII
FeII
CII]
CIII
FeII
FeII*
FeII
MgII
MgI
[OII]
[NeIII]
Hδ
Hγ
Balmer lines
Hβ
[OIII]
Ca II K
G-band
MgI
He I
Na I D
Na I
[OI]
[NII]
Hα
[SII]
Telluric lines
Wavelength (Angstrom)
ISM emission lines
ISM absorption lines
Stellar lines
Balmer recombination lines
(absorption and emission)

Fig. 3.1: (figure on facing page) Composite spectrum of distant galaxies from UV-to-Optical rest frames. The first upper two panels show the composite spectrum of 74 star forming galaxies at $z \sim 2$ from Talia *et al.* (2012). Low- and high-ionization interstellar absorption lines, indicated by green dashed lines, are the most prominent features. All absorption features do not arise exclusively from ionized metals from the ISM, some of them are weak photospheric lines which are labeled in violet. Weaker features such as stellar emission from P Cygni profiles and fine-structure emission (Si II* and Fe II*) are also visible. The lower two panels show optical spectra of a distant star-forming galaxy from the IMAGES survey (Rodrigues *et al.*, 2012b). In the visible, emission lines from the ISM are the most prominent features. Emission lines from metals in the ISM are shown by red dot lines and H I recombination lines are indicated in blue. The grey area is a region affected by strong telluric $O_2$ absorption at 6300Å while the faint absorption lines at wavelength redder than 6500Å are telluric lines.

lines, such as O IV 1407Å, C IV 1550Å, and Mg II 2799Å. The numerous absorption lines are described in Sec. 3.7 as well as their use to probe heavy element abundances in the cold and warm ISM and IGM.

#### 3.1.1.2 *Visible spectrum*

In the spectra of distant galaxies, the most noteworthy features are emission lines. These lines are emitted by low density gas photo-ionized by UV photons from nearby O and B type stars in H II regions. Emission lines are classified according to the radiative process which induced the transition[1]:

- Recombination lines, such as the Hydrogen lines from the Balmer series (H$\alpha$, H$\beta$ etc., transition to level $n = 2$), or Helium lines such as He I lines at 3888.6Å, 4471.5Å, 5876Å. They are produced during the radiative cascade occurring after the recombination of a free electron with an ion. He II line at 4886Å requires a particularly strong ionization flux, which often reveals Wolf–Rayet stars.
- Forbidden lines, which are the majority of lines emitted by metals in the visible. By convention they are noted in brackets, e.g., [O II], to distinguish them from permitted lines. For the majority of ions, the radiative decay from collisionally-excited transitions is forbidden: collisional de-excitation is more probable than spontaneous emission. However in the case of H II regions, the extremely low density of the gas ($N_e < 10^4\text{cm}^3$) make collisional de-excitation less likely than spontaneous emission from collisionally excited levels. Forbidden lines therefore correspond to the spontaneous emission from metastable levels populated by collisions. They are the main coolant of the ISM, converting thermal energy from collisions into radiation. Ratios of forbidden lines are used as diagnostics to measure the temperature, density, and abundance of ions in the ISM of distant galaxies (see Sec. 3.6.4).

[1]The fundamental physics of the radiation mechanism of the ISM can be found in the graduated textbook by Kwok (2007).

At visible wavelengths, the strongest stellar features are the Hydrogen recombination lines from the Balmer series arising from the photospheres of A- and F-stars, and the transition lines from the ionized metals embedded in F- and G-star atmospheres (Ca II K, Mg I, Na I, Fe). Molecular bands, such as the G band at 4300Å, from CH molecules and TiO bands, are also detected in the optical continuum of distant galaxies. In galaxies dominated by old stars, a break can be seen at wavelengths below 4000Å, dubbed the *4000 Å break*. This break corresponds to a drop in opacity in the spectra of cool stars produced by a plethora of metallic lines originating from their atmospheres. In galaxies with recent star formation, a break occurs at wavelengths under 3646Å, called the *Balmer Break* (see the slope inversion in Fig. 3.1). This break is produced by bound-free transitions between energetic photons and atoms of hydrogen in the first excited state ($n = 2$). The strength of these stellar features — absorption lines and breaks — help us estimate the age and metallicity of a galaxy's stellar population. Contrarily to the UV spectrum which exhibits plethora of absorption lines from the cold and warm ISM, only few of them lie in the optical spectrum, e.g., Na I D at 5890, 5896Å, and are often blended with lines produced by the stars.

### 3.1.2 *Integrated spectroscopy of distant targets*

The observational limitations when observing distant galaxies can be easily summarized as: *distant galaxies are faint, barely resolved and their spectra are redshifted.* The first statement — *distant galaxies are faint* — is the central problem when observing distant galaxies. The observational strategy has to be optimized in order to maximize signal-to-noise (S/N). This requires a trade-off between the S/N, the spectral resolution and spectral coverage (Sec. 3.3.2), but also to minimize the impact of possible sources of noise such as the sky background or instrument defects (flat residuals, scattered light), and to maximize the number of photons collected from the target. Even then, the observer is often confronted with the issue low S/N detections and measurements of spectral features. In this regime, false detections or measurements of line fluxes dominated by noise become critical (Sec. 3.4).

The second point — *distant galaxies are merely resolved* — implies that the spectra collected are often integrated (see spatially resolved spectroscopy in the next chapter). When observing a distant galaxy with a slit or fiber spectrograph, all the light from the stars and the gas is convolved by the spatial and spectral resolutions and blurred into a single integrated spectrum. The task of the spectroscopist is to strip the integrated spectrum layer by layer, in order to constrain the properties of the ionized (Secs. 3.6 and 3.7) and neutral gas (Sec. 3.7), and of the stellar population (Sec. 3.8). Overlaid onto the intrinsic properties of their building blocks, galaxies have morphological and kinematic properties that imprint specific features on the spectrum. For instance, the velocity dispersion of the stars in the line-of-sight will increase the width of the stellar absorption lines, and big clumps of star forming regions in motion following complex kinematics could induce multi-Gaussian emission line profiles.

Finally, due to the expansion of the Universe the *spectra of distant galaxies are redshifted.* As shown in Fig. 3.2, diagnostic spectral features used in the optical are shifted towards the red and into near-infrared (NIR) wavelengths where they become heavily affected by strong sky features. For the most distant galaxies, only rest-frame UV spectra are accessible from ground-based optical spectrographs. UV light is often significantly dimmed by dust, thus only yielding an obscured and incomplete view of distant starbursts.

## 3.2 Basics of spectroscopy

A spectrograph[2] is an instrument able to measure the intensity of light $I_\lambda$ as a function of wavelength $\lambda$. The central component of the spectrograph is the disperser element, which can be gratings, VPH (holographic) gratings or grisms, see Fig. 3.3. The main characteristics of a spectrograph are the resolving power and the wavelength coverage. The resolving power $R$ gives the ability of a spectrograph to separate two adjacent spectral lines separated by $\Delta\lambda$ at the central

[2] An extended review on spectroscopy can be found in "*Astronomical Spectroscopy*" by Massey and Hanson (2010).

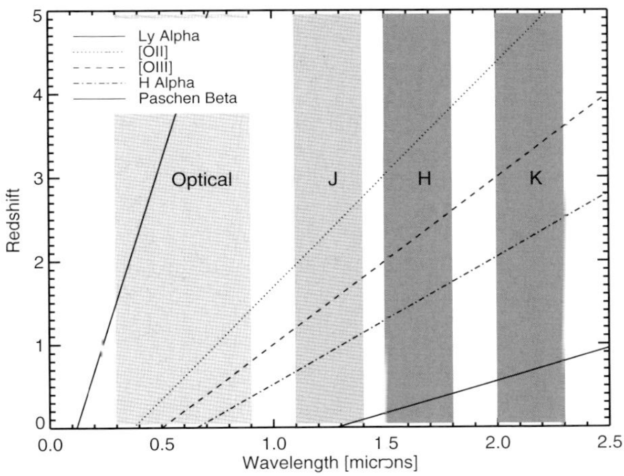


Fig. 3.2: Diagram illustrating the redshifts at which several important emission lines are observable from the ground in the optical (3500–9500Å,) and through the J H K NIR atmospheric windows. Adapted from Moustakas et al. (2006).

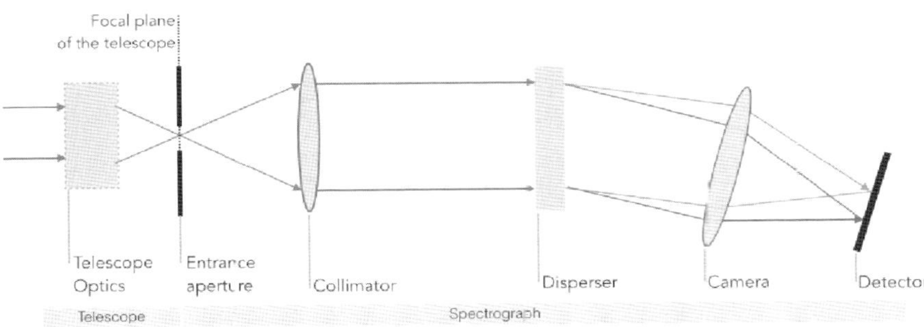


Fig. 3.3: Schematic optical-path of a spectrograph. The light source is selected by the aperture in the focal plane of the telescope. The divergent light beam passes through a collimator in order to enter in the disperser with parallel light. The light is dispersed by the disperser element and then focus on the detector by the camera.

wavelength $\lambda_c$. It is a dimensionless quantity and it is defined has:

$$R = \frac{\lambda_c}{\Delta\lambda} = \frac{c}{\Delta v}. \tag{3.1}$$

The spectral resolution depends primarily on the disperser element. It is also inversely proportional to the width of the aperture for fiber and slit spectroscopy or to the actual size of the observed source for slit-less (or open-grism) spectroscopy.

The wavelength coverage is the range of wavelengths that can be observed in a single setup. The wavelength coverage is set by the grating dispersion, and is limited by the detector size. In grism or grating spectrographs, several orders can overlap in the same locus of the detector. The region free of order contamination is called the free spectral range. It diminishes with increasing diffraction orders. To avoid the overlapping of orders, higher orders may be suppressed using blocking filters.

### 3.2.1 *Multi-object spectrographs*

In order to maximize the number of spectra observed in one exposure, distant galaxies are usually observed using multi-object spectrographs (MOS). The high density of sources in the field of view of the telescope allows us to simultaneously acquire the spectra of several tens or hundreds of galaxies. MOS are split into two categories depending on how the light from the targets is transported to the spectrograph, i.e., multi-slits or fiber-fed spectrographs.

In multi-slit spectrographs the light from the targets is sampled in the image plane of the telescope by a mask having multiple slits. The observer designs the mask in order to have the slits at the object positions. This can be quite challenging because slits cannot overlap in the spectral direction, since this would result in a blending of the spectra at the detector level. These instruments are optimal to observe faint targets because of their high global throughput and their capability to subtract accurately the sky background. Moreover, the pre-imagery mode allows the observer to take an image of the field which can then be used to design a mask with accurate slit position. This relaxes the need for an external catalog with accurate astrometry. One of the main drawbacks is that the spectral coverage varies from slit to slit and depends on the location of the slit in the direction parallel to the dispersion axis. Observers need to take a particular care of the variation of the spectral coverage when designing masks. In practice, the variation of the spectral coverage reduces the field-of view available in multi-slit spectrographs.

In fiber-fed spectrographs, the light from the objects is collected in the image plane by optical fibers and transported into the spectrograph where they are arranged into a "pseudo-slit" row. The fibers are positioned in the focal plane by a robot. Fiber-fed instruments have the advantage to be very flexible in the positioning of the targets, and to provide the same spectral coverage regardless of the target location in the field. The only constraint is the minimal separation between two fibers. Fiber-fed spectrographs are well-suited for observations in crowed fields. On the other hand, the current generation of fiber instruments have lower throughput and sky subtraction efficiency than slit instruments and are thus less competitive for faint targets. However, the next generation of Fiber-fed instruments will achieve global throughput and sky-subtraction accuracy comparable to slit spectrographs. Because they offer extra versatility compared to slit spectrographs, their use in extra-galactic studies is expected to increase.

### 3.2.2 *Slit-less grism spectrographs*

The optical path of slit-less grism spectrographs is identical to conventional multi-slit spectrographs, but there is no slits or masks. The focal plane image is therefore a stack of monochromatic images of all objects in the field-of-view. This observation mode is only used in space telescopes, such as HST and GALEX, for which observations are not limited by the atmosphere (Pasquali *et al.*, 2006). The higher multiplex capabilities and the low levels of sky background (observation are often RON limited) make space-based slit-less spectroscopy a powerful tool for measuring redshifts of galaxies in high density fields. However, the spectral resolution depends mostly on the sizes of the objects that define virtual slits. Then slit-less spectroscopy is more efficient to observe compact emission lines objects and is expected to be severely biased against large galaxies with moderate or without emission lines. It has also the drawback to be heavily affected by the overlapping of object spectra in crowded fields and by order overlapping. The contamination between objects can be partly disentangled by observing the same field with different grism angles.

Slit-less spectroscopy can be very useful to complement redshift surveys by providing magnitude or mass limited samples as defined in Chapter 1. However selection effects described above have to be controlled.

### 3.2.3 *NIR spectrographs*

Above $z = 1$ most emission lines used at intermediate redshifts to study the ionized gas are redshifted into the NIR bands J, H, and K bands (see Fig. 3.2). For this reason, the construction of NIR spectrographs has been promoted in the past decade to study the properties of galaxies at $z \sim 2 - 3$ and to detect distant galaxies beyond $z > 5$. At the time of writing, multi-object spectrographs in the NIR domain are relatively new, or in commissioning phases, or under construction: Subaru/FMOS, Gemini-N/FLAMINGOS-2, VLT/KMOS, GTC/EMIR, VLT/MOONS.

There are two main differences between NIR and optical spectroscopic observations. The first difference is the detector. NIR instruments do not use CCDs but infrared detectors with a different architecture (e.g., read-out capabilities, array defects), see Chapter 2. The second point is the sky background. NIR observations are strongly affected by strong sky lines with fluctuating intensities. To overcome the effect of the sky background, observational strategies such as beam-switching or nodding strategies are systematically used in NIR spectroscopy. A complete description of sky subtraction strategies is described in the next section.

## 3.3 Preparing observations

When studying distant galaxies, the observer is confronted with the compromise between spectral resolution and wavelength coverage on one hand and on the other hand between S/N and exposure time. The S/N varies with the resolution depending on the object and aperture sizes as well as on the actual width of the emission lines. As

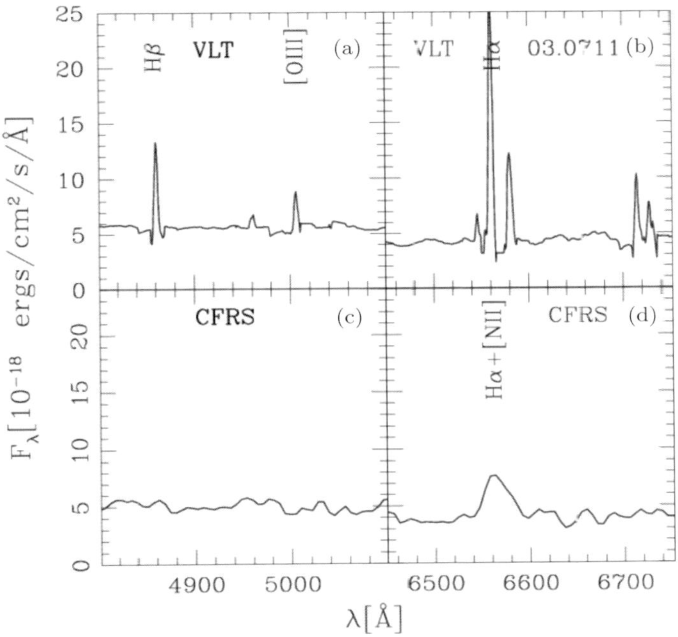


Fig. 3.4: (a), (b) Rest-frame spectra of CFRS03.0711 from the VLT observation around H$\beta$ and H$\alpha$ wavelengths; (c), (d) the corresponding low-resolution CFRS spectra. At red wavelength the H$\alpha$ line is blended with the N II $\lambda\lambda$ 6584,6548 lines Credit: Liang *et al.* (2004), A&A, 417, 905, reproduced with permission ©ESO.

a common rule, as far as an emission line is unresolved, increasing the resolution always improve its S/N: such an effect is well illustrated in Fig. 3.4. In this sub-section we review the minimal spectral resolution and S/N ratio required for several science cases.

### 3.3.1 *The choice of spectral resolution*

Low spectral resolution $R \sim 300$ and low S/N are sufficient for redshift surveys in the optical domain, for which the only requisite is to detect the position of a few lines. Such low resolution is however insufficient to measure other physical quantities, such as star formation rate (SFR), metallicities, extinction, stellar populations. The example *Effect of low spectra resolution on the estimated quantities* illustrates the dramatic impact of using low resolution spectra on the estimates of galaxy properties (Liang *et al.*, 2004). The majority of optical and UV diagnostics will require $R > 1000$ in order to properly measure emission line fluxes as well as absorptions lines. Kinematic studies of the ISM from absorption lines also require high spectral resolution, $R > 5000$, in order to de-blend velocities of close components. Precise measurements of abundances from the absorption lines in the UV can be performed for bright continuum sources such as QSOs and require $R > 20000$ in order to resolve the lines and reject saturated lines, see Sec. 3.7 and Fig. 3.29.

**Example 3.1: *Effect of low spectral resolution on the estimated quantities***

Figure 3.4 illustrates the dramatic effect of low spectral resolution on the detection of emission lines in a $z \sim 0.6$ galaxy. A low $R = 300$ (bottom panel) and a moderate $R = 1200$ (top panel) spectral resolution spectrum of a distant galaxy are shown. The emission lines are hidden or diluted in the low resolution spectra, while there are clearly detected in the moderate resolution spectrum. The H$\beta$ line is undetectable in low resolution spectrum because it is completely diluted by the underlying H$\beta$ absorption. This dilution of lines in the low spectral resolution observations have strong repercussions on the estimation of physical properties such as extinction, SFR, and metallicities, and even first order analysis such as spectral classification. Liang *et al.* (2004) have shown that spectral resolution over 1,000 and S/N $> 10$ are mandatory in order to retrieve reliably these quantities (Liang *et al.*, 2004; Hammer *et al.*, 2005b; Hammer, 2005a).

### 3.3.2 *Optimization of the S/N ratio*

#### 3.3.2.1 *Limits on S/N and error propagation*

Table 3.2 summarizes the lower limit on the S/N per element of resolution required to detect and measure spectral features. Prediction of the S/N can be done using an exposure time calculator (ETC).

Table 3.2: Lower limits on S/N required to detect and measure spectral features.

| Spectral features | S/N | Remarks |
|---|---|---|
| **Emission lines** | | *(Profile S/N)* |
| Position | >5 | Rola and Pelat (1994) |
| Flux | >10 | c.f Sec. 3.5.2 |
| Flux of Balmer line | >15 | *Need absorption correction* |
| **Absorption line** | | *(S/N per element of resolution)* |
| Position | >5 | |
| | | Fox *et al.* (2005) |
| **Stellar population** | | *(S/N per element of resolution)* |
| Lick indices | >25 | Wei *et al.* (2011) |
| Full spectra fitting | >20 | Cid Fernandes *et al.* (2005) |

For emission lines the S/N is the profile S/N, defined in Sec. 3.5.3. For absorption lines and stellar population, the limits are for detection S/N per element of resolution, as also defined in Sec. 3.5.3.

In many cases the physical quantities we aim to estimate involve the measurement of several lines, such as line flux ratios or equivalent width ratios. The uncertainty on the final quantities depends on the propagation of the measurement uncertainties of the individual lines. The precision on the final quantity is thus limited by that of the faintest lines. For example, gas metallicities estimated using the strong line ratio $R_{23}$ (see Sec. 3.6.5) depend on the fluxes of four lines — [O II]$\lambda$3727Å, H$\beta$ and [O III]$\lambda\lambda$4959Å, 5007 Å — and on the amount of extinction estimated from the ratio of two Balmer lines. If the ratio H$\gamma$/H$\beta$ is used for the estimated extinction instead of the H$\alpha$/H$\beta$ because of observational constraints, the error budget will be dominated by the large uncertainties on the flux of the faint H$\gamma$ line. The uncertainties associated with the H$\gamma$ line will first propagate to the estimate of the extinction, then to the extinction-corrected fluxes, and hence to the estimated SFR or metallicity.

#### 3.3.2.2 *Aperture losses*

The first step to maximize the S/N is to prevent aperture loss. At a given wavelength, if the aperture is wider than the dimension of the object (intrinsic dimension convolved by the seeing), the aperture loss is negligible. However, increasing the aperture does not necessary imply an increase of the S/N of the spectrum. The optimal slit width for a point source is $\sim 1.2 \times seeing$, while it is $\sim 1.4 \times seeing$ for the diameter of a fiber. A smaller aperture would lose part of the object flux and thus implies a smaller S/N. For wider apertures, the S/N decreases because the signal from the object remains constant (no aperture loss) but the noise contribution from the sky increases. In slit spectrographs, tuning the width of the slit is also constrained by the requirements on the spectral resolution. A wider slit would also decrease the spectral resolution. There is thus a trade-off between the slit width, the mean seeing during observation- and the spectral resolution when preparing the observations. As a simple rule it is not optimal to observe a galaxy with a slit (fiber) width larger than the observed target diameter.

#### 3.3.2.3 *Airmass*

When crossing the atmosphere, the light from a celestial source is dimmed by the atmosphere. The amount of attenuation is proportional to the optical length crossed by the light ray. The airmass (AM) expresses the optical length as a function of the position of the source in the sky, given by its zenith angle $za$. Airmass values are normalized to 1 for observation at zenith, and can be approximated by $AM = 1/cos(za)$ for $za \leq 60$–$70°$. To minimize the attenuation by the atmosphere, observations at low AM are highly recommended for observing faint targets.

The attenuation by the atmosphere, respectively for magnitudes and flux, is given by:

$$\delta m = m_{obs} - m_0 = AM\, k(\lambda) \tag{3.2a}$$

$$F_{obs}/F_0 = 10^{[-0.4 \text{ AM } k(\lambda)]}, \tag{3.2b}$$

where $k(\lambda)$ is the atmospheric extinction (or absorption) and $\delta$m is the magnitude difference between the observed object ($M_{obs}$) and its magnitude as it would be measured without the atmosphere ($M_0$). The airmass of a target can be checked by using visibility plot (altitude or airmass vs time) for the observing nights.

#### 3.3.2.4 *Atmospheric refraction*

The atmospheric refraction affects a spectrum in two ways: a chromatic effect and a field differential effect. The chromatic effect is due to the differential refraction of the light in the atmosphere, the light is dispersed at different angles as a function of wavelength due to the variation of the refracting index. This effect introduces an image elongation, which rotates with time as the hour angle changes, and impacts the quality of the spectrophotometry. The differential refraction has a huge impact when observing at blue wavelengths and high airmass and can lead to dramatic slit losses (Filippenko, 1982). For example, the image of an object observed at airmass 1.5 centered in a slit at 5500Å, is shifted upwards by 1.3 arcsec at 3500Å, see Fig. 3.5. This chromatic effect is constant in the field and can be corrected by an atmospheric diffraction corrector (ADC). For spectrographs without an ADC, the differential atmospheric diffraction leads to a lost of light at the entrance of the slit which fraction depend on wavelength, and the width and orientation of the slit. In long slit

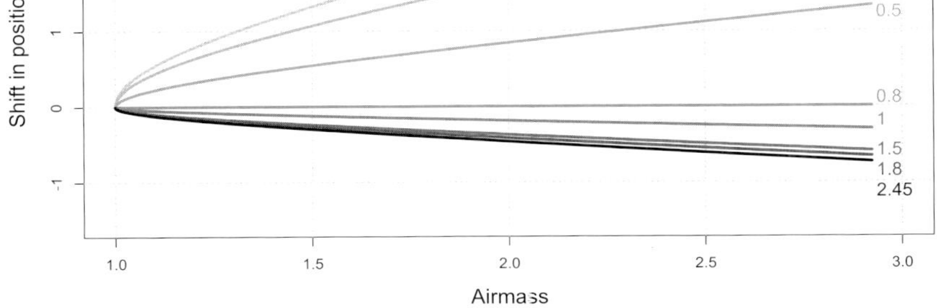


Fig. 3.5: Shift in position (arcsec) due to the atmospheric differential refraction as function of airmass. The shift was estimated relatively to the reference position of 8000Å, and for the temperature and pressure conditions of the Paranal site ($P = 557.25$ mm Hg and $T = 11.5°\mathrm{C}$).

spectroscopy this problem is overcome by orienting the slit along the parallactic angle.[3]

The atmosphere also induces a monochromatic distortion in the field of view, which depends mainly on airmass (stronger at high airmass). The variation of airmass across the field-of-view induces a field differential refraction. This differential effect is a source of slit losses of the order of 40%. In large field-of-view spectrograph, the objects at the edges of the field can be moved out of their slits after a few hours. With Fiber-fed spectrographs, observations must be taken shortly after the configuration of the fibers in the field. Since the airmass and local coordinates are known, the position of the fibers can be fine-tuned to correct for the atmospheric diffraction at the time of observation. In slit spectrographs with wide fields of view (e.g., VLT/VIMOS), this effect can be minimized by observing close to the meridian and at low airmass (Cuby *et al.*, 1998).

### 3.3.3 *Minimizing the impact of the sky background*

The sky background is the main enemy of extragalactic observations. Distant galaxies are extremely faint and can be several magnitudes fainter that the sky background. A good strategy to subtract the sky is thus crucial for a successful observation.

#### 3.3.3.1 *Sources of sky background*

Sky background is usually defined as all the light arriving to the detector that does not proceed from neither the observed astronomical source nor the telescope/instrument. The sky background is composed by a mixture of radiation produced by several sources, such as the moonlight, the Airglow (the nonthermal emission from the upper layers of the atmosphere), the zodiacal light, the starlight from unresolved sources, and the thermal emission from the atmosphere. These radiations are attenuated and scattered by the Earth atmosphere, following the atmospheric extinction law. Figure 3.6 shows

[3]To minimize the differential refraction the slit should be perpendicular to the horizon. When the object is on the meridian, the position angle of the slit should be 0° (or 180°).

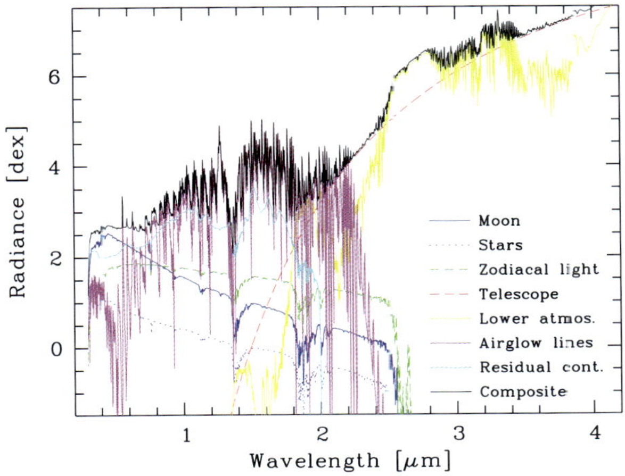


Fig. 3.6: Components of the Paranal sky model in logarithmic radiance units for wavelengths between 0.3 and 4.2 $\mu$m, during a night with the moon over the horizon. The scattered light and airglow components were only computed up to the K band because of their negligible importance at longer wavelengths. Credit: Noll *et al.* (2012), A&A, 543, 92, reproduced with permission ©ESO.

the contribution of the each background components as a function of the wavelength.

The moonlight is potentially the main source of sky background at blue wavelengths. Moonlight increases the night sky brightness in the blue bands (at new and full moon, $\mu_B$= 22.4 mag/arcsec$^2$ and 19.5 mag/arcsec$^2$, respectively), which may decrease the S/N of observation by up to a factor 4. In general, for deep imagery and/or spectroscopy of faint sources, dark nights are requested especially for UV or blue observations. Observations in the red part of the spectrum and NIR are less affected by the moonlight and can be observed in grey time at large angular distance of the moon. Above 1.5 $\mu$m the contribution of the moon to the sky background becomes progressively negligible and observations at these wavelengths can be carried at any phase.

In dark nights, the sky spectrum is dominated by the contribution of the airglow light. It corresponds to the light radiated by electronally or/and vibration-rotationally excited atoms and molecules located in the upper part of the atmosphere. Strong emission lines

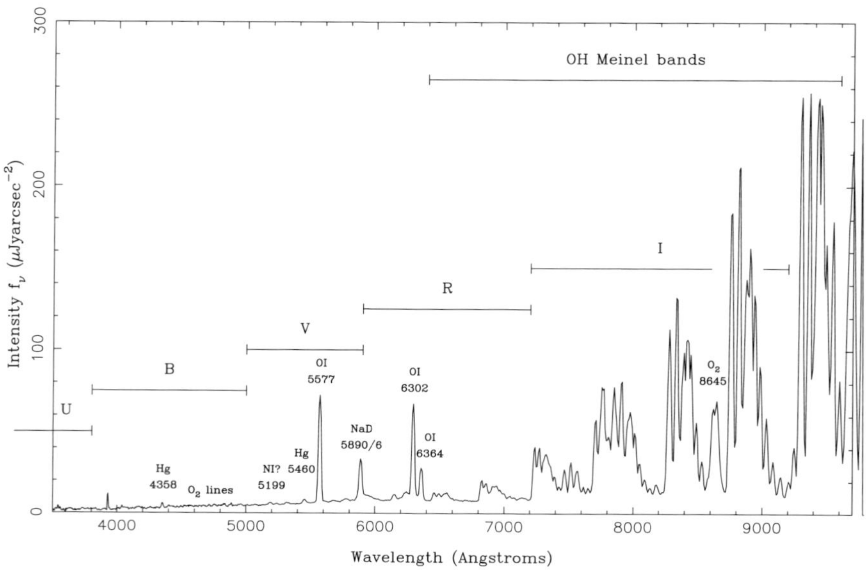


Fig. 3.7: Night spectrum at La Palma observatory. The principal emission lines and bands of the airglow are indicated. Reprinted from Benn and Ellison (1998) with permission from Elsevier.

from the airglow appear between 0.557 and 2.62 $\mu$ and are several orders of magnitude above the sky emission from other sources, see Fig. 3.7. The NIR observations are far more affected by the earth atmosphere that those in optical, because OH emission lines are particularly strong and numerous at these wavelengths. In addition, the absolute intensities of the sky lines and the continuum show temporal and spatial variations. The brightness of the airglow has a smooth diurnal variation with higher values registered at the beginning of the night (Patat, 2008). In addition to these long-scale variations, the airglow can vary randomly, by up to a few 10%, spatially and temporally. Such short-period variations, of the order of few minutes to an hour, originate from the passage of a density and temperature perturbations through the upper atmosphere. In the NIR (1–1.8 $\mu$m), the fluctuation of the OH emission lines can reach 5–10% and the sky continuum can fluctuate by 1% (Puech *et al.*, 2012).

#### 3.3.3.2 *Observational strategies*

OH lines can be the dominant source of noise in fully processed data. Notice that even with an extremely careful sky subtraction, the science signal in the regions of strong sky emission has very low S/N and therefore only spectral features falling between sky lines are generally used. Following are some strategies to optimize the sky subtraction before the data reduction:

(1) *Avoid emission lines*: If the redshift of the source is known, the observer should preferentially observe targets for which the object lines do not fall on strong sky lines or molecular absorption bands. Atlases of sky emission lines can be found in Hanuschik (2003) for the optical spectrum ($\lambda < 0.92\ \mu$m) and in Rousselot *et al.* (2000) for the NIR.
(2) *Minimal spectral resolution*: At redder wavelengths, the fraction of regions with strong OH bands increases. The quality of the sky subtraction in this region will depend on the spectral resolution of the observations. At low resolution, the OH emission lines will be heavily blended, increasing dramatically the values of the apparent amplitude of the sky continuum and creating large spectral regions with low S/N. A spectral resolution higher than $R > 3000$ is minimal when observing at $\lambda > 7000$Å, in order to have regions of the spectrum free of strong sky lines that are large enough (see Fig. 3.8).
(3) *Nodding in the IR*: NIR observations are often done in nodding mode for slit spectrographs or beam-switching mode for fibers to correct for the intense and rapidly changing OH lines. An object is observed at two positions along the slits (Positions A and B) or in ON and OFF position in the case of fibers (ON is the position with the object in the fiber and OFF the offset position with the sky). The science exposure is split into several nodding cycles, generally defined as A–B–B–A or ON–OFF–ON–OFF. The exposure time in each individual position is set to have the maximum number of counts but to avoid saturating the detector in regions of strong OH lines.

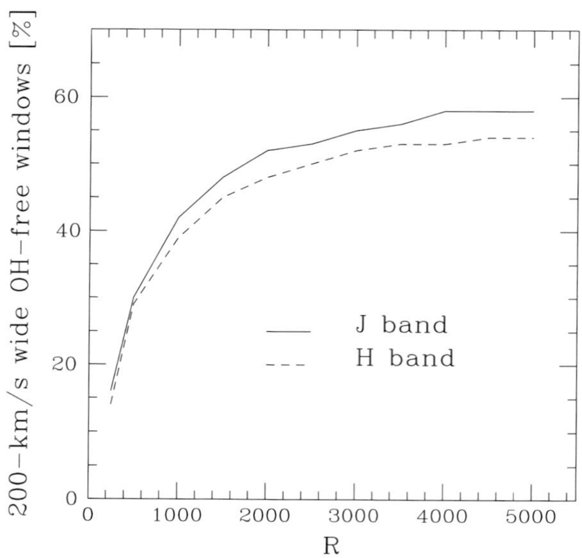


Fig. 3.8: Percentage of the spectral band available between OH sky lines in the J (solid line) and H (dash line) bands as a function of spectral resolution. Only regions providing a 200 km/s wide window have been counted.

(4) *Sky subtraction in Fiber-fed spectrographs*: It is a critical issue because the sky cannot be evaluated in the immediate vicinity of the object as in slit spectrographs. In the optical domain, the sky is usually subtracted using fibers distributed over the whole field-of-view that sample the sky during the science exposure. Because the value of the sky varies about 5–10% in the field-of-view, the sky sampled by a sky fiber can be different from those in the object. For this reason, the position of the sky fibers should be optimized according to the target positions. A more accurate sky subtraction can be obtained by placing the sky fibers close to the object fibers and using the closest sky fibers to subtract the sky in a science fiber (Rodrigues *et al.*, 2010, 2012a). The most accurate method is to observe alternatively the object and the sky with a couple of two fibers (cross-beam switching mode, see Fig. 3.9). This strategies allow us to reach a accuracy of 0.6% on the sky continuum subtraction Yang *et al.* (2013).

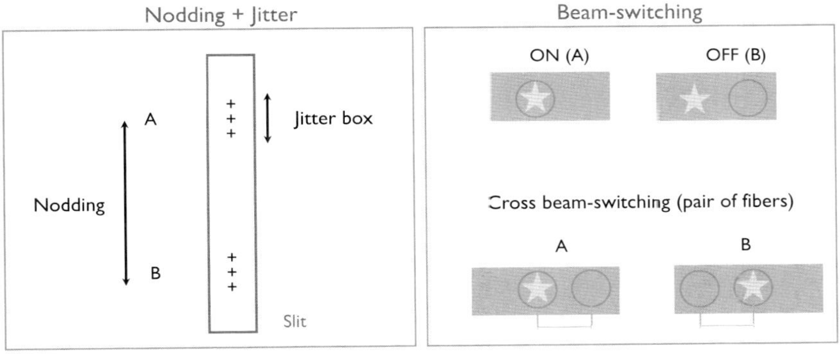


Fig. 3.9: Observational strategies for subtracting the sky in spectroscopic observations. The three strategies illustrated here are based on a sequence that alternatively observe the object and a sky position. The name of this sequence depend on the type of aperture (see text): nodding for slit spectrograph, where the object is moved from a position A to B along the slit; and beam switching for fiber-fed instruments where the aperture fiber probe alternatively the object and then the sky. In NIR observation, small changes on the position of the object in the slit are add between each exposure to move the position of the spectra in the detector and correct its response.

(5) *Nod & Shuffle mode in multi-slits*: This technique is used when the responses of the instrument and of the sky need to be subtracted with extreme accuracy. It allows the object and the sky to be observed using exactly the same optical path, including the same detector pixels (see details in the Fig. 1 of Glazebrook and Bland-Hawthorn, 2001 and also in Cuillandre *et al.*, 2003).

#### 3.3.3.3 *Water vapor column constrain*

Molecules in the atmosphere, such as $H_2O$ and $O_2$, $CO_2$, CO, $CH_4$, $O_3$, absorb part of the light, which results in absorption lines and bands in the near infrared, known as telluric lines. The amount of absorption created by the atmosphere molecules is proportional to the concentration of these species at the time of observation. In the NIR, the telluric absorption lines are dominated by the water molecules. While the concentration of $CO_2$, CO, $CH_4$, and $O_3$ varies over large timescales (seasonal cycles), or punctual phenomena (like volcano eruption), the amount of water molecules in gaseous phase, defined as the amount of precipitable water vapor (PWV),[4] varies rapidly with meteorological conditions.

Putting constraint on the maximal amount of water vapor tolerated for a given observation is thus especially important in NIR observations to optimize the final S/N. In wet conditions, extragalactic lines can vanish under strong bands of telluric absorption. In addition, high amounts of water vapor also increase the intensity of emission bands from water molecules and thus decreases the S/N of the observations, see Fig. 3.10. For example, in J band the S/N ratio of a Ly$\alpha$ line from a $z = 8.63$ galaxy will be divided by two when observing in wet conditions (PWV $\sim 7$ mm) compared to dry conditions (PWV $< 3$ mm). Moreover, the 2.85–3.25 $\mu$m region contains numerous strong $H_2O$ lines, and observations in this band should thus be carried in the driest conditions with PWV $< 3$ mm.

## 3.4 Data reduction

### 3.4.1 *Main steps*

The first stages of data reduction for spectra are common with images and consist in removing the defects, mostly originating from the detector and spectrograph: bias, bad pixel, flat-field, scattered-light, and cosmic rays. In multi-slit spectrographs, the individual 2D spectra associated to each slit are then trimmed and reduced individually in a way similar to long-slit spectroscopic data. After the wavelength calibration and sky subtraction steps, the 1D spectra are extracted from the 2D image in the regions of interest. In general, distant galaxies have relatively low continuum S/N, which makes the trace of the spectrum difficult to follow over the full wavelength range. 1D spectra are therefore extracted along the optical distortion of the instrument and within a fix aperture with a size that optimizes the final S/N. Data from multi-fiber spectrographs are reduced in a similar way with the exception that the flat-field correction, wavelength calibration, and sky subtraction are done directly on the 1D spectra. The

[4]The PWV is expressed in mm and is equivalent to the liquid precipitation that would be obtain if if all the water vapor in the column is condensed.

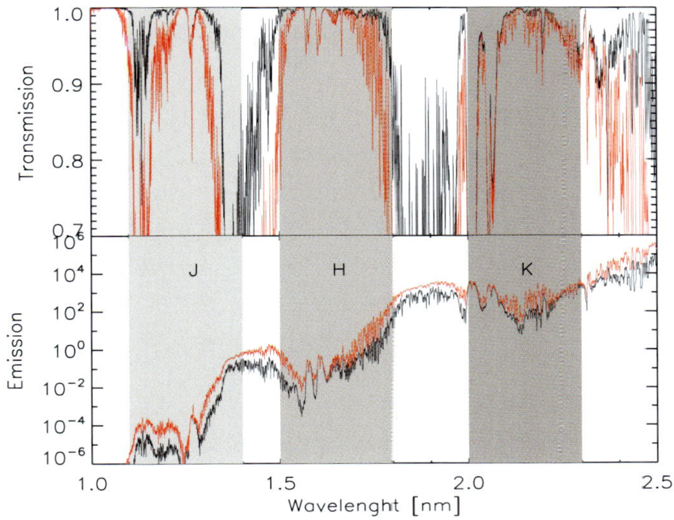


Fig. 3.10: Transmission (upper panel) and thermal emission (lower panel), in photons/s/$\mu$m/m$^2$/arcsec$^2$, of the atmosphere in the NIR bands, for a water vapor column of 0.5 mm (black) and 10 mm (red). The strong decrease of the atmosphere transmission between the photometric bands are produced by molecular bands. The molecular emission produced by the same species are added to the thermal emission of the atmosphere.

extraction uses the trace measured in high S/N flat-field frames. We encourage the readers to continue the data reduction until the end, by flux calibrating their spectra and, in the case of NIR observations, by correcting for the telluric lines.

#### 3.4.1.1 *Flux calibration*

Flux calibration requires additional telescope time to observe spectrophotometric standards and it is a delicate step of data reduction. For this reason, some observers have opted not to flux calibrate their observations. They restrict their study to the use of equivalent width of emission lines. This method introduces unnecessary large uncertainties on the estimated physical properties, which are much larger than the uncertainties of the flux calibration (see Ex. *The importance of being flux calibrated*). It is thus highly recommended to flux calibrate observations.

**Example 3.2: *The importance of being flux calibrated***

In the optical, the use of uncalibrated spectra has two big drawbacks: the underlying absorption of emission lines cannot be simply corrected and the emission lines cannot be corrected from extinction. Not correcting for these two effects can rapidly lead to large uncontrolled systematics that are stronger than errors introduced by the flux calibration. Some authors have tried to overcome this issue by correcting the measured flux of Balmer lines by adding a constant equivalent width $W_\lambda$ to take into account the underlying absorption. However, the intensity of the Balmer absorption lines vary from a galaxy to another because it depends on the star formation history. This method is thus not recommended since it adds $\sim +0.2$ dex systematic uncertainties on the measurements. Figure 3.11 illustrates the effects of estimating metallicities from $W_\lambda$ compared to flux method (Liang *et al.*, 2004).

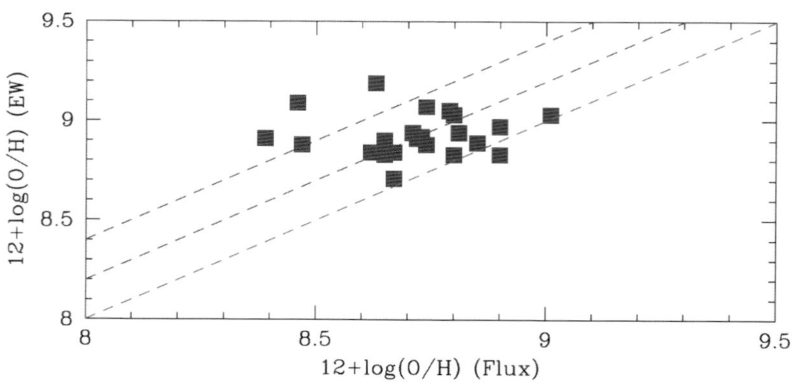


Fig. 3.11: Comparison of metal abundances derived from extinction corrected flux ratio and those derived following the equivalent width method as described in Kobulnicky and Kewley (2004). Credit: Liang *et al.* (2006a), A&A, 447, 113, reproduced with permission ©ESO.

#### 3.4.1.2 *Telluric correction*

After the standard data reduction, a NIR spectrum may appear quite disappointing as it remains full of absorption spikes. These spikes are the telluric lines which have to be corrected before making any further use of the spectrum.

These lines can be corrected by observing a telluric standard star, whose normalized spectrum will be divided to the object spectrum, see Ex. *How to correct a spectrum from telluric absorption*. The intensity of the telluric absorption lines varies with the relative concentration of the molecules, and therefore vary with time and airmass. To minimize the residuals due to the variability of the lines, the telluric standard has to be observed right before or after the science exposure. The airmass of the standard star should be as close as possible to that of the object, for warranting a small residual error on the correction of unsaturated features (Kenworthy and Hanson, 2004). The standard star should be located in the same quadrant of the sky than the target and within 30 to 60 min before or after the science exposure. When the total exposure time is longer than one hour, it is recommended to divide the total exposure in several blocks of one hour each and observe a telluric standard between each block. To avoid introducing lines intrinsic to the telluric star during the telluric correction, the telluric standard should be a star with very few intrinsic stellar absorption lines such as main sequence A or B stars, or stars with well known stellar lines such as G2V and A0V stars (respectively Sun and Vega like stars).

**Example 3.3: *How to correct a spectrum from telluric absorption lines***

There are a number of softwares that can be used to correct telluric lines, such as in Vacca *et al.* (2003), MAKEE (Keck observatory), or TELLURIC in IRAF. The main steps are the following:

- The continuum and the photospheric lines of the standard star are removed. The continuum is normalized by dividing

*(Continued)*

**Example 3.3: *Continued***

the spectrum with a black body function at the temperature of the star.

- The resulting spectrum is scaled in wavelength by adjusting the lines with the telluric lines in the object spectrum.
- Since the standard star has been observed at sightly different airmass, the flux of the telluric spectrum $F_T$ needs to be scaled to the same airmass that the object. As we know that the atmospheric extinction (expressed in magnitude) of the atmosphere is proportional to the airmass (see Eq. 3.2), this can be done by applying an airmass correction to the telluric spectrum:

$$F_T(\text{scaled}) = (F_{\text{T}})^{(AM_{\text{object}}/AM_{\text{star}})} \tag{3.3}$$

- The object spectrum is then divided by the flux-scaled telluric spectrum. The previous steps may be repeated, if needed, by adjusting the airmass and wavelength scales until the residuals have been minimized.

A new generation of telluric correction tools which limits the need for telluric standard observations are in development by several groups. For example, MOLECFIT from ESO recovers the column densities of the atmospheric constituents at the time of the observation from a small set of tropospheric and stratospheric telluric features in the spectra of the astronomical target (Smette *et al.*, 2015). It then derives a correction function to remove the telluric features.

The correction of telluric features is critical for studies of UV absorption lines of distant galaxies. At high redshift, the metallic lines from the cold and warm ISM are redshifted from the UV to the NIR. The telluric lines can be misleadingly identified as metallic lines if not properly removed.

### 3.4.2 Characterize the quality of the data reduction

In the low S/N regime, a significant contribution of the total noise comes from the instrument itself. The global response of the instrument is removed from the signal during the data reduction. However, this may leave residuals and could introduce additional noise. We describe hereafter some quality control parameters that can be used to characterize the quality of the reduced spectrum.

#### 3.4.2.1 *Effective spectral resolution and wavelength calibration accuracy*

The final spectral resolution of a spectrum can be calculated by measuring the full half width maximum (FWHM) of sky emission lines. As the line spread function (LSF, the spectral equivalent of the PSF in the spectral dimension) can vary with wavelength, several sky lines distributed along the spectrum should be used.

The accuracy of the wavelength calibration can be checked by comparing the measured position of strong sky lines with their tabulated wavelengths.The precision of the wavelength calibration has to be estimated during the data reduction process. It is given by the residual of the least square fitting of the wavelength solution. Note that bad accuracy on the wavelength calibration can be due to an improper correction of the optical and spectral distortions during the data reduction, which should be redone if necessary. A good wavelength calibration provides uncertainties smaller than a tenth of the spectral resolution element. The accuracy and precision of the wavelength calibration are crucial to determine accurate redshifts and above all, to distinguish variations in the emission/absorption line positions that are due to different physical processes.

#### 3.4.2.2 *S/N ratio of detection*

The sources of noise in a spectrum are the same as those for imagery (see Sec. 2.3.4). The difference with imagery is that the spectral dependence of several sources of noise is projected into the spectral dimension of the 2D frame. The S/N of a spectrum as a function of wavelength[5] and spatial position($x$) can be expressed as:

$$\mathrm{S/N}(x,\lambda) = \frac{O(x,\lambda) \times \sqrt{ndit}}{\sqrt{O(x,\lambda) + S(x,\lambda) + RON^2 + R(x,\lambda)}}, \tag{3.4}$$

where $O$ is the signal from the object, $S$ is the signal from the Sky, $RON$ the read-out noise, and $R$ are the residuals after the data reduction (see Sec. 2.3.4). Observations of distant galaxies are always done in photon-noise limited regime, where the $RON$ is negligible, and the noise generally follows a Poison distribution (dominated either by the Sky or Object). A feature is considered as a detection with 98% of confidence for S/N > 5. At the wavelength of strong sky lines, for example in the NIR, the signal-to-noise is sky-dominated and readily decreases to values lower than the detection threshold. The portion of signal within these strong sky lines cannot be used for the analysis.

#### 3.4.2.3 *Spectrophotometry and Aperture correction*

The accuracy of the flux calibration has to be checked by comparing the flux measured in the spectrum to those given by aperture photometry in several bands. A fast method can be through the use of AB magnitude, see Ex. *Fast method to control spectrophotometry using independent photometry.*

**Example 3.4: *Fast method to control spectrophotometry using independent photometry***

The method is based on the use of $f_\nu$ that is directly related to the AB magnitude, independently of the wavelength (see Sec. 1.2.2). First convert your spectrum into $f_\nu$, then extract the values from the AB photometry that could be immediately superposed to the spectrum at the right wavelength (see Fig. 3.12). Because spectroscopy is made through an aperture one

(*Continued*)

[5] The S/N of a spectrum is usually expressed per element of resolution, instead of S/N per pixel. A resolution element is usually sampled by at least 2 pixels in the detector (Nyquist–Shannon rule).

**Example 3.4: *Continued***

should expect higher photometric $f_\nu$ values than those from spectrophotometry. The aperture correction at the $n$th $\lambda$ value will be simply $C^n_{\text{aper}} = f_\nu(\text{spectroscopy})/f_\nu(\text{photometry})$. If the spectral flux calibration is correct, the aperture correction is simply the median of the $C^n_{\text{aper}}$ values.

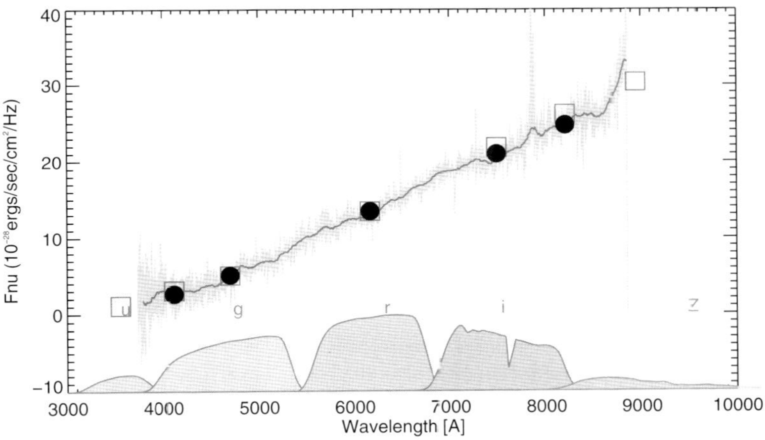


Fig. 3.12: Assessment of the spectrophotometric accuracy of a spectrum of a distant galaxy. The flux-calibrated spectra (light grey line) are compared with the AB photometry from SDSS (open squares). The solid black line corresponds to a heavy smoothing of the spectrum and the solid circles are the summed fluxes on the spectrum. Since the flux calibration is more problematic at the edges of the spectra, the magnitude in two intermediate pseudo-bands ub and iz ($\lambda_c \sim 4118$Å, 8206Å) have been estimated, so as to be used as visual guidelines.

In most cases, discrepancy between spectroscopy and photometry may occur at the edges of the spectrum, which is often due to the response of the spectrograph. In such cases, one can restrict the analysis to the region that is not affected by the instrumental response. As a last resort, and when the number of photometric bands covering the spectrum is larger than 3, a correction function can be estimated using photometry as a reference. Indeed, fluxes measured from the photometry are not affected by edge effect and are in general measured with a much higher S/N. The fluxes from the set of points $(\lambda^{n_{\rm central}}; C^{n}_{\rm aper}/\langle C_{\rm aper}\rangle)$ are fitted by a low order polynomial which is divided to the original spectrum.

## 3.5 Emission lines

In the UV-optical spectrum, the ISM properties are revealed by the presence of strong emission lines produced by the partially ionized gas at $10^4$ K $< T <$ $10^6$ K. This gas is mainly photo-ionized by radiation of young stars in H II regions and/or by an AGN. In the Milky Way, the ionized phase accounts for 20% of the disk ISM. At higher redshift, the relative fraction of ionized gas is expected to increase, powered by the radiation of strong starburst regions. In a partially-ionized gas the strength of the lines emitted by atomic transitions are ruled by: (1) the properties of the ionizing radiation, which is derived from the spectral energy distribution emitted by the ionizing source, and the number of ionizing photons per atom defined as the *ionization parameter* U; (2) the chemical composition of the medium, and (3) the density and temperature of the ionized gas. Emission line fluxes and their ratios can be used as diagnostics of the ionized gas, e.g., extinction, detection of an AGN, SFR, temperature, density, and metallicity.

### 3.5.1 *Emission line properties*

Emission lines are defined by three properties: the central wavelength $\lambda_c$, the line width profile ($\sigma$), and the flux ($F_\lambda$). Each one of these properties provides insight into the state of the emitting gas: its velocity, the temperature of the gas, and the abundance of the emitting species.

The *central wavelength* $\lambda_c$ of a line is determined by the energy of the atomic transition responsible for the emission which are known with great precision from laboratory experiments (see Appendix A). The position of a line with respect to its reference value is directly linked to the velocity of the gas cloud from which the line is emitted. When observing distant galaxies, the lines are redshifted by a factor $(1 + z)$ (see Chapter 1).

The *line profile width* is linked to the velocity distribution of the emitting particles. Theoretically, an emission line has a Lorentzian distribution with a natural width of a few m/s, resulting from the quantum uncertainty on the energy of the atomic transition. In practice, the emission lines measured on galaxy spectra have widths of the order of hundreds of km/s and their profile frequently follows a Gaussian distribution. The broadening of the lines is mainly due to the thermal broadening i.e., to the velocity dispersion of the emitting gas particles in a galaxy. However, profiles can differ from a Gaussian one if regions have different emissivities. For example, the well-know "double-horn" profile typical of rotating disks is essentially due to the low emissivity of the central bulge region. Distant galaxies can exhibit complex morphologies and star forming regions with different velocities. In this case the resulting emission line can be composed of multiple Gaussians with different velocity shifts and/or dispersions (see Fig. 3.13). The kinematics of the gas is described in Chapter 4. Finally, the observed line profile is the convolution of the intrinsic profile with the spectral response of the instrument that has, in general, a Gaussian profile with width $\sigma_{\mathrm{inst}}$.[6] In the case of Gaussian profiles, the intrinsic width is therefore:

$$\sigma_0 = \sqrt{\sigma^2_{\mathrm{obs}} - \sigma^2_{\mathrm{inst}}}. \tag{3.5}$$

The *flux of a line* may depend on the abundance of the excited species that produce the line, which in turn depends on the abundance of the element, the temperature, density, and ionization state of the gas. The flux of an emission line is defined as the area below the line and above the continuum level. The intensity of a line can also be characterized by the *equivalent width* ($W_\lambda$), defined as the width of a rectangle that would have the same level as the continuum but the same total flux (area) as the line. By convention, the $W_\lambda$ is negative for emission lines.[7] $W_\lambda$ is given in continuum unit and it can therefore be measured in spectra which are not flux calibrated.

[6] $\sigma_{\mathrm{inst}}$ or the LSF can be estimated from the FWHM of the very narrow emission sky lines.

[7] Historically, an equivalent width is a measurement of the flux absorbed relatively to the continuum. Emission lines have thus negative values of $W_\lambda$, but the sign is often left aside.

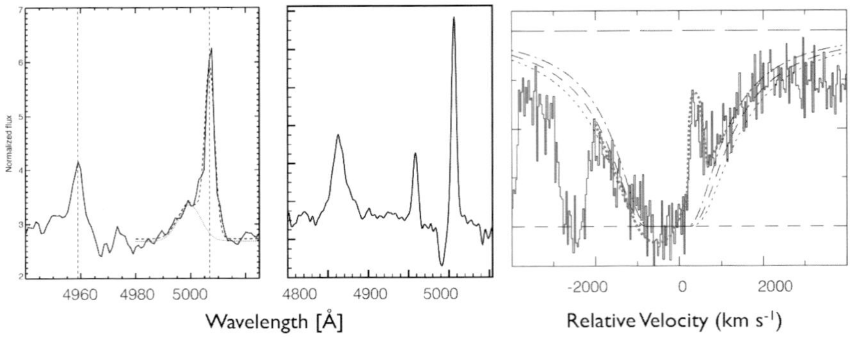


Fig. 3.13: Emission lines can have a variety of line profiles ranging from multiple Gaussian distributions or asymmetric profiles (left panel), broad lines produced by AGN (center) to P-Cygni profiles (right, dashed profile) due to large scale motions of the emitting gas or dust scattering, e.g., multiple components of the ionized gas with different velocities, expanding shells around star-forming regions. Right panel from Pettini *et al.* (2002), ©AAS, reproduced with permission.

### 3.5.2 *Proper methods for measuring emission lines*

Before measuring the width or flux of the lines, the observed spectrum has to be set to the rest-frame. This can be done by dividing the wavelength scale by $(1 + z)$ and multiplying the fluxes $F_\lambda$ by $(1 + z)$. Notice that $F_\nu$ should be divided by $(1 + z)$.

At low and moderate resolution, emission lines often display profiles compatible with a Gaussian distribution. The line is measured by fitting a simple Gaussian model. In galaxy spectra, lines lie above the continuum produced by the stars, and a continuum component is added to the Gaussian model, usually a low polynomial order (a constant or a line). The flux of the line is measured by integrating the spectra between $\pm 3\sigma$ of the central wavelength of the fitted Gaussian (see Fig. 3.14).

For helium and hydrogen recombination lines, such as the Balmer lines, the underlying continuum is more complex because the emission lines are superimposed on the absorption lines produced by the same transition in stellar photospheres (see Fig. 3.15 and Example *Correct from underlying absorption*). Not accounting for the underlying absorption when measuring the emission lines leads to the

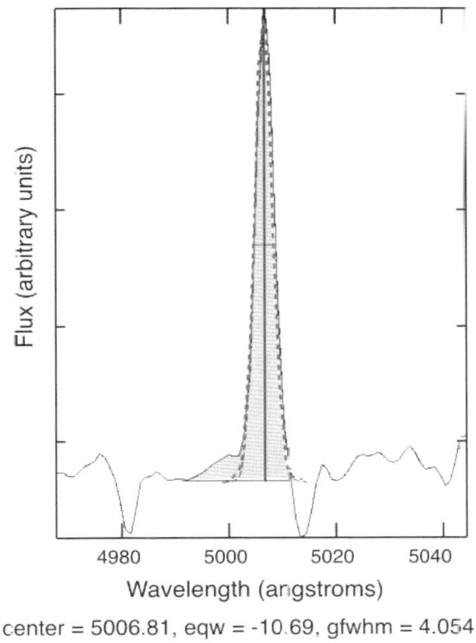


Fig. 3.14: Measure of the [O III]5007 emission line in a $z = 0.6$ galaxy spectrum, using IRAF. The line is composed of a main component at the systematic velocity of the galaxy (dashed line), and a blue shifted broad component. The total emission flux from both components can be measured with IRAF by integrating the spectra in the vicinity of the two components (filled area).

important underestimation of their fluxes and thus induces a bias for all the derived properties (Liang *et al.*, 2004).

### Example 3.5: *Correcting from underlying absorption*

*The most efficient method to subtract the stellar component consists in fitting a synthetic spectrum from a template library to the spectrum of each galaxy. The* STARLIGHT *software is taken for illustration but the steps are valid for all full spectrum fitting*

*(Continued)*

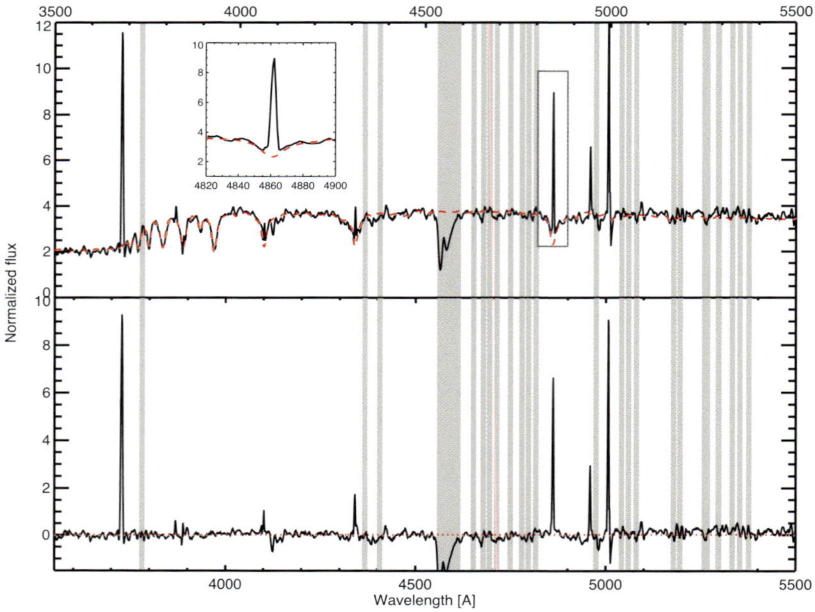


Fig. 3.15: Correcting from underlying absorption in a spectrum of a $z = 0.66$ galaxy. *Up*: The continuum of the rest-frame spectrum (black line) has been fitted using a stellar library and the STARLIGHT software. The contribution from the stellar absorption is plotted in red line. The insert box is a zoom on the H$\beta$ line. The emission line and the underlying absorption are clearly visible in the H$\beta$ line. The regions affected by sky lines and the strong $O_2$ absorption feature from the atmosphere (at 7600Å, in the observed frames) are indicated by grey areas and have been masked during the fit. *Down*: Final spectrum following subtraction of the stellar continuum, from which the flux of the emission lines can be properly measured.

## Example 3.5: *Continued*

*softwares. For correcting the underlying absorption in Balmer emission lines the steps are*:

(1) *Adjust the spectral resolution between the templates and the observed spectrum.* A majority of softwares automatically match the spectral resolution between the spectral templates and the input spectrum. This works reasonably well in

*(Continued)*

**Example 3.5: *Continued***

high S/N spectra and when the difference between the two spectral resolution is not too large (Cid Fernandes *et al.*, 2005). However, for S/N in the continuum under ten, it is highly recommended to adjust the template spectral resolution prior to that of the observed spectra, using a Gaussian deconvolution. Always use templates with higher spectral resolution than the observations. Mismatch on the resolution induces bias in the fluxes of Balmer lines.

(2) *Mask emission lines in the spectrum to be fitted.* Regions associated with emission lines have to be excluded from the fit. In STARLIGHT this can be done by setting the weight flag to 0 in the input file. Special attention has to be taken when excluding the emission in Balmer lines: the wings of the absorption line, not contaminated by the emission line, should not be masked since these features give strong constraints on the fit. Strong sky emission or absorption also need to be excluded, see the grey vertical lines in Fig. 3.15.

(3) *Create a library base.* The measurement of the fluxes does not strongly depend on the choice of template library. In general, the stellar evolutionary population models are used as base templates in the literature because of their completeness in the $Z$-age grid and their wide wavelength coverage, but empirical stellar or star cluster libraries can also be used. The only requisite is that the templates spectra cover the wavelength range of the input spectrum and that the spectral resolution is higher than the observed spectrum.

(4) *Fit and quality control.* The fitting algorithm will find the best mix of stellar populations from the template base that adjusts the observed spectrum.[8]The quality of the fit is

(*Continued*)

[8]At this point, we strongly advise against using the fit results to formulate any conclusion about star formation history, extinction, or stellar mass. The reader is referred to Sec. 3.8 for a complete discussion on stellar population decomposition.

**Example 3.5: *Continued***

evaluated by the $\chi^2$ value and a visual inspection of critical features, such as the wings of the Balmer lines, the 4000Å, break and the G band, as well as the global slope of the synthetic spectrum. If the slopes of the observed and synthetic spectra do not match, check the spectrophotometric quality of the input spectrum or wherever a nebular continuum should be added in the template base for extremely young objects.

(5) *Subtract the synthetic spectrum to the observed spectrum.* The synthetic spectrum is subtracted directly to the observed spectrum and emission lines can be measured in the continuum-subtracted spectrum. The continuum of the final spectrum should be flat and close to 0. In particular, check the quality of the fit close to emission lines. A negative continuum means that the absorption lines have been underestimated, while a bump suggests that the underlying absorption has been overestimated. Should one of these problems arise, check that: (a) the spectral resolution has been correctly matched between the observed spectrum and the templates; (b) the Balmer emission lines are correctly masked; (c) if the problem remains this can be due to the template base used, e.g., a young stellar template may be missing.

The estimation of the synthetic spectrum is sensitive to the wavelength window available for the fit. For example, the spectral window between [O II]$\lambda$3727 and [O III]$\lambda$5007 has several strong absorption lines (e.g., Balmer lines and Ca lines) and the 4000Å, break, which give strong constraints on the fit. The underlying absorption in Balmer lines can thus be accurately constrained when fitting this wavelength interval.

There are many softwares designed to automatically measure emission lines in large number of spectra and for which the stellar continuum subtraction is included. They are generally based on the assumption that emission lines have simple Gaussian profiles. Since this is not always the case and that at increasing redshifts, galaxies have more complex and disturbed kinematics (Yang *et al.*, 2008), such measurements can lead to systematics (~10%). Moreover the Poisson noise can affect the profile of faint emission lines, which may become difficult to detect with automatic algorithms. It is thus recommended to manually measure the lines, or at least to manually check the fit accuracy during this process.

### 3.5.3 *Low S/N regime: measurement bias*

In most distant galaxy spectra, some of the diagnostic lines are faint and with low S/N. Here, we show why this problem has to be treated and provide a solution for emission line ratios as well as limits on undetected emissions. The *profile signal-to-noise*, can be defined as the median S/N in a line:

$$\mathrm{S/N_{line}} = \frac{F_{\mathrm{line}}}{\sigma_c \times \sqrt{N}}, \tag{3.6}$$

where $F_{line}$ is the line flux, $\sigma_c$ is the noise of the pseudo-continuum, $N$ the number of resolution elements over which the line spreads. If the line has a Gaussian profile, $N$ is equal to six times the emission line width.

The statistical significance is generally evaluated as a function of the equivalent fluctuation of a standard Gaussian stochastic variable. A S/N above 3 implies that the line is detected within a $3\sigma$ significance level, and thus with a probability above 96%. However, the $3\,\sigma$ significance level criterion assumes that the detected signal lies above a known background, affected with pure Gaussian noise. These assumptions are not valid in the case of galaxy spectra, since the continuum is not known *a priori*, and it is not constant as a function of wavelength due to the underlying stellar populations. Even after subtracting the stellar continuum (see Ex. *Correct from underlying absorption*), the background still has residual signals arising from uncertainties in the flux calibration and templates mismatches.

In an ideal case of Poisson noise and a Gaussian line shape with unknown background, Rola and Pelat (1994) demonstrated that the integrated flux and S/N of a narrow emission line can be described by a lognormal probability distribution. This has a strong impact in the low S/N regime:

- The fluxes from lines having observed S/N under 5 are systematically overestimated, because *"At very low S/N an emission line can be detected with the help of some chance positive flux fluctuation, but never if it is a negative fluctuation. Most of the time, the observed line intensity is the true line intensity plus a noise intensity under the line, thus the bias"*;
- Observed S/N over 4–5 (respectively $2\sigma$ and $3\sigma$ confidence level) are required to exclude false detections;
- As for color in imagery (see Chapter 2, Sec. 2.4.6.1), the resulting noise of the logarithm of a line ratio follows a normal distribution (see Appendix in Zheng *et al.*, 2004).

Estimating the *upper limit* of the flux of an undetected line can be very useful, including to set limits on line flux ratios. Figure 3.16 shows how this can be done by assuming that the undetected line has a Gaussian line-profile with a width equals to the mean width ($<\delta_v>$) of other detected lines, and a height equals to the pseudo-continuum noise $\sigma_c$. The line-integrated flux is then $F = \sqrt{2\pi} \cdot \sigma_c \cdot <\delta_v>$, where $\sqrt{2\pi}$ is coming from the integration of the Gaussian profile, $\sigma_c$ is in flux units (either $f_\nu$ or $f_\lambda$), and $<\delta_v>$ is either in Hertz or Angstrom, respectively.

## 3.6 Emission diagnostics of the ionized ISM

This section provides guidelines to measure physical quantities characterizing the ionized ISM from optical spectra, see summary in Table 3.1. It focuses on diagnostics relevant to distant galaxies, given the available lines and their S/N ratio. Three properties of the ISM are discussed: (1) Dust extinction; (2) Ionization source: photoionization from young stars, old stellar populations, active galactic nuclei, or collisional ionization from shocks; (3) Chemical composition of the gas.

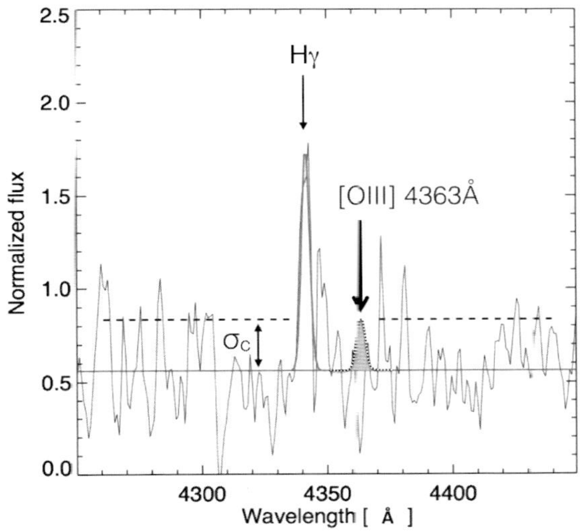


Fig. 3.16: Portion of a FORS2 spectrum of a $z \sim 0.6$ galaxy, which is shown at rest-frame. The goal is to estimate a limit on the [O III]$\lambda$4363Å line while $H\gamma$ is detected and can be used to estimate the expected width of a line ($<\delta_v>$). One has to evaluate the noise in the pseudo-continuum ($\sigma_c$), and from the Gaussian surface (grey area) one may derive the maximal line flux possessed by a line at 4363Å above the continuum (see the Gaussian with a dotted line) represented by the horizontal full line.

The reader is referred to Osterbrock (1989) and the review *What can emission lines tell us*? by Stasińska (2007) for a detailed description of photo-ionization modeling and emission lines diagnostics.

### 3.6.1 *Dust extinction*

Interstellar dust grains are the solid-state component of the ISM and the main source of opacity. Although dust grains account for only 1% of the ISM mass and ~0.1% of the total baryonic mass in our Galaxy, the impact of dust on the luminosity of galaxies is huge. In our Galaxy, around 40% of the UV starlight is absorbed by dust and re-emitted in the FIR. In distant starbursts, dust drives the global shape of the spectral energy distribution. More than 75% of the global SFR density at $z \leq 1$ is hidden by dust in the UV-optical (Flores *et al.*, 1999).

#### 3.6.1.1 *Extinction laws*

The main effect of dust upon light is to absorb and scatter the light. It can be modeled as a single point-like source behind a dust screen. Under this assumption, the observed intensity $I^{\mathrm{obs}}(\lambda)$ of a light source crossing a dust layer is:

$$I_\lambda^{\mathrm{obs}} = I_\lambda^0 \, e^{-\tau_\lambda}, \tag{3.7}$$

where $I_\lambda^0$ is the intrinsic intensity and $\tau_\lambda$ is the optical depth of the dust layer. Dust grains absorb the incoming photons with different efficiencies depending on their energy. Blue wavelengths are more attenuated than the reddest wavelengths, leading to an effect known as *reddening*.

Dust extinction is usually expressed as total extinction $A_\lambda$ in magnitude units. $A_\lambda$ is defined as the decrease in magnitude produced by a layer of dust with optical depth $\tau_\lambda$:

$$A_\lambda = 2.5 \log\left(e^{\tau_\lambda}\right) = 1.086 \, \tau_\lambda \tag{3.8}$$

so that the attenuation of a source can be written as:

$$I_\lambda^{\mathrm{obs}} = I_\lambda^0 \, 10^{-0.4 \times A_\lambda}. \tag{3.9}$$

The total extinction can typically be split into two components, an extinction amplitude at a given reference wavelength (e.g., $A_V$) and the extinction law that gives the normalized reddening as a function of wavelength. Figure 3.17 shows the normalized-to-$A_V$ extinction law[9] ($A_\lambda/A_V$) for three interstellar extinction curves scaled from the Milky Way and from local dwarf starbursts Cardelli *et al.*, 1989.

It is generally useful to estimate the extinction through intensity ratios, generally using a reference intensity $I_{\mathrm{ref}}$ that is either at V band or H$\beta$ wavelengths for photometric and spectroscopic data, respectively. The two-components behavior of the extinction is better

[9]Other parametric forms of the extinction laws are used and are strictly equivalent, e.g., $E(\lambda - V)/E(B - V)$, where $E(\lambda - V) = A(\lambda) - A(V)$ (and $E(B - V) = A_B - A_V$, see Fitzpatrick, 1999) or $\kappa(\lambda) = A_\lambda/E(B - V)$ (Calzetti *et al.*, 2000).

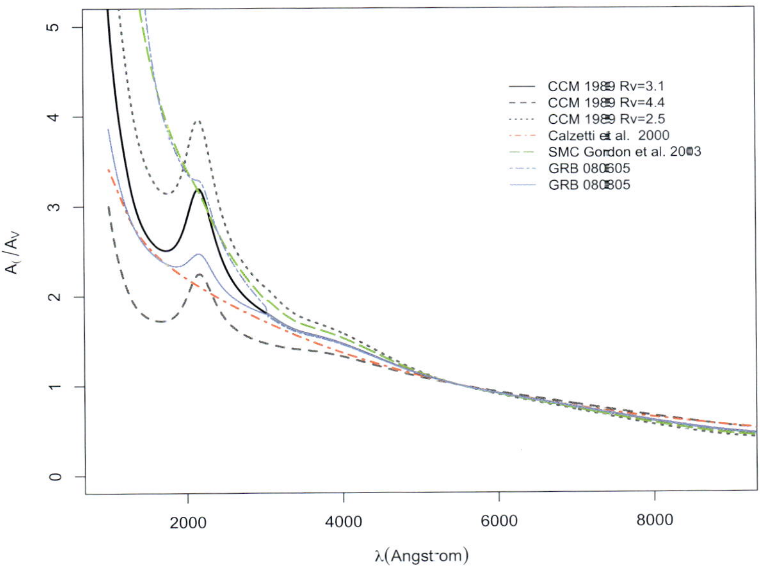


Fig. 3.17: Interstellar extinction laws ($A_\lambda/A_V = f(\lambda)/f(V)$) from Cardelli *et al.* (1989) for three values of $R_V = 2.5$, 3.1 (average Milky Way), 4.4, the average SMC (Gordon *et al.*, 2003) and local dwarf (Calzetti *et al.*, 2000). In dense clouds, such as the Orion nebulae, $R_V$ shows higher values from 4.4 to 5.5 (Osterbrock, 1989).

illustrated by:

$$\frac{I_\lambda^{\rm obs}}{I_{\rm ref}^{\rm obs}} = \frac{I_\lambda^0}{I_{\rm ref}^0} 10^{-c_{\rm ext}\times(f(\lambda)-f({\rm ref}))}, \tag{3.10}$$

where the logarithmic extinction coefficient ($c_{\rm ext}$) normalizes the extinction amount and the function $f(\lambda)$ gives the shape of the wavelength dependence of the extinction law (Osterbrock, 1989). They can be related through: $A_{\rm ref} = -2.5\, c_{\rm ext}\, f({\rm ref})$. Table 3.3 shows the $A_\lambda/A_V$ ($= f(\lambda)/f(V) = f(\lambda)$) values at the reference wavelengths (V band), which can be used for simple evaluations of the extinction.

The Milky Way-type extinction law can be described through a parametric form using the total-to-selective ratio $R_V$ (e.g., Cardelli *et al.*, 1989; Fitzpatrick, 1999), which is directly linked to the curve slope at 0.55 $\mu$m (V):

$$R_V = A_V/(A_B - A_V) = A_V/E(B-V). \tag{3.11}$$

Besides this, the Calzetti *et al.* (2000) curve lacks a 0.2175 $\mu$m bump, sharing this property with e.g., that of the SMC (Osterbrock, 1989).

Table 3.3: Values of the extinction curve at different wavelengths and for several extinction laws.

| $\lambda$ | CCM | | | SMC | Calzetti |
|---|---|---|---|---|---|
| (Å) | $R_V = 3.1$ | $R_V = 4.5$ | $R_V = 2.5$ | $R_V = 2.74$ | $R_V = 4.05$ |
| $L\alpha$ | 3.51 | 2.11 | 4.64 | 6.61 | 2.96 |
| C IV | 2.60 | 1.74 | 3.31 | 4.56 | 2.50 |
| C III] | 2.62 | 1.83 | 3.27 | 3.58 | 2.25 |
| Mg II | 1.95 | 1.54 | 2.28 | 2.16 | 1.79 |
| 2800 | 1.95 | 1.54 | 2.28 | 2.16 | 1.79 |
| [O II] | 1.54 | 1.37 | 1.68 | 1.61 | 1.45 |
| $H\delta$ | 1.43 | 1.30 | 1.54 | 1.49 | 1.33 |
| $H\gamma$ | 1.35 | 1.24 | 1.43 | 1.39 | 1.27 |
| B | 1.31 | 1.22 | 1.38 | 1.35 | 1.24 |
| H$\beta$ | 1.16 | 1.12 | 1.20 | 1.18 | 1.14 |
| [O III] | 1.12 | 1.09 | 1.15 | 1.13 | 1.10 |
| V | 1.00 | 1.00 | 1.00 | 1.00 | 1.00 |
| [O I] | 0.86 | 0.88 | 0.84 | 0.85 | 0.86 |
| H$\alpha$ | 0.82 | 0.84 | 0.80 | 0.81 | 0.82 |
| 24 $\mu m$ | 0.10 | 0.11 | 0.09 | 0.09 | 0.00 |

Values of the extinction curve expressed as $f(\lambda) = A(\lambda)/A(V)$ at different wavelengths and for several extinction laws: three $R_V$ values of the Cardelli *et al.* (1989) law, the average SMC (Gordon *et al.*, 2003) and the starburst extinction law from Calzetti *et al.* (2000). Values at other wavelengths can be interpolated assuming $f(\lambda) \sim 1/\lambda$, while at very large wavelengths, $f(\lambda) \sim 0$.

The detailed shape of the extinction law depends on dust grain size and composition, both properties which are difficult to evaluate for distant and faint galaxies.

### 3.6.1.2 *Balmer decrement*

In optical spectroscopy, dust extinction can be evaluated using the Balmer decrement method. Ratios between Balmer lines (H$\alpha$, H$\beta$, H$\gamma$) are nearly constant in the typical range of conditions encountered in H II regions (Osterbrock, 1989). The offset between the theoretical ratio $R_0 = I^0_{\lambda_1}/I^0_{\lambda_2}$ and the observed ratio $R_{\rm obs} = I^{\rm obs}_{\lambda_1}/I^{\rm obs}_{\lambda_2}$ of two Balmer lines is then attributed to dust extinction.[10] The theoretical ratios are given in Table 3.4 assuming a Case B recombination[11] typical of H II regions.

[10]It is mandatory to verify beforehand that measurements of Balmer lines have been properly corrected for absorption, see Ex. *Correct from underlying absorption*.

Table 3.4: Theoretical Balmer ratios.

| $T_e$ | H$\alpha$/H$\beta$ | H$\gamma$/H$\beta$ | H$\alpha$/H$\gamma$ |
|---|---|---|---|
| 5000 K | 3.10 | 0.458 | 6.76 |
| 10000 K | 2.86 | 0.469 | 6.09 |

Theoretical Balmer ratios assuming a Case B recombination and an electron density of $N_e = 10^4 cm^{-2}$ for two values of temperature[12] (Osterbrock, 1989).

Balmer decrements are commonly defined relatively to the H$\beta$ line using Eq. (3.10).

#### 3.6.1.3 *Global balance between IR and UV-H$\alpha$ SFR*

Extinction can be evaluated from the balance between the infrared and the UV/H$\alpha$ luminosities to which SFRs are proportional. Therefore the amount of extinction can be estimated from the ratio between the SFR estimated in the UV-optical and the SFR$_{\rm IR}$, and by applying Eq. (3.10):

$$c_{\rm ext}(\rm IR, Balmer) = \log\left(\frac{SFR_{IR}}{SFR_{H\alpha_{nc}}}\right)/(f(IR) - f(H\alpha)), \qquad (3.12)$$

$$c_{\rm ext}(\rm IR, UV) = \log\left(\frac{SFR_{IR}}{SFR_{2800\mathring{A}}}\right)/(f(IR) - f(UV)), \qquad (3.13)$$

where SFR$_{\rm H\alpha_{nc}}$ is the SFR estimated from the luminosity of the observed H$\alpha$ line (corrected for aperture effects but not for dust extinction) and SFR$_{\rm UV}$ the SFR estimated from the UV continuum luminosity, which should be taken below 2800Å to avoid contamination by intermediate old stars. f(IR) and f(UV) values can be interpolated from Table 3.3. These diagnostics have the advantage to be global, i.e., accounting for all the emitting regions of a galaxy conversely to that from the pure Balmer ratio. In fact, in dust-enshrouded regions both Balmer lines could be fully extinguished

[11] It assumes that the nebulae is optically thick to Lyman-series photons which are re-emitted as high-order lines of the Balmer series.

[12] The temperature of an H II region depends on the metal abundance, see Sec. 3.6.4: it is inversely proportional to the metallicity. High temperatures are appropriate for H II regions having low metallicities, while $T_e \sim 5000$ K are a better fit to the condition of metal-rich nebulae. Nevertheless the uncertainties on the electron temperature has little impact on the estimation of dust extinction. A variation in the electron temperature by a factor of two would result in ~0.1 mag difference in the estimation of $A_V$.

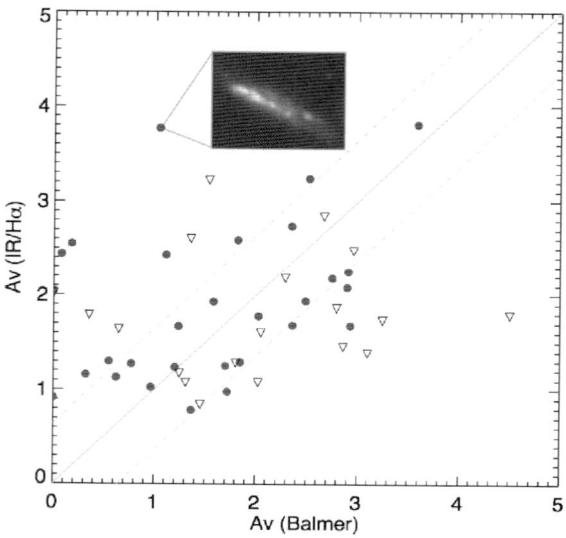


Fig. 3.18: Relation between the extinction $A_V$ values derived by Balmer decrement and by $SFR_{IR}$ and $SFR_{H\alpha\,nc}$ ratio. The objects with infrared detections (24 $\mu$m MIPS) are plotted as filled circles while objects with upper limit estimations of the infrared flux are plotted as triangles. The two dashed lines refer to the results with $\pm 0.64$ mag. The insert shows the image of one of the edge-on galaxies for which the two methods give discrepant values.

leading to an underestimation of the extinction as it could occur, e.g., in an edge-on galaxy (see Fig. 3.18). Limitations of the global balance method are linked to the difference between estimates of the SFR, which is instantaneous ($\tau \sim$10 Myr) in the case of Balmer lines, while in IR or UV it is caused by $\sim$100 Myr timescale events (see also chapter 2). At low SFRs, the IR light could also be contaminated by cirrus (Kennicutt, 1998).

#### 3.6.1.4 *Which extinction law for distant galaxies?*

Considerable progress has been made in the understanding of the dust extinction laws in galaxies. For a simple geometry such as disk galaxies, Charlot and Fall (2000) have adopted a two-component dust distribution, one for giant blue stars embedded into birth clouds and the second for other stars lying in the ISM. This description is superior to that of a dust screen, and points out that extinction laws intimately depend on what we are measuring. For example, extinction from Balmer line ratios is expected to be different then that from continuum observations (such as in Eq. (3.13)) and this depends on geometry (inclination for disk galaxies), star formation activity, metal abundance, etc. For disk-dominated galaxies, Wild *et al.* (2011) provided useful guidelines for evaluating the extinction.

For distant galaxies the choice of extinction law is far more speculative, since their morphologies are more complex, and we are lacking a model for their spatial distribution, emission, and radiation transfer. Important progress has been made in identifying the presence or the absence of the 0.2175 $\mu$m bump in distant ($z = 1$–$2$) galaxies (Noll *et al.*, 2009; Buat *et al.*, 2011) by using transfer models between FIR and UV. However the observations of power-law, very bright GRB afterglows provided for the first time accurate extinction laws for individual distant galaxies (Zafar *et al.* 2012 and references therein). All of them show a strong 0.2175 $\mu$m bump as well as significant variations from one case to another. Because GRBs are associated with low mass galaxies, this ended former expectations from the Local Group that the bump may have been associated with massive galaxies (e.g., the MW) and not to dwarfs (e.g., the SMC). It bestows on the extinction law in distant galaxies a considerable variance, which mostly affects studies in the rest-frame UV (see Fig. 3.17).

#### 3.6.1.5 *Extinction for correcting the (UV) continuum*

Determination of the global extinction through Eq. (3.13) is mandatory to correctly extract quantities linked to the continuum (see also Sec. 2.6.2). Figure 3.17 illustrates that the latter is very dependent on the extinction law and at the time of writing it is very difficult to propose robust criteria for helping with this choice. The standard galactic extinction with $R_V = 3.1$ or alternatively the Calzetti *et al.* (1994) law are often assumed. These are highly speculative and one has to account for the systematics this may bring to spectroscopic estimates, e.g., when considering the wide range of extinction laws. It is also mandatory to avoid determining properties of the young stellar population at rest-frame wavelength near 0.2175 $\mu$m. Attempts to use the UV slope (see Sec. 2.6.2) are not compelling because besides the impact of the 0.2175 $\mu$m bump, it requires a precise knowledge of stellar ages, star and medium abundances and overall distribution. To note that simple prescriptions may help in many instance, such as with disk morphologies (Wild *et al.*, 2011; see Sec. 2.6.2) or luminous compact galaxies for which the overall stars and gas are likely bathing in approximately the same medium, see Ex. *How to choose the extinction law in a compact distant galaxy.*

#### 3.6.1.6 *Extinction for correcting optical emission lines*

Finally and perhaps more importantly, for calculating line ratio estimates e.g., for metal abundances or diagnostic diagrams, one should adopt the Balmer decrement estimate under the assumption that it probes the same medium from which the lines are emitted. Notice that in Fig. 3.17, from [O II]$\lambda$3727Å to the optical/NIR lines, the different extinction curves show relatively modest variations.

**Example 3.6: *How to choose the extinction law in a compact distant galaxy***

Let us compare extinction laws for the specific case of a redshifted compact starburst with multi-wavelength observations. The object is a LIRG at $z \sim 0.4$ and photometry from UV to IR, deep imagery from Ultra Deep Field HST and optical spectrum (VLT/FORS2) are available (from Hammer *et al.*, 2009). According to the Balmer decrement, the extinction is $c_{\mathrm{ext}} = 1.0$ independently of the assumed extinction law (e.g., $R_V = 5$ also called Orion law or the standard extinction law). A discrepancy appears when estimating extinction using the IR-to-UV SFR balance: $c_{\mathrm{ext}} = 0.54$ and 0.8 when using the standard extinction and the Orion extinction law, respectively. This favors the Orion extinction law for this object. This is furthermore supported by the $(b - z)$ color from HST images. Without correction, the

(*Continued*)

**Example 3.6: *Continued***

$(b-z)$ is equal to 0.78 which appears consistent with the color of an Sbc galaxy, whereas the galaxy (a compact LIRG) is in fact a very strong starburst. Applying the Orion extinction curve to the overall color of the galaxy resolves most of the discrepancy due to the gap between Sbc and pure starburst colors $(b-z) = 0.78$. A similar exercise assuming the standard extinction curve would provide a de-reddened color bluer than that of a pure starburst, which is not plausible.

### 3.6.2 *Active galactic nuclei*

Galaxies, and perhaps all massive galaxies, can host *active galactic nuclei* (AGN), i.e., compact central regions producing large amounts of ionizing radiation with a nonstellar origin. Most diagnostics used to determine physical quantities of the ISM are based on the assumption that the gas is exclusively photo-ionized by massive stars. To avoid misleading inferences on distant galaxy properties, it is mandatory to test for the presence of an AGN.

In the optical, AGN hosts are classified into three categories according to their spectral features: Seyfert I, Seyfert II, and LINERs. *Seyfert I* can be easily identified due to their characteristically broad Balmer lines with dispersion velocities $\sigma > 800$ km/s, see central panel of Fig 3.13. The presence of multiple ionized species such as [N V], [O VI], [Fe VII], [Fe X] in their spectra reveals a high energy ionization source incompatible with the energy produced by stellar radiation. *Seyfert II* are characterized by very bright collisionally excited lines. They can be isolated from star formation using diagnostic diagrams based on emission line ratios. "Low Ionization Narrow Emission line" (LINERs) spectra are characterized by intense collisional lines, abnormally faint Hydrogen recombination lines, and emissions from low ionized species such as [O I]$\lambda$6300Å, or [S II]$\lambda\lambda$6716, 6731Å. LINERs are not ubiquitous AGN because similar low ionization emissions can be produced by late evolutionary stages of "retired" stars (hot evolved low-mass stars, Stasińska *et al.*, 2008) or by shocks

(Heckman, 1980), the latter can be important in distant galaxies in which feedback or outflows are often expected or observed. Because it is the most abundant population of AGNs (e.g., in the SDSS), LINERs should not be ignored in galaxy evolution studies.

#### 3.6.2.1 *Diagnostic diagrams*

The most definite test for the presence of an AGN comes from the hardness of the X-ray emission, or if present, from the shape of the radio emission spectrum. However X-ray observation satellites generally do not provide sufficient resolution and sensitivity to access these distant galaxy diagnostics, which are instead restricted to a few bright AGN.

Diagnostic diagrams link two lines ratios: (1) a ratio tracing the mean level of ionization and temperature, such as [O III]/H$\beta$; (2) a ratio sensitive to high energy photons produced by the AGN, e.g., [N II]/H$\alpha$, [S II]/H$\alpha$, or [O I]/H$\alpha$. Diagnostic diagrams are divided into several regions that separate star-forming/AGN and Seyfert/LINERs, and allow us to classify galaxies according to their location (Osterbrock, 1989).

The *Baldwin–Phillips–Terlevich diagram* (BPT) (Baldwin *et al.*, 1981), also called the O3N2 diagnostic, is the most widely used method. It is defined as the line ratio of [O III]$\lambda\lambda$4959, 5007/H$\beta$ vs. [N II]$\lambda$ 6584/H$\alpha$. Galaxies fall into two well-defined branches, see Fig. 3.19: a left branch, which is populated by star-forming galaxies and a right branch attributed to AGN dominated galaxies (with high [N II]$\lambda$ 6584/H$\alpha$).

The limits of the Fig. 3.19 have been modeled (see Kauffmann *et al.*, 2003; Kewley *et al.*, 2006) for local SDSS galaxies. More violent star formation can be expected in distant, star-forming galaxies which may affect these limits, though apparently this occurs mostly above $z = 1.5$ (see Kewley *et al.*, 2013). This is an important limitation for the BPT method. Readers might use Figs. 1 and 2 of Kewley *et al.* (2013) to make their own AGN/starburst classification of high redshift galaxies, though a prerequisite is the thorough estimation of the ionization parameter and the ISM density.

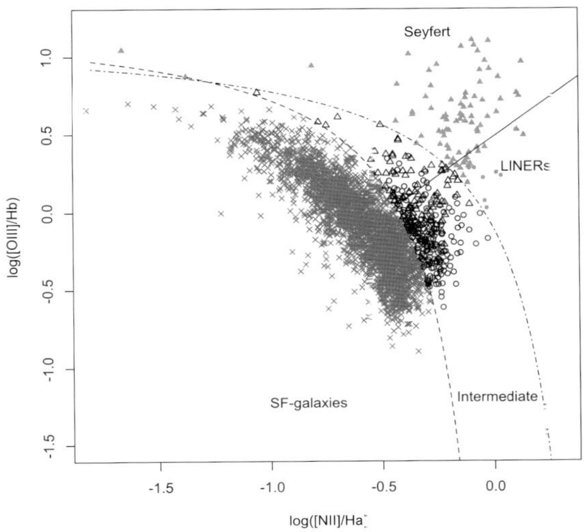


Fig. 3.19: BPT diagram for a sample of local galaxies (drawn from the GAMA survey, see Rodrigues *et al.*, 2016): Seyfert (triangles), LINERs (circles) and star-forming galaxies (crosses). Targets above the dot-dashed line ($\log[\text{O III}]/\text{H}\beta = 0.61/(\log[\text{N II}]/\text{H}\alpha - 0.47) + 1.19$) are robust AGN/LINERs hosts (solid symbols) according to the theoretical separation of Kewley *et al.* (2001). This separation corresponds to the theoretical limit when assuming only stellar photo-ionizing radiation. The dashed line ($\log[\text{O III}]/\text{H}\beta = 0.61/(\log[\text{N II}]/\text{H}\alpha - 0.05) + 1.3$) corresponds to the empirical limit between star-forming galaxies and AGN/LINERs from Kauffmann *et al.* (2003b). Galaxies falling between the two lines are composite galaxies (open symbols), for which the AGN contribution is expected to be within 3–10% in e.g., the H$\beta$ line flux (Kauffmann *et al.*, 2003b). Below the line, galaxies are classified as pure star-forming galaxies (crosses). The solid line shows the separation between Seyfert and LINERs.

At intermediate redshift, the [N II] and $H\alpha$ lines may not fall in the visible spectrum. In this case the diagnostic diagram based on [O II]$\lambda$ 3727/H$\beta$ vs. [O III]$\lambda\lambda$4959, 5007/H$\beta$ ratio can be used, see Fig. 3.20. The delimitation of the maximum starburst line in the [O II]/H$\beta$ vs. [O III]/H$\alpha$ diagram is the limit ratio predicted by photo-ionization models for the maximal stellar temperature of 60000 K, which is the maximum temperature for the most massive stars (Tresse *et al.*, 1996; Hammer *et al.*, 1997a). The [O II]/H$\beta$ ratio involves two lines separated by more than 1000Å, and is thus sensitive to extinction. It is mandatory to correct the flux of emission lines for dust extinction before using this diagnostic. As an example a galaxy with an extinction $A_V = 2.0$ and a line ratio log [O II]3727/H$\beta$ = 0.7

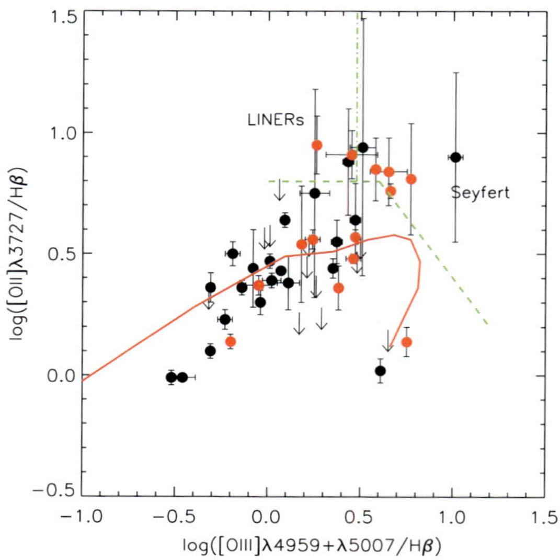


Fig. 3.20: Diagnostic diagram [O II]$\lambda$3727/H$\beta$ vs. [O III]$\lambda\lambda$4959, 5007/H$\beta$ for a sample of $z \sim 0.6$ galaxies. The objects undergoing shocks, $log$([Ne II]$\lambda$3868/[O II]$\lambda$3727) $> -1.3$, are plotted in red dots. The solid line shows the theoretical sequence from McCall *et al.* (1985) for extra-galactic H I regions. The dashed line shows the photo-ionization limit for a stellar temperature of 60000 K and empirically delimits the Seyfert 2 and LINER from the H II regions. The dot-dashed line shows the demarcation between Seyfert 2 and LINERs from Osterbrock (1989). Credit: Rodrigues *et al.* (2008), A&A, 492, 371, reproduced with permission ©ESO.

uncorrected from extinction gives a corrected ratio of 0.96. This galaxy would be identified as a star-forming galaxy if not corrected for dust extinction, while it should be classified as an AGN.

At high redshift, the presence of a hard photo-ionization radiation field from an AGN can be detected in the rest-frame UV spectrum from the ratio of collisional emission lines with the Ly$\alpha$ recombination line. AGNs have typical ratios of C IV/Ly$\alpha$ $\sim$ 0.25 and C III]/Ly$\alpha$ $\sim$0.125 (Steidel *et al.*, 2002), while starburst galaxies have ratios under 0.05. These ratios are comparable to those of local Seyfert 2 galaxies (Ferland and Osterbrock, 1986), though the method certainly requires further tests/modeling before being routinely used.

Because of the above limitations, alternative diagnostic diagrams have been suggested, e.g., using the empirical correlation between

[N II]/H$\alpha$ and the stellar mass (the Mass-Excitation diagram with [O III]/H$\beta$ ratio and $M_{stellar}$, see Juneau *et al.*, 2011) or with color (the Color-Excitation diagram with [O III]/H$\beta$ ratio and the rest-frame $U-B$ color, see Yan *et al.*, 2011). It is not advised to use these relations for various galaxy populations, because they use empirical and not universal correlations and also because the stellar mass is a quantity relatively poorly constrained (see Chapter 2).

#### 3.6.2.2 *Complementary diagnostics*

The presence of other excited lines can also hint at the presence of an AGN. For example, Seyfert galaxies produce a broader range of ionization states than the radiation from young stars, which can be detected by the presence of both low ionization species (e.g., [Ne III]$\lambda$3868Å, [O I]$\lambda$6300Å) and high ionization elements ([C IV], [Ne V]). In the UV, the strong N V$\lambda\lambda$1238,1242 and He II$\lambda$1640 lines may indicate the presence of an AGN. Low values of the ratio C III]/C IV also indicate hard photo-ionization radiation field from AGN.

### 3.6.3 *Star formation rate (SFRs)*

#### 3.6.3.1 *SFR from the H$\alpha$ line*

The SFR estimates based on emission lines rely on measuring the starlight that has been reprocessed by interstellar gas. In the case B recombination regime, the nebulae is thick: the Lyman photons are absorbed by the ISM Hydrogen atoms and re-emitted during a recombination cascade into Balmer photons. The luminosity of the Balmer lines is proportional to the amount of ionizing photons from young and massive stars. The conversion between the number of ionizing photons and the SFR is done by assuming an initial mass function (IMF) that gives the distribution in mass of the stars born in a same cloud (see Sec. 2.6.5). The SFR is very sensitive to the high mass end of the IMF (30–40 $M_\odot$). As the lifetime of these stars is around ~3–10 Myr, the SFR estimated from the Balmer lines is an *instantaneous* measurement of the SFR when compared to the

UV and IR SFR (100 Myr timescale, see Sec. 2.6.5). In practice, the H$\alpha$ luminosity is calibrated against the SFR using photo-ionization models and evolutionary synthesis models, with a prescribed age mix, chemical composition, and IMF, see review Kennicutt and Evans (2012). The calibration from Kennicutt and Evans (2012) gives:

$$\mathrm{SFR}_{\mathrm{H}\alpha}(\mathrm{M}_{\odot}\mathrm{yr}^{-1}) = 5.37 \times 10^{-42} L_{\mathrm{H}\alpha}\ (\mathrm{erg\,s}^{-1}). \tag{3.14}$$

The H$\alpha$ luminosity is calculated from the dust-corrected line flux by:

$$\log L_{\mathrm{H}\alpha}\ (\mathrm{erg\,s}^{-1}) = 50.078 + 2\log D_L + \log F_{\mathrm{H}\alpha}, \tag{3.15}$$

where $D_L$ is the luminosity distance in Mpc, $F_{\mathrm{H}\alpha}$ is the flux in the H$\alpha$ line corrected for extinction in $\mathrm{erg\,s^{-1}\,cm^{-2}}$ and the luminosity of the line is $L_{\mathrm{H}\alpha}$ in $\mathrm{erg\,s^{-1}}$. $SFR_{\mathrm{H}\alpha}$ is very sensitive to uncertainties on aperture correction and on extinction correction. Because it is very sensitive to the very top-end of the IMF, Eq. 3.14 cannot be applied on low mass galaxies for which the IMF is certainly less top-heavy (see, e.g., Kroupa *et al.*, 2013 and references therein).

At intermediate redshift, the H$\alpha$ line is redshifted outside the optical window. In such a case, the SFR can be estimated from the higher order Balmer lines such as H$\beta$ line and H$\gamma$, though they are more affected by uncertainties on the dust extinction and by low S/N (see Sec. (3.5.2)). There is unfortunately no alternative line (see Ex. *Can we estimate SFR from the forbidden* [O II] *line*?) for robustly measuring the SFR.

**Example 3.7: *Can we estimate SFR from the* [O II] *line?***

In high-$z$ galaxy spectra, the Balmer recombination lines fall outside the optical window and the high order Balmer lines (H$\gamma$, H$\beta$) can be too faint to derive reliable SFR. Kennicutt (1992) proposed to use the [O II]$\lambda$3727Å line, to derive the SFR. This line has the advantage to be strong in galaxy spectra and to be detected in the optical range up to $z \sim 1.5$. However, the [O II] line intensity is not directly proportional to the SFR, because their intensity mainly depends on, e.g., metallicity and

(*Continued*)

**Example 3.7: *Continued***

ionization parameter. Kennicutt (1992) claimed to have found a correlation between the $W_\lambda$([O II]) and $W_\lambda$(H$\alpha$) in a sample of local galaxies, making possible the implementation of an indirect calibration based on the [O II] luminosity. This line has been used in the literature to derive the SFR in high-$z$ surveys. However, Hammer *et al.* (1997b) and later Jansen *et al.* (2001) showed that the [O II] luminosity is not a good tracer of the SFR. Using a larger survey of intermediate-$z$ and local galaxies, they found a large scatter between the $W_\lambda$([O II]) and $W_\lambda$(H$\alpha$) that invalidates the Kennicutt calibration. It reveals that the later is based on a sample of nearby spiral galaxies having very similar ISM properties (dust, ionization parameters) and stellar populations. Besides the fact that it does not trace the SFR, using [O II] cumulates numerous problems: (1) The UV, [O II] line is very sensitive to extinction. (2) The [O II] line intensity depends strongly on the metallicity of the gas and is extremely weak at low metallicities. (3) The line is sensitive to the ionization parameter. It is therefore not recommended to estimate SFR using this emission line.

#### 3.6.3.2 *SFR from the Ly$\alpha$ line*

At $z > 4$ the only emission line accessible to optical instruments is the Ly$\alpha$ line. This line arises from the recombination of Hydrogen in the photo-ionized medium and in principle it could hence be used as a proxy to the SFR. In practice, the Ly$\alpha$ SFR is derived from the Ly$\alpha$ flux corrected for dust reddening (e.g., dust content estimated from the UV slope, which is largely debated, see Chapter 2). However, the SFR(Ly$\alpha$) are found to be inconsistent with other diagnostics by large factors (2 to 6, see Ajiki *et al.*, 2003; Atek *et al.*, 2008). The discrepancy is due to the resonant nature of the Ly$\alpha$ photons and therefore to the complex radiative transfer processes occurring in the Ly$\alpha$ line. When crossing the foreground ISM, the Ly$\alpha$ photons are highly scattered and/or absorbed by the neutral Hydrogen gas and dust grains (see discussion on the Ly$\alpha$ line in Sec. 3.7.3.1), and the line intensity depends on the H I content, the geometry and kinematics, and the dust content and its distribution (Kunth *et al.*, 1998; Atek *et al.*, 2008; Hayes *et al.*, 2011). The escape fraction $f_e$ of Ly$\alpha$ photons, defined as the fraction of the number of Ly$\alpha$ photons emitted over the number of photons emerging from the ISM, can span a wide range of values, e.g., between 0 to 100% in a sample of $z \sim 3$ Lyman break galaxies (Kornei *et al.*, 2010; Atek *et al.*, 2009). Thus the SFR is:

$$\mathrm{SFR}_{\mathrm{Ly}\alpha}[\mathrm{M}_\odot \cdot \mathrm{yr}^{-1}] = \frac{1}{f_e} \times 9.1 \times 10^{-43}\, L(\mathrm{Ly}\alpha[\mathrm{erg.s}^{-1}]), \quad (3.16)$$

assuming the SFR calibration of Kennicutt (1998) and a Salpeter (1955) IMF, following Atek *et al.* (2009), and which considerably depends on $f_e$. The best method to retrieve $f_e$ is from the ratio of Ly$\alpha$ to the extinction-corrected H$\alpha$ flux (Atek *et al.*, 2009), implying that Eq. (3.16) is rather useless in practice.

### 3.6.4 *Direct measurement of the gas metallicity ($T_e$ method)*

Oxygen accounts for about half of the metals in the ISM (Calura and Matteucci, 2004) and emits relatively intense lines at visible wavelengths. In H II regions, the abundance of heavy elements (called metallicity) in the gas phase is therefore primarily probed using the relative abundance of oxygen compared to hydrogen. The oxygen abundance in the gas is by convention expressed as the logarithmic quantity A(O) = log (O/H) + 12, where O/H is the fraction of oxygen relatively to hydrogen and in which the value for hydrogen is 12.

Metallicities can be directly measured from the temperature and the density of a gas of free electrons, defined as the electron temperature $T_e$ and electron density $N_e$, respectively. The ions[13] $O^+$, $O^{++}$, and $N^+$ are the primary coolants of H II regions, as they convert the

[13]For spectral lines the neutral elements are noted I, and single ionized species are noted with II. For instance, the transition emitted by the doubled-ionized oxygen $O^{++}$ is noted O III.

kinetic energy of free electrons (induced by collisions) into radiation arising from forbidden transitions. The ionic abundances of Oxygen and Nitrogen drive the temperature of the gas. To compute the total abundance of an element, the ionic abundance of all the ionization stages of that element need to be summed. For instance with Oxygen, the total metallicity is given by the sum of the two main stages of low and high ionization, respectively:

$$\frac{\mathrm{O}}{\mathrm{H}} = \frac{\mathrm{O}^{+}}{\mathrm{H}^{+}} + \frac{\mathrm{O}^{++}}{\mathrm{H}^{+}}. \qquad (3.17)$$

Currently one does compute the abundance of oxygen only from ionized species because the corresponding neutral species are strongly linked by charge exchange (Stasińska *et al.*, 2012). The contribution of $O^{3+}$ is assumed to be negligible in most H II regions, but this is perhaps not the case in distant galaxies or very recent and strong star formation galaxies, with e.g., an expected presence of hot Wolf Rayet stars (such as in GRB hosts, see Hammer *et al.*, 2006). The contribution of $O^{3+}$ can be assessed by the UV line [O IV] $\lambda$1661, and the contribution of higher ionized species can be assessed through [O V] $\lambda$1213 and [O VI] $\lambda$1031.

The temperature $t_2$ in the low-ionization zone can be estimated through the following ratios: [O II] I(3726 + 3729)/I(7320 + 7330), [N II] I(6548 + 6583)/I(5755), and [S II] I(6716 + 6731)/I(4068 + 4076). The temperature $t_3$ in the zone of double ionized species, such as $O^{++}$ can be calculated from the [O III]$\lambda 4959, 5007/\lambda 4363$ ratio. The electron density can be estimated from two transitions arising from two levels of a similar ionized specie having different radiative transition probabilities. The ratio of these levels depends mainly on the electron density, via the collisional de-excitation term, and to a second order on $T_e$ (Osterbrock, 1989). The most commonly used ones are [S II]$\lambda$6716/6731Å, [O II]$\lambda$3726/3729Å, and [N I]$\lambda$5200/5198Å. Since the [S II] and [N I] doublets are rapidly redshifted into the near infrared, the intense [O II] doublet is usually used for distant galaxies. For high redshift galaxies, one may alternatively use the C III]1907,1909, C III]1883,1892 and N IV1483,1487 doublets.

For redshifted galaxies, it is often difficult to detect all the lines needed to calculate the temperature, density, and abundance for each ionic stage. If not, one may for example assume that the electron temperature (or density) in a given ionization stage can be calculated from the adjacent stage using photo-ionization models (Garnett, 1992). In practice, the metallicity from electron temperature can be derived using routines such as *stsdas/nebular.zones and .abund under IRAF*, see Ex. *Metallicities from effective temperature.*

**Example 3.8: *Metallicities from effective temperature***

This example uses the observations of a low-$z$ gamma ray burst host from Hammer *et al.* (2006), obtained with VLT/FORS2. The galaxy spectrum extends the 3700Å to 7000Å, rest-frame and has a spectral resolution of $R = 1500$.

The first step is to accurately calculate the extinction from the Balmer lines, while carefully accounting for underlying absorptions (see Sec. 3.5.2). If Balmer emissions are very intense, it becomes mandatory to calculate the extinction from two Balmer ratios, and then to derive the underlying absorption assuming a constant absorption under each Balmer line (Izotov *et al.*, 1994). Subsequently, each measurable line is corrected for extinction using Eq. (3.9).

The second step consists in calculating the electron temperature and density in the different zones (neutral or single, double or multiple ionized species). In theory, the electron density can be calculated from the [O II] or [S II] doublet. However, the spectral resolution of the spectrum is not high enough to resolve the [O II] doublet. $N_e$ is thus computed from the [S II]6716/[S II]6731 ratio, equal to 1.26, using the *stsdas/nebular.temden* and assuming as a first guess $T_e = 10000$ K.

```
cl> temden dens 1.26 atom=sulfur spectrum=2
assume=10000.
```

(*Continued*)

**Example 3.8: *Continued***

```
Density ratio [\ion{S}{2}]: I(6716)/I(6731) =   1.26
Density: 164 /cm^3
```

The temperature of the double ionized species can be computed from the [O III] lines. The [O III]$\lambda$4959, 5007/$\lambda$4363Å, ratio is equal to 161.

```
cl> temden temp 161. atom=oxygen spectrum=3 assume=164
Temperature ratio  O  III: I(4959+5007)/I(4363) = 161.
Temperature = 10915.6 K
```

$T_e$([O II]) cannot be estimated directly because the [O II]$\lambda$7320+7330Å, fall outside the spectral coverage. $T_e$([N II]) could not be computed either, because the faint [N II]$\lambda$5755Å, line is not detected. However, the temperature in the zone of single ionized ions, noted $t_2$, can be estimated from $T_e$([O III]) using (Garnett, 1992):

$$t_2 = 0.7 \times t_3 + 0.3(K) \tag{3.18}$$

where $t_2 = T_e([\text{O II}])/10^4$ and $t_3 = T_e([\text{O III}])/10^4$ are respectively the temperatures of the single and double ionized zones. This gives $T_e$([O II]) = 10640 K. Finally, the abundances of each ionic species are computed from the collisional lines of the ions and Hydrogen recombination lines, using the *stsdas/nebular.abund* task. The *nebular.abund* routine requires inputting the lines fluxes in a table format.

Alternatively and for pedagogical purposes, the ionic abundances could be computed using the equations published by Izotov *et al.* (2006) (see their Sec. 3.1) for the determination of the oxygen abundances in H II regions for a five-level atom:

$$12 + \log(\text{O}^{++}/\text{H}^+) = \log \frac{I_{[\text{O III}]4959,5007}}{I_{\text{H}\beta}} + 6.2 + \frac{1.251}{t_3} - 0.55 \log(t_3) - 0.014\, t_3. \tag{3.19}$$

(*Continued*)

**Example 3.8: *Continued***

Line measurements give $I_{\mathrm{H}\beta} = 25.5$, $I_{[\mathrm{O\ III}]4959} = 49.7$ and $I_{[\mathrm{O\ III}]5007} = 201.20$. This leads to an ionic abundance equal to $12 + \log(\mathrm{O}^{++}/\mathrm{H}^{+}) = 8.30$. The $\mathrm{O}^{+}/\mathrm{H}^{+}$ ratio can be calculated using the following equation:

$$\begin{aligned} 12 + \log(\mathrm{O}^{+}/\mathrm{H}^{+}) = \log \frac{I_{[\mathrm{O\,II}]3727,3729}}{I_{\mathrm{H}\beta}} + 5.961 + \frac{1.676}{t_2} \\ -0.40 \log t_2 - 0.034 t_2 + \log(1 + 1.35 x_2), \end{aligned} \tag{3.20}$$

where $x_2 = 10^{-4} n_e t_2^{-1/2}$. The unresolved [O II] doublet has an intensity of $I_{[\mathrm{O\,II}]3727} = 44.0$. This gives a ionic abundance equal to $12 + \log(\mathrm{O}^{+}/\mathrm{H}^{+}) = 8.24$.

The total abundance of oxygen is given by Eq. (3.17). This galaxy has a nearly solar abundance with $12 + \log(\mathrm{O/H}) = 8.57$. The accuracy of the metallicity in this example is limited by the uncertainties of the photo-ionization model from which the $T_e$([O II]) was estimated and which propagate into the $(\mathrm{O}^{+}/\mathrm{H}^{+})$ abundance.

#### 3.6.4.1 *Limitations of the $T_e$ method for distant galaxies*

The direct method provides a physical determination of abundances, and the associated uncertainties are easy to estimate from the flux uncertainties. As for any (1D) spectroscopic measurement, the temperature and metallicity computed from the direct method are dominated by the brightest H II regions.

For most distant galaxy spectra, the metallicity cannot be measured from the direct method based on $T_e$ because the auroral lines, such as [O III]λ4363Å, are too faint. At high metallicities $12 + \log(\mathrm{O/H}) \sim 8.6$, the gas cools down through transitions emitting in the far infrared. This induces a decrease of the [O III]λ4363Å, line that becomes immeasurably weak. Other line ratios can be used to measure $T_e$ in metal rich galaxies such as the [O II]$\lambda$3726,3729/$\lambda$7320,7330Å, ratio, e.g., Liang *et al.* (2007). However this could be of no help for distant galaxies since these lines are again redshifted in the IR. Another alternative is to use UV diagnostics for high-z galaxies. A relation may exist between temperature and the [O III]$\lambda$1661,1666/[O III]$\lambda$5007Å, ratio (Villar-Martín *et al.*, 2004), though this ratio is extremely sensitive to dust extinction and modeling.

### 3.6.5 *Indirect metallicity estimates from strong line ratios*

For faint or high metal-rich galaxies, empirical calibrations between ratios of strong emission lines and metallicity have been established. There is a large number of strong line ratios in the literature and the selection of the "best" strong-line method is still in debate. Strong-line parameters are calibrated using an empirical method (comparison with $T_e$ measurements) or a theoretical method based on photo-ionization models (see review from Kewley and Dopita 2002).

#### 3.6.5.1 *The $R_{23}$ parameter*

$R_{23}$ is the most frequently used strong-line parameter in the local Universe and at intermediate redshifts. The ratio was defined by Pagel *et al.* (1992) as:

$$\mathrm{R}_{23} = ([\mathrm{O\,II}]\lambda 3727 + [\mathrm{O\,III}]\lambda\lambda 4959, 5007)/\mathrm{H}\beta. \qquad (3.21)$$

It however cannot be used alone because:

- For a given value of $R_{23}$ there are two values of O/H: a low and high metallicity branch (Kobulnicky and Kewley, 2004), see Fig. 3.21. An additional criterion is then required for choosing the right metallicity branch (see Ex. *Estimating metallicities using $R_{23}$*);
- The ratio depends on the ionizing potential, which requires an additional quantity sensitive to the ionization parameter, e.g., the excitation parameter $P$ defined as

$$P = [\mathrm{O\,III}]\lambda 5007/([\mathrm{O\,II}]\lambda 3727 + [\mathrm{O\,III}]\lambda 5007), \qquad (3.22)$$

  which is calibrated through theoretical photo-ionization models (Pilyugin, 2000; Pilyugin and Thuan, 2005).

**Example 3.9: *Estimating metallicities using* $\mathrm{R}_{23}$**

The $R_{23}$ ratio has been widely used in studies with redshift ranges $0 < z < 1$ and $z > 3$. Figure 3.21 evidences that it is double valued with $12 + \log(\mathrm{O/H})$: at low metallicity $R_{23}$ scales with metal abundance, but at higher metallicity ($12 + \log(\mathrm{O/H}) \sim 8.3$) gas cooling occurs through metallic lines and $R_{23}$ decreases with metallicity. The selection between these low or high metallicity branches can be done using another line ratio, such as $\mathrm{N_2}$ or $\mathrm{O_3O_2}$. The main steps to estimate metallicities from $R_{23}$ are therefore:

- Subtract the stellar continuum (Sec. 3.5.2) and measure the flux of the Balmer, [O II]$\lambda$3727Å and [O III]$\lambda\lambda$4959Å, 5007Å lines.
- Estimate the amount of extinction $A_V$ using the Balmer decrement ($\mathrm{H}\alpha/\mathrm{H}\beta$ or $\mathrm{H}\gamma/\mathrm{H}\beta$) and cross-check with the extinction estimated through the SFRs ratios.
- Correct the emission lines fluxes for extinction.
- Calculate the $R_{23}$ ratio from the extinction-corrected fluxes.
- Select the metallicity branch of the $R_{23}$-log O/H relation using another metallicity estimator, for example $\mathrm{N_2}$.

#### 3.6.5.2 *Other strong-line ratios*

Other estimates of metal abundances have been assessed through the use of the $\mathrm{O_3N_2}$ ratio, [O III]$\lambda$5007/[N II]$\lambda$6584 by Alloin *et al.* (1979), the $\mathrm{N_2}$ ratio, [N II]$\lambda$6584/H$\alpha$ by Storchi-Bergmann *et al.* (1994), and the $\mathrm{N_2O_2}$ ratio, [N II]$\lambda$6584/[O II]$\lambda$3727 by Dopita *et al.* (2000). Their validity relatively to the $R_{23}$ method have been discussed in Yin *et al.* (2007) and in Kewley and Ellison (2008). However they depend on the log(N/O) abundance ratios which renders them quite inaccurate for estimating metallicity, what's more they saturate at high metallicity (see Ex. *Limitation of the* $N_2$ *parameter*).

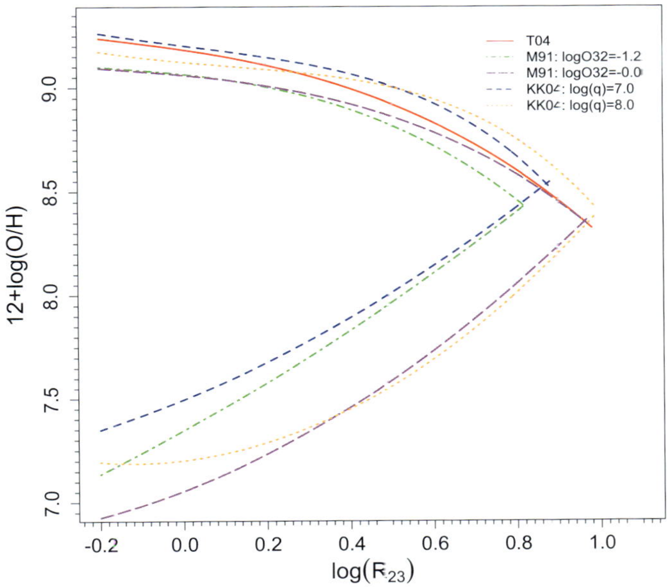


Fig. 3.21: The plot shows the degeneracy between the $R_{23}$ parameter and $12 + \log(\mathrm{O/H})$. For a given $R_{23}$ two solutions are possible: a low and a high metallicity branch. The turn-off between the two branches is around $12 + \log(\mathrm{O/H}) = 8.4$. Several calibrations of the $R_{23}$ strong line parameter are plotted: Tremonti *et al.* (2004a) (solid red curve); Kobulnicky and Kewley (2004) (short-dashed blue and dotted orange curve, corresponding to $\log(U) = -3.5$ and $-2.5$, respectively); and McGaugh (1991) (dot-dashed green and long-dashed purple curve, corresponding to $\log(\mathrm{O}_{32}) = -1.0$ and 0.0, respectively). $\mathrm{O}_{32}$ is defined as the ratio O III5007/O II and $U$ is the ionization parameter (Dopita and Evans, 1986). Adapted from Moustakas *et al.* (2011).

## Example 3.10: *Limitations of the* $\mathrm{N}_2$ *parameter*

The $N_2$ ratio has been intensively used in the literature as a metallicity estimator for high redshift studies, $1 < z < 3$. This parameter has the advantage of linking two closely separated lines in wavelength, H$\alpha$ and [N II]$\lambda$6584Å, and is hence not affected by extinction. It can be observed within small spectral windows such as the atmospheric bands in the near infrared. The main drawback of the $\mathrm{N}_2$ parameter is that it saturates at $12 + \log(\mathrm{O/H}) > 8.8$ and becomes insensitive to metal

(*Continued*)

**Example 3.10: *Continued***

abundance. Notice that $N_2$ is not a direct estimator of the oxygen abundance but of N/H. Since the N/O ratio depends on the star formation history of each galaxies, the estimation of O/H from N/H is quite uncertain (Liang *et al.*, 2006b; Stasińska *et al.*, 2008). However, it is perfectly suited to perform a first guess of the metallicity and, for example to select the branch of the $R_{23}$ diagnostic to be used.

#### 3.6.5.3 *Which strong-line ratio to use?*

The SDSS provides large samples of nearby galaxies to test the validity of the different diagnostics, as compared to a direct method. Figure 3.22 evidences the large scatter resulting from the $N_2$ and $O_3N_2$ methods when compared to the $R_{23}$ method. It is important to bear in mind that the latter method systematically offsets O/H by about +0.2 dex (Yin *et al.*, 2007) when compared to the direct method values (see Fig. 3.22, top-left panel). Most of this offset is corrected once the ionization parameter has been accounted for (see Fig. 3 of Yin *et al.*, 2007). The bottom panels of Fig. 3.22 shows the same relations for distant galaxies (or their local analogs), which appear similar than those of the local Universe. Then O/H values obtained from the $R_{23}$ method are better correlated to direct O/H estimates after accounting for the systematic offset of +0.2 dex (see dashed lines in the right panels of Fig. 3.22), while other methods provide larger scatters.

The most reliable estimate using strong line ratios is thus likely provided by the $R_{23}$ method if applied after accounting for the ionization parameter and with an additional estimate to identify the metal branch of the $R_{23}$–O/H relation, (López-Sánchez *et al.*, 2012; Shapley *et al.*, 2014). Moreover, Kewley and Ellison (2008) have shown that the relative metallicities between galaxies can be reliably estimated within ∼0.15 dex when using the same metallicity calibration. Several authors have pointed out that H II regions in $z > 1$ galaxies could have different physical conditions than their

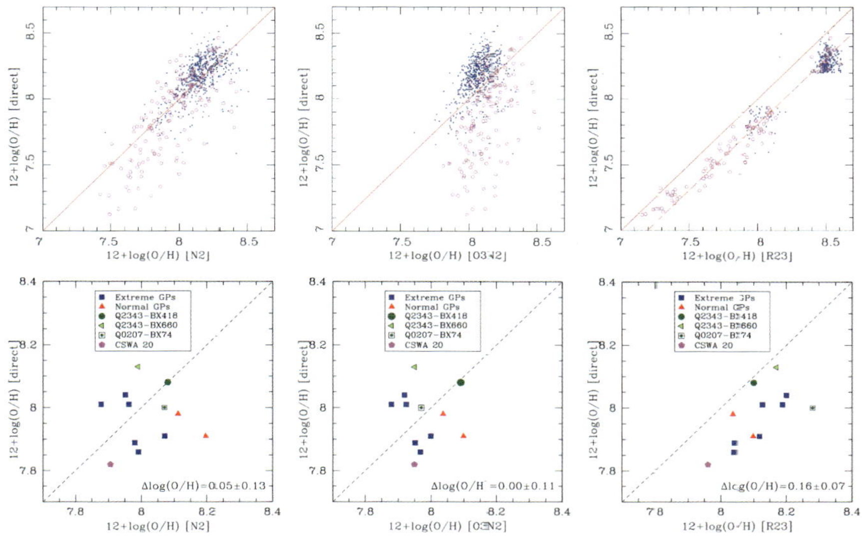


Fig. 3.22: Correlation between the O/H abundances derived from the direct $T_e$ method with those derived from the indirect $N_2$, $O_3N_2$, and $R_{23}$ methods. In the top panels are shown results from local galaxies (blue points from the SDSS, pink circles from the literature. Credit: Yin *et al.* (2007), A&A, 462, 535, reproduced with permission ©ESO. In the bottom panel are shown the same relations for a few distant galaxies and local analogues, see Steidel *et al.* (2014) for more details (©AAS, reproduced with permission).

local counterparts — harder ionizing spectra from low metallicity stars, higher ionization parameters — and are more readily affected by shocks and AGN. However there is no evidence for a change in the physics at high redshift, as it is shown by e.g., Fig. 3.22. It is hence recommended, when estimating the evolution of metal abundances, to use the same line ratios and calibrations at different redshifts to minimize systematics.[14]

## 3.7 Absorption line diagnostics of ISM & IGM

Absorption lines result from the absorption of an incident background light by the atoms of a foreground cold gas cloud.

[14]In order to compare the metallicity of galaxies measured from different calibrations, several authors have derived conversions between metallicity calibrations (e.g., Liang *et al.*, 2006b and Kewley and Ellison, 2008).

The incident photons are absorbed by bound–bound atomic transitions in the atoms/ions composing the cloud. The excited atoms decay back to their original state by re-emitting photons or cascades of photons. The light is scattered out off the observed line of sight and the resulting spectrum appears with absorption lines at the frequencies of the transitions. At high $z$, it allows to directly study the ISM of star-forming galaxies (see Fig. 3.23, top panel) as well as the line of sight ISM/CGM/IGM that is illuminated by a background source (e.g., quasar or gamma ray bursts, see Fig. 3.23, bottom panel). Absorption line studies in UV-optical are unique to probe column densities, abundances and masses of the cold, warm and warm-hot phases in the ISM, CGM, and IGM.

### 3.7.1 *Measuring absorption lines*

For a given transition, the absorption line profile follows:

$$I_\lambda = I_\lambda^{\text{cont}} e^{-\tau_\lambda} \tag{3.23}$$

in which $I_\lambda^{\text{cont}}$ is the intensity of the continuum at the central wavelength of the transition $\lambda_0$ and $\tau_\lambda$ is the optical depth as a function of wavelength. The optical depth is given by:

$$\tau_\lambda = N\, a_0\, \Phi(\lambda, \gamma, b), \tag{3.24}$$

where the number of ions/atoms per unit area is defined as the column density $N$, $a_0$ is the absorption cross-section coefficient of the given transition and $\Phi(\lambda, \gamma, b)$ is a broadening function that gives the probability of photons to be absorbed in the wavelength interval $\lambda, \lambda + \mathrm{d}\lambda$. The broadening function can be decomposed into two parts:

- A Lorentzian profile arising from the intrinsic uncertainty on the transition energy and is called *natural broadening*. The width of the natural broadening is given by $\gamma$, the natural damping constant of the transition ($\sim$10 m/s);
- A Gaussian profile of width $b$, produced by thermal and nonthermal processes in the ISM. The thermal broadening, also called *Doppler broadening* ranges from <20 km/s in cold clouds to several

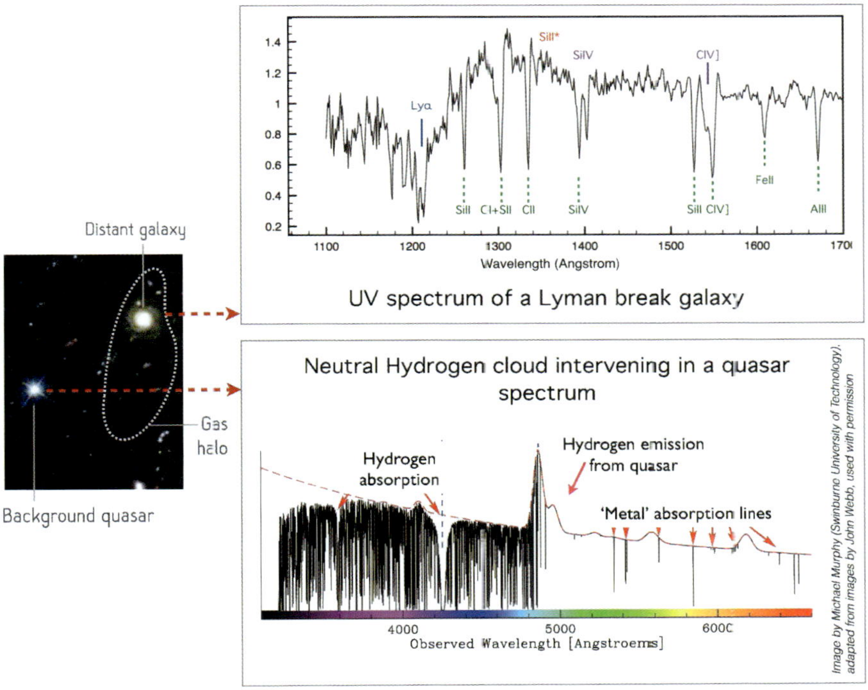


Fig. 3.23: The left panel shows a distant star-forming galaxy and its gaseous halo which is illuminated by a background quasar. The upper panel shows the UV rest-frame spectrum of a distant star-forming galaxy (adapted from Talia *et al.*, 2012). The species producing the absorption lines span a wide range of ionization states, from low-ionization ions like O I, C II, and Si II, to highly-ionized ions such as C IV. This variety of ionization states is due to the presence of multiple phases in the ISM (components at different density and temperature) in the optical disk. The bottom panel is a quasar spectrum (in red) showing strong absorption lines resulting from intervening H I gas cloud (black), image by Michael Murphy (Swinburne University of Technology) used with permission.The plethora of narrow lines at wavelengths bluer than the Ly$\alpha$ emission of the quasar is the so-called "Ly$\alpha$ forest" produced by the multitude of small clouds of H I along the line of sight. The strong absorption line, indicated with a vertical blue line, is the Ly$\alpha$ absorption of a Damped Lyman absorber, i.e., a large and dense cloud of H I gas.

100 km/s in the warm gas ejected by outflows. Nonthermal processes can be dominant contributors, such as those caused by large scale motions induced e.g., by rotation, mergers, or prominent outflows.

Absorption lines often have Voigt profiles, resulting from the convolution between a Gaussian and Lorentzian profile. At high spectral resolution, the lines are resolved and they can be fitted with a Voigt profile parametrized by $\lambda_0$, the column density $N$, and Doppler broadening $b$. Multi-component Voigt profiles can be fitted to observations using dedicated software such as VPFIT or the routines FITLYMAN in MIDAS (Fontana and Ballester, 1995).

#### 3.7.1.1 *Equivalent width*

Resolving the profile of absorption lines requires very high spectral resolution ($R > 20000$), unattainable for most distant galaxies. To overcome this limitation, the line strength is estimated by its equivalent width which measures the fraction of energy removed from the spectrum by the line (see Ex. *Measure an equivalent width*). The equivalent width is defined as:

$$W_\lambda = \int_{\text{line}} \frac{I_\lambda^{\text{cont}} - I_\lambda}{I_\lambda^{\text{cont}}} \mathrm{d}\lambda, \tag{3.25}$$

where $I_\lambda^{\text{cont}}$ is the intensity of the continuum and $I_\lambda$ is the intensity profile of the line. This quantity is invariant with the resolution of the observation. The equivalent width can be written as a function of the optical depth $\tau_\lambda$:

$$W_\lambda = \int_{-\infty}^{+\infty} (1 - e^{-\tau_\lambda}) \mathrm{d}\lambda. \tag{3.26}$$

**Example 3.11: *Measuring an equivalent width***

$W_\lambda$ can be measured using iterative routines such as the IRAF task SPLOT or automatic codes. In general, the main source of uncertainty on $W_\lambda$ measurements is the evaluation of the continuum. The continuum at both sides of an absorption line can be evaluated using a low order spline function, a polynomial fit, a broad-band smoothing filter or using wavelet decomposition (Starck *et al.*, 1997).

(*Continued*)

**Example 3.11: *Continued***

The *upper limit* on $W_\lambda$ for an undetected line above a S/N threshold can be estimated by:

$$W_{S/N} = \frac{FWHM}{S/N}, \tag{3.27}$$

where FWHM is the typical full width at half maximum of the detected absorption lines and S/N the signal-to-noise ratio of the continuum.

3.7.1.2 *COG*

Combining Eqs. (3.24) and (3.26), we can see that the equivalent width of a transition depends on the column density $N$ of the emitting atoms/ions. Figure 3.24 shows the dependence of $\log W_\lambda$ as a function of column density, dubbed COG, for the Ly$\alpha$ line. The COG is divided into three regions that correspond to different regimes of the optical depth:

- *Linear regime* — At low column densities, the medium is optically thin ($\tau_\lambda \ll 1$). The line is unsaturated (see region 1 of Fig. 3.24). For the limit $\tau_\lambda \ll 1$, the simplification of Eq. (3.26) shows that $W_\lambda$ increases linearly with $N$, independently of the broadening parameter $b$,
- *Flat regime* — At higher column densities, the medium is optically thick ($\tau_\lambda > 1$) and the line intensity reaches zero level (saturation). All photons are already absorbed and a further increase of the column density cannot increase their number. This region is known as the flat regime since $W_\lambda$ is almost insensitive to $N$ and depends primarily on the broadening parameter $b$ (see region 2).
- *Square root regime* — At very high densities ($\tau_\lambda \gg 1$), the saturated core of the lines increases further, up to a point at which it extends beyond the Doppler core creating prominent wings. The wings are always unsaturated and therefore exhibit absorptions that are proportional to column density. $W_\lambda$ is becoming insensitive to the velocity dispersion of the absorber. This

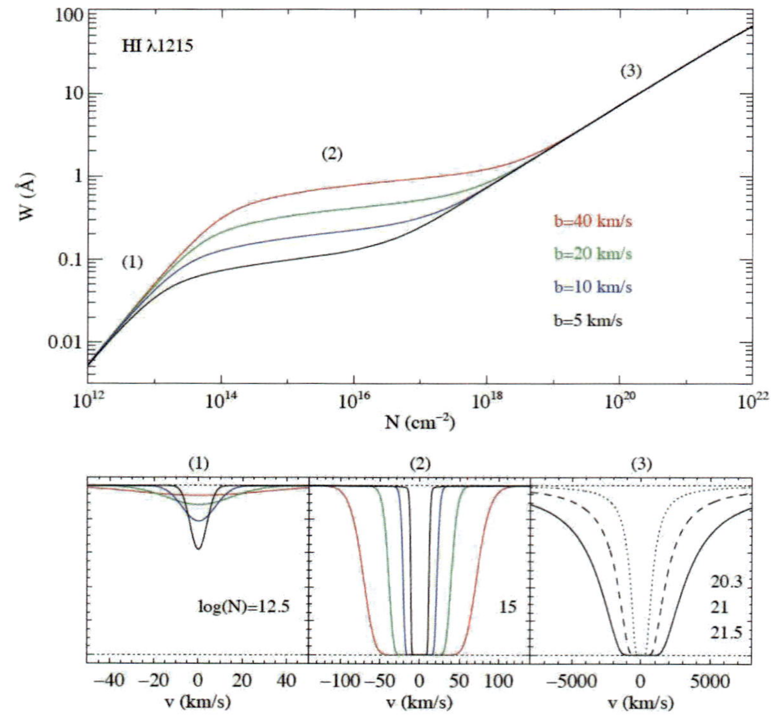


Fig. 3.24: COG of H I λ1215Å, (Lyman-$\alpha$): Equivalent width $W_\lambda$ as a function of column density $N$(H I). The different regimes of the COG are indicated: (1) Linear regime at low $N$(H I), where $W_\lambda$ does not depend on the broadening $b$; (2) Saturated regime, the $W_\lambda$ is sensitive to the broadening and is not proportional to $N$(H I); (3) Square root regime, the $N$(H I) of damped Lyman alpha systems can be measured from their equivalent width. Reproduced from Noterdaeme (2008), with kind permission.

profile is called a *damped* wing profile and is usually detected for Ly$\alpha$ and Mg II in environments with $N > 10^{19}\,\mathrm{cm}^2$. This regime is called the damping or square root regime because $W_\lambda \propto \sqrt{N}$.

### 3.7.2 *Back-illuminated neutral Hydrogen gas*

Photons with energies close to the Ly$\alpha$ line are effectively absorbed by neutral Hydrogen, even at small column densities. Ly$\alpha$ systems are clouds of neutral Hydrogen detected when they intersect the line of sight of background sources. The Hydrogen atoms from the clouds absorb the background light at wavelengths $\lambda_{\mathrm{Ly}\alpha} \times (1 + z_{\mathrm{cloud}})$, see Ex. *The Ly$\alpha$ forest and the The Gunn–Peterson trough*. Ly$\alpha$ systems are classified according to the column density of neutral Hydrogen $N(\mathrm{H\ I})$. Systems with $N(\mathrm{H\ I}) < 1.6 \times 10^{17}\mathrm{cm}^{-2}$ are called Ly$\alpha$ forest absorbers (Rauch, 1998), Lyman limit systems are absorbers with $1.6 \times 10^{17} < N(\mathrm{H\ I}) < 2 \times 10^{20}\mathrm{cm}^{-2}$ (Péroux *et al.*, 2003) and Damped Ly$\alpha$ absorbers have $N(\mathrm{H\ I}) > 2 \times 10^{20}\mathrm{cm}^{-2}$. Notice also that sub-DLAs are defined with $10^{19} < N(\mathrm{H\ I}) < 2 \times 10^{20}\mathrm{cm}^{-2}$.

**Example 3.12: *The Ly$\alpha$ forest and the Gunn–Peterson trough***

The plethora of absorption lines produced by the foreground clouds of neutral gas in the IGM in the line-of-sight of quasars is called the Ly$\alpha$ forest. The number of interposed systems visible in the spectra of quasars increases up to $z = 6$. Above this redshift the intergalactic medium has so much neutral Hydrogen that all the light from the background source at wavelengths smaller than Ly$\alpha$ is absorbed by the IGM, see Fig. 3.25. This effect is called the Gunn–Peterson trough.

#### 3.7.2.1 *Damped Lyman Absorbers*

Damped Lyman Absorbers (DLAs, see the extended review on DLA by Wolfe *et al.*, 2005) are the densest Ly$\alpha$ systems, with column densities comparable to that found in disks of local galaxies. They are particularly interesting systems to investigate the chemical evolution up to high redshift because many chemical elements (as O, Na, Mg, Si, Ar, Zn, Cr, Fe) can be observed in their UV spectrum. Their simple ionization structures allow very precise measurements of metal abundances. Notice that absorption line studies are very efficient for measuring abundances and can detect very low gas column densities. For instance, an $W_\lambda \sim 100\,\mathrm{m\AA}$, of the Zn II transition at 2026Å, corresponds to a column density of only $N \sim 6 \times 10^{12}\,\mathrm{cm}^{-2}$.

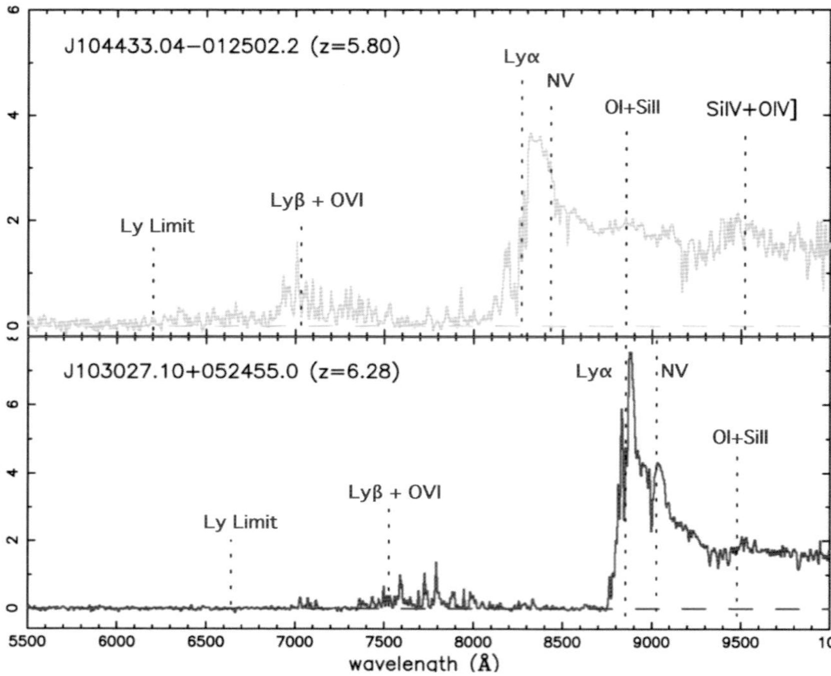


Fig. 3.25: Optical spectra of $z = 5.8$ and $z = 6.2$ quasars observed with Keck/ESI in the observed frame, adapted from Becker *et al.* (2001), ©AAS, reproduced with permission. In $z = 6.2$ quasars, the flux of the quasar drops at wavelengths smaller than Ly$\alpha$ due to the absorption by the neutral Hydrogen in the intergalactic medium.

#### 3.7.2.2 *The H* I *column density in DLAs*

DLAs are easily recognizable by the damped profile of the Ly$\alpha$ line. The H I column density of these systems can be measured accurately from the equivalent width of the Ly$\alpha$ line. This damped Ly$\alpha$ line is commonly associated to DLAs but can also be detected in Lyman Break Galaxies (e.g., Fig. 3.26).

In high resolution spectra ($R > 20000$), the Ly$\alpha$ line can be resolved (FWHM $< 25\,\mathrm{km\,s^{-1}}$) and $N$(H I) is directly measured by fitting the line with a Voigt profile (see Fig. 3.26 and the review by Wolfe *et al.*, 2005 for details on Voigt profile fitting). At lower resolution, the density can be measured from the $W_\lambda$ and the COG of the Ly$\alpha$ line. Since damped lines are produced in the square root

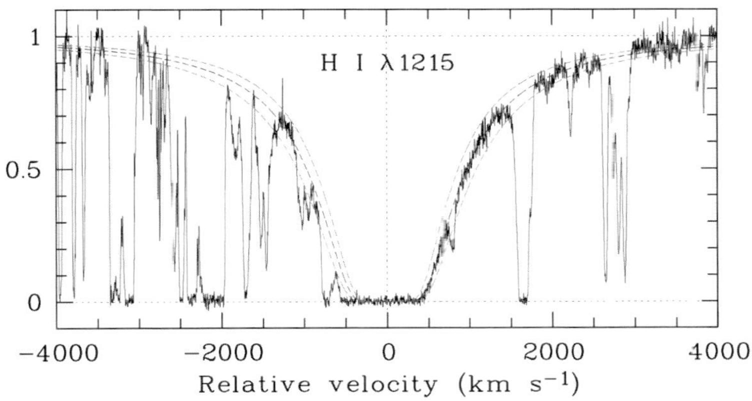


Fig. 3.26: Fit of damped Lyman-$\alpha$ line at $z_{abs} = 2.431$ towards Q 2343 + 125. The derived neutral Hydrogen column density is $\log N(\mathrm{H\,I}) = 20.40 \pm 0.07$. The dashed curves correspond to the best-fitted Voigt profile and the associated uncertainties (upper and lower curve). Credit: Noterdaeme *et al.* (2007), A&A, 474, 393, reproduced with permission ©ESO.

regime of the COG, it follows that:

$$N(\mathrm{H\ I}) = 1.88 \times 10^{18}\, W_{\AA}^{2}(\mathrm{Ly}\alpha)\ (\mathrm{cm}^{-2}). \tag{3.28}$$

For reference, a damped Ly$\alpha$ line with an equivalent width of 10Å, corresponds to a column density of $1.88 \times 10^{20}$ cm$^{-2}$.

#### 3.7.2.3 *Abundances and metallicities in* H I *clouds*

Because of their high $N(\mathrm{H\ I})$ values, the gas in DLAs is optically thick to the ionizing radiation and the self-shielding makes the gas predominantly neutral. The ionization structure of the gas is therefore relatively simple and the column densities of H I and of metals can be inferred from the strengths of the resonance lines of their dominant level of ionization. The abundance [X/H] of an element X is usually defined relatively to the abundance of Hydrogen and normalized to the solar abundances:

$$[\mathrm{X/H}]_{\mathrm{total}} = \log\left(\frac{N(\mathrm{X})}{N(\mathrm{H})}\right) - \log\left(\frac{N(\mathrm{X})}{N(\mathrm{H})}\right)_{\odot}, \tag{3.29}$$

in which $\log\left(N(\mathrm{X})/N(\mathrm{H})\right)_{\odot}$ is the reference solar abundance of X over Hydrogen (e.g., Asplund *et al.*, 2009). In DLAs, metallicities

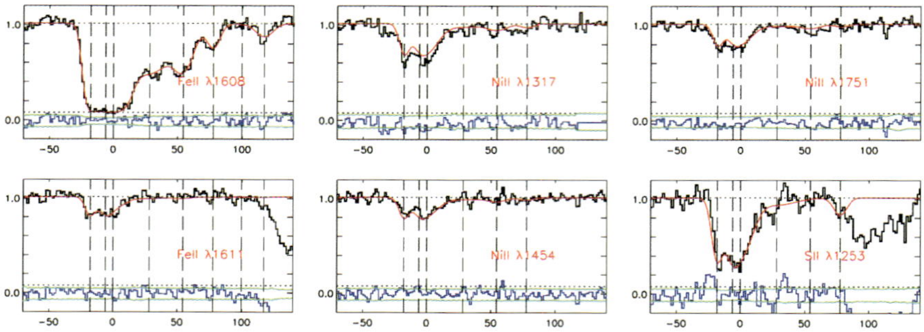


Fig. 3.27: Line fitting of several metallic UV lines in high spectral resolution observations of a DLA (black solid line). Best-fitting Voigt profiles are overplotted in red. Several velocity components have been fitted, and the position of the component are indicated by vertical dashed line. The uncertainties in flux and residuals from the fit are shown at the bottom in green and blue respectively. From Dutta *et al.* (2014), MNRAS, 440, 307, reproduced with permission of Oxford University Press on behalf of the Royal Astronomical Society.

are measured from the absorption lines of reference elements that are not or hardly depleted into dust grains such as Zn, S, or Si, conversely to Fe.

At high spectral resolution ($R = 40000$) the interstellar lines can be resolved into several velocity components, corresponding to the absorption of the background light of the different clouds forming the ISM. The abundances are measured directly by fitting theoretical Voigt profiles to the observed absorption lines via $\chi^2$ minimization. Line fitting can determine the redshifts, column densities, and broadening parameters ($b$ values) of the absorption components, see Fig. 3.27. To reproduce the observed profiles, the theoretical profiles are convolved with the instrumental LSF. The estimation of the Voigt parameters can be improved by fitting simultaneously all the transition from the same element. Simultaneous fitting of all transitions is highly recommended in the case of complex lines for which lines from different velocity clouds are blended.

At moderate resolution ($R = 20000$), the lines are unresolved and the abundance of the ionic species can be grossly estimated (uncertainties $\sim 0.4\,\mathrm{dex}$) from the COG analysis described above for H I. Equation (3.24) can be generalized for a transition $i \rightarrow j$ of an element in the quantum level $i$ as:

$$\tau_\nu = \frac{\pi e^2}{m_e c} f_{ij} N_i \Phi_\nu, \tag{3.30}$$

in which $f_{ij}$ is the oscillator strength, $N_i$ the column density of the ion, $e$ and $m_e$ are respectively the charge and mass of the electron, and $c$ is the light speed. The COG analysis of the abundances can provide accurate measurements if lines are resolved and unsaturated. The structure of the COG of other ions are similar to that of Ly$\alpha$($\lambda$1215Å): a linear region where lines are unsaturated, a flat region where the lines are saturated, and the square root region corresponding to damped wings lines. Metals (except Mg) are generally not abundant enough to produce damping wing profiles. Their absorption lines are thus unsaturated or saturated, and if the regime is known, the column density of a specie can be estimated from one of its transitions using the COG. In the unsaturated region of the COG the equivalent width is proportional to $N_i f_{ij} \lambda^2$, where $\lambda$ is the central wavelength of the transition. In practice, to avoid hidden saturation, several transitions of the same ion are fitted simultaneously in the COG, see Fig. 3.28.

Column densities can only be estimated from $W_\lambda$ if the line is unsaturated or damped. However, at low spectral resolution, the line profile is heavily smeared by the instrumental response and saturated lines can be misidentify as unsaturated lines, e.g., see Fig. 3.29. That is why high resolution should always be favored, even at a cost of S/N. To verify in which region of the COG a line falls, the best way is to compare the equivalent width of two transitions from the same ion having different oscillator strengths (follow Example *How to check the saturation of a line in low resolution spectra*). If it is not possible to check the saturation of a line, it is better to assume that the line is saturated which provides only a lower limit of column density.

Precise measurement of abundances can only be achieved with Voigt profile fitting in high spectral resolution observations. COG provide reliable estimates provided that the nonsaturation of the lines has been verified and that $N_a$ and $b$ have been constrained by several transitions. If the saturation of the lines could not be assessed

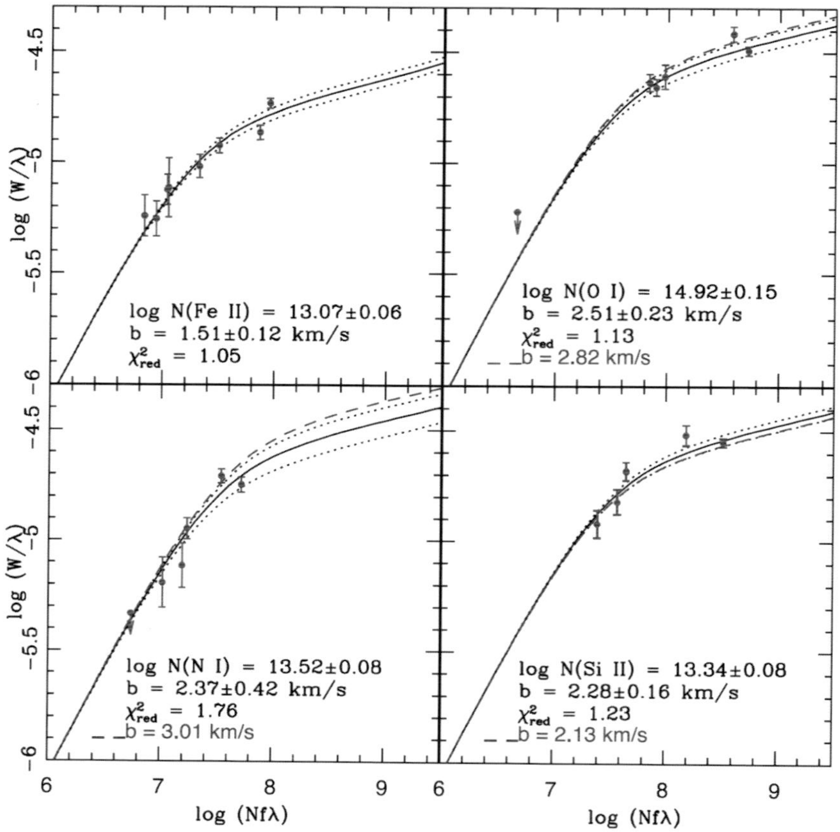


Fig. 3.28: Curves of growth for the ions detected in the DLA at $z_{abs} = 2.34006$ towards J0035-0918. The solid curves are obtained using the estimated $b$, while the dotted lines show the $1\sigma$ error range. The dashed curves are obtained from $b$ under the condition of pure thermal broadening. The points with errorbars mark the observed rest-frame equivalent widths and the column densities calculated. From Dutta *et al.* (2014), MNRAS, 440, 307, reproduced with permission of Oxford University Press on behalf of the Royal Astronomical Society.

due to the lack of lines, then COG only provide a lower limit of the abundance. Absorption lines are often composed of a blend of lines arising from atoms/ions located in clouds with different velocities, which further limits the interpretation using a single profile, especially at low spectral resolution, see Fig. 3.29.

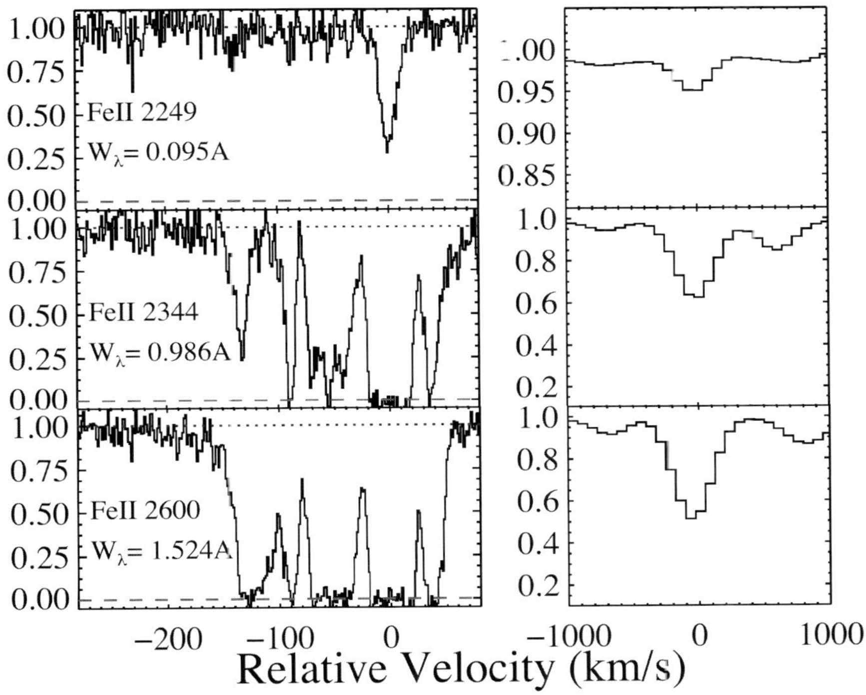


Fig. 3.29: The effect of spectral resolution on the profile of absorption lines. The left panel shows the line profile of three Fe II transitions observed in the ISM foreground of GRB 051111 from a high resolution spectrum (Keck/HIRES $R \sim 50000$). The right panel represents the same lines smoothed to lower resolution ($R \sim 1200$). The complexity of line profiles and depths of the Fe II 2344 and 2600 lines are totally blurred in the low resolution spectra (from Prochaska (2006), ©AAS, reproduced with permission).

**Example 3.13: *How to check the saturation of a line***

The two transitions of Si II $\lambda1260$, $\lambda1526$ have a theoretical $W_\lambda$ ratio equal to $W_0(\text{Si II}\lambda1260)/W_0(\lambda1526) \sim 5$ if the lines are in the linear regime of the CCG. An observed ratio with $1 \leq W_0(\text{Si II}\lambda1260)/W_0(\lambda1526) << 5$ indicates that the lines are saturated and that the column density of $\text{Si}^{++}$ cannot be estimated from these lines without knowing the $b$ parameter.

#### 3.7.2.4 *Chemical patterns*

The plethora of metallic transitions in the UV allows us to measure abundances for numerous elements. The study of abundance ratios between different elements yields constraints on the chemical evolution, dust content, and star formation history of DLAs (Prochaska *et al.*, 2003; Centurión *et al.*, 2003; Vladilo *et al.*, 2006; Dessauges-Zavadsky *et al.*, 2007).

The pattern of abundance, defined as the abundance ratio of elements over Fe, provide insight on the chemical enrichment of DLAs. In particular, the ratio between $\alpha$-elements and Iron-group elements put constraints on the typical timescale of star formation. $\alpha$-elements, such as Si and Mg, are synthesized in Type II Supernova and are released into the ISM at timescales shorter than 100 Myr. On the contrary, Fe-peak elements and nitrogen are produced in SN Ia explosions and enrich the surrounding medium with a delay of 1 Gyr after the episode of star formation. Relative abundances can thus be used as cosmic clocks if they involve two elements formed on different timescales.

However, the above abundances account only for the amount of metals in the gas phase. To determine the whole abundance of an element, the fraction of metals locked into dust grains needs to be estimated. Notice that the effect of chemical evolution and dust depletion are degenerated and are thus difficult to disentangle. The reader is referred to Vladilo *et al.* (2006) for a detailed description on the procedure for extracting the dust-phase column density from the abundance pattern, for example, the rest-frame extinction $A_V$ scales with the dust-phase column density of iron. Because Zn is mostly not depleted, one can measure its true column density from the gas phase. Adding a specific condition for the abundance of the Zn/Fe ratio (e.g., a solar value that seems realistic exhibiting not-too-under abundant DLAs) helps to evaluate the Fe abundance as well as its fraction in dust. The derived relationship between $A_V$ and dust-depleted $Fe$ in $z \sim 2$ DLAs is similar to that of the MW or of the SMC, providing further confidence to this method.

#### 3.7.2.5 *Ionization correction*

In DLAs, the bulk of the Hydrogen is supposed to be neutral while the other elements are assumed to be in majority in a given ionization state. The total abundance of an element can thus be approximated as the column density of the dominant specie. For example, elements such as silicon, iron, zinc, sulphur, nickel, titanium, and chromium, the singly ionized ions is the dominant specie. The abundance of $N$(Si) is thus approximated as being equal to $N$(Si II). This assumption is not always valid, in particular in sub-DLAs (densities under $N$(H I)$< 2 \times 10^{20}$cm$^{-2}$), in which the gas is not totally self-shielded and part of the Hydrogen is ionized. Several element ionization stages are hence present in the cloud. In this case, the abundance measured from the dominant ionized species have to be corrected for secondary stages of ionization. The ionization correction is computed using photo-ionization models (such as CLOUDY[15]) depending on the intensity and shape of the spectrum of the ionization radiation (Vladilo *et al.*, 2001; Dessauges-Zavadsky *et al.*, 2003; Milutinovic *et al.*, 2010). The level of ionization, and therefore the need for ionization corrections, can be assessed by comparing the abundance of adjacent ions of a given element, for example $N(\mathrm{Fe}^{++})/N(\mathrm{Fe}^{+})$.

### 3.7.3 *Absorption lines from the ISM of distant starbursts*

Spectra of star-forming galaxies are a composite mixture of multiple sight lines in which background sources are stars that are not uniformly shielded by the multi-phase gas clouds. The strongest absorption features visible in the UV-rest frame of distant galaxies are shown in Fig. 3.1 (see also Appendix A).

In theory, the diagnostics described above for DLAs are applicable to star-forming galaxies. In practice, the interstellar medium of star-forming galaxies is much more complex then that in DLAs

[15]See `http://www.nublado.org/`.

(Pettini, 2006). The ISM of galaxies have complex ionization structures as revealed by the detection of species with a wide range of ionization stages. Low ionization species are detected similarly as in DLAs, but so are high-ionization species, such as Si IV and C IV transitions, which are associated with the warm phase of the ISM ($T \sim 2 \times 10^5$ K). Ionization corrections are potentially large in such multi-phase medium. In addition, some lines can have P cygni profiles (e.g., C II$\lambda$1334, C IV$\lambda$1550.8Å) due to the presence of several components located in different regions of the galaxy which can be seen in emission or in absorption. All these difficulties complicate the estimation of metallicities of distant galaxies from absorption lines.

#### 3.7.3.1 *Ly$\alpha$ properties of distant starbursts*

The Ly$\alpha$ emission line is emitted by the photo-ionized gas excited by hot and massive stars,[16] and it is the strongest feature visible in distant star forming galaxies (Partridge and Peebles, 1967; Schaerer, 2003). It is widely used to detect high-$z$ galaxies (see Chapter 1), to estimate SFRs and to probe the ionization state of the intergalactic medium (IGM) at the early stage of the re-ionization epoch. Due to its resonant nature, the Ly$\alpha$ photons can scatter several times in the neutral ISM or/and IGM before escaping the medium or before being absorbed by the dust grains. This complex radiative transfer system generates a large variety of Ly$\alpha$ line profiles as observed in distant galaxies, with damped wings, asymmetric line profiles, P Cygni lines, double-peak or pure symmetric emission profiles (see Fig. 3.30). The final line profile depends on the distribution and geometry of the H I ISM (coverage of H I gas over the emitting source, clumpiness of the medium), its kinematics (presence of out-falling or in-falling shells), and the dust content and its distribution. Even for relatively simple profiles such as Ly$\alpha$ damped lines, it depends on a large number of properties of the ISM, which prevents the direct extraction of parameters such as NH I or the velocity of the gas cloud. For example, the

[16]Other sources of emission cannot be excluded in the early Universe, such as radiative shocks caused by large-scale outflows of gas and/or cooling radiation from the gas heated during a gravitational collapse of the gas.

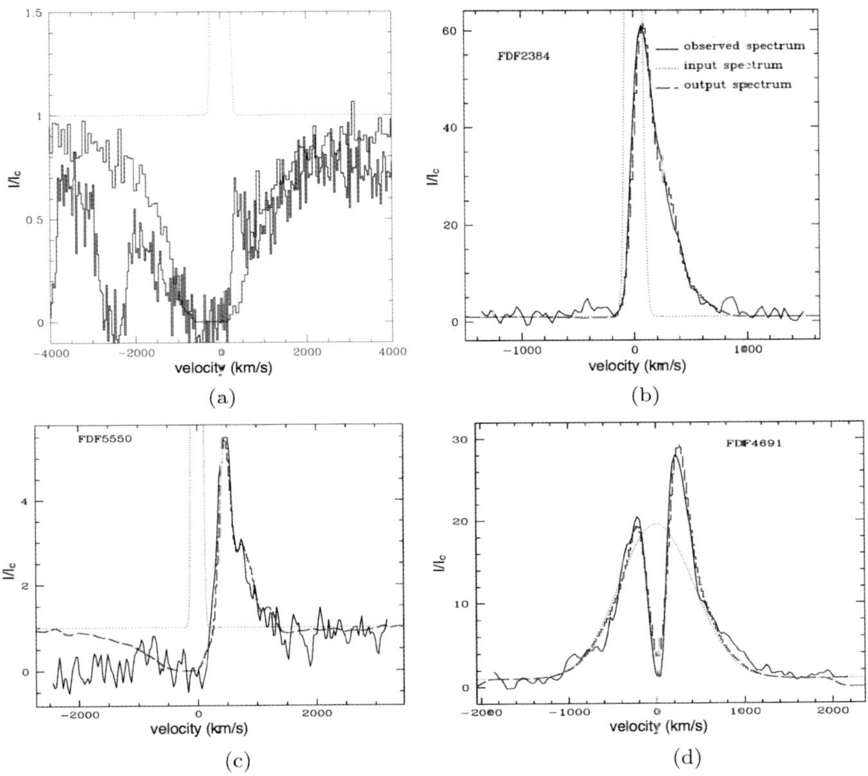


Fig. 3.30: Variety of Ly$\alpha$ lines observed in the spectra of distant galaxies. The observed spectra are shown as black solid lines, model fits as blue dashed curves, and the intrinsic (input) profile as red dotted lines. All spectra are normalized to unity in the red part of the spectrum(positive velocities). Panel (a) shows an example of Voigt profile Ly$\alpha$ from the (lensed) LBG MS 1512-cB58. The observed absorption profile is found to result naturally from the observed amount of dust and the relatively high H I column density $N$(H I). Radiation transfer effects and suppression by dust transforms strong intrinsic Ly$\alpha$ emissions with $W_\lambda$(Ly$\alpha$) $> 60$Å, into the observed faint superposed Ly$\alpha$ emission peak. Panels (b) and (c) show examples of asymmetric emission line profiles from two strong Ly$\alpha$ emitters, FDF2384 and FDF5550 respectively. The expanding shell model reproduces well the faint and broad asymmetric Ly$\alpha$ emission lines. Panel (d) Illustrates a double peaked profile, seen in the galaxy FDF4691. This profile can be modeled with a static or almost static shell. Credit: Schaerer and Verhamme (2008), A&A, 480, 369, and Verhamme *et al.* (2008), A&A, 491, 89, reproduced with permission ©ESO.

standard Ly$\alpha$ absorption line fitting, as applied to DLAs, may lead to an underestimation of the true Hydrogen column density under certain geometrical conditions (Verhamme *et al.*, 2006). Intense efforts are carried out to model the Ly$\alpha$ lines with 3D radiative transfer models, so as to allocate line profiles to particular configurations of the ISM against the emitting source (Verhamme *et al.*, 2006; Schaerer and Verhamme, 2008).

#### 3.7.3.2 *Gas motions*

The absorption lines in the UV spectrum can be used to investigate the presence of galactic-scale gas flows. In the presence of galaxy-wide outflows triggered by violent star formation activity, the lines from the ISM are systematically shifted relative to the systemic velocity (Steidel *et al.*, 1996). The systemic velocity is measured from stellar photospheric lines (Shapley *et al.*, 2003), or from emission lines (Pettini *et al.*, 2001). In Lyman break galaxies, the interstellar absorption lines and the Ly$\alpha$ emission lines are often separated by several hundred $\mathrm{km\,s^{-1}}$ from the systemic velocity, with velocity shifts spanning $-800$ km/s $< v <$ $+200$ km/s (see Fig. 3.31). This may depend on the orientation of the object. For example if it is disk-dominated, the detection of winds are favored in face-on systems (the outflow is generally perpendicular to the disk plane, therefore into our line-of-sight), while radial in-fall of gas would be easier to detect in edge-on systems.

An important concern about spectral resolution is that it should be sufficient to identify the unsaturated nature of absorption lines. It is indeed recommended to perform a kinematic analysis, such as dispersion measurements, only on lines (generally low ionization) with:

$$0.1 < \min(I_\lambda/I_\lambda^{\mathrm{cont}}) < 0.6, \tag{3.31}$$

where $I_\lambda$ is the observed intensity and $I_\lambda^{\mathrm{cont}}$ is the continuum near the transition. This criterion precludes the inclusion of weak lines and also those that are heavily saturated (i.e., $I_\lambda/I_\lambda^{\mathrm{cont}}$) and focuses the analysis on the bulk of the gas while maintaining a good S/N ratio (Prochaska and Wolfe, 2010). However, should the observer be interested in characterizing the statistical kinematic properties of a galaxy population, this criterion cannot be used to exclude galaxies

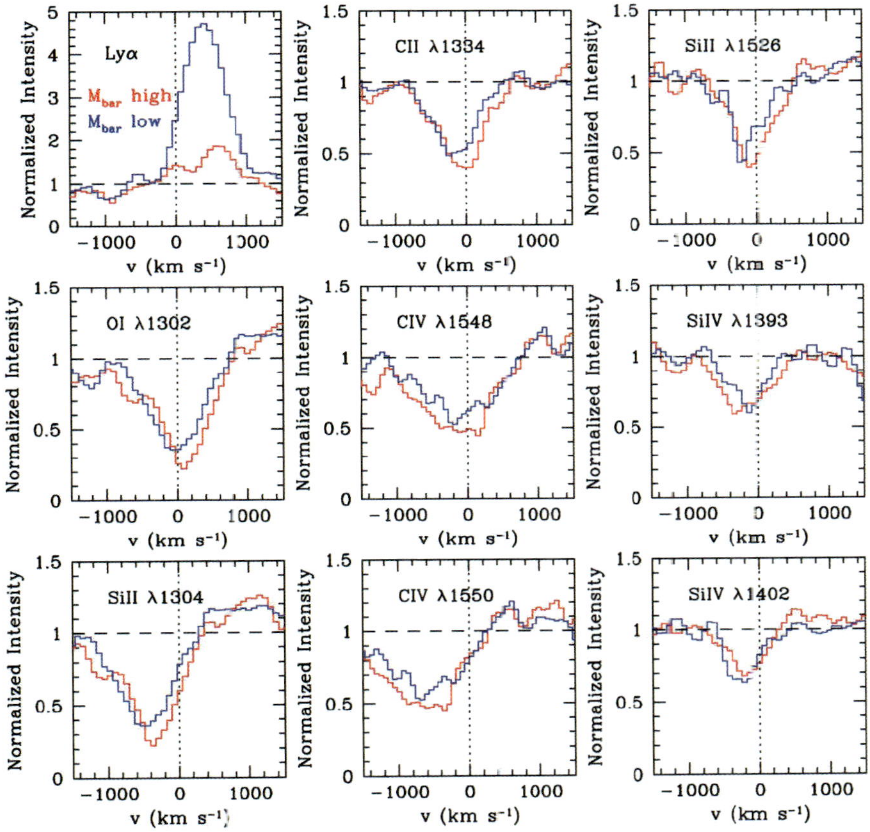


Fig. 3.31: Line profile of the Ly$\alpha$ and inter-stellar lines compared to the systemic velocity (dotted vertical lines) measured from the stellar photosphere absorption lines for two composite spectra of $z \sim 2-3$ galaxies. The composite from objects with $M_{\rm bar} > 3.7\times10^{10}$ M$_\odot$ is plotted in red, and that from objects with $M_{\rm bar} < 3.7\times10^{10}$ M$_\odot$ in blue. Note the differences in the profiles of the low ions (e.g., Si II, O I, C II) for $v \sim 0$, where the high baryonic mass objects appear to have stronger low-ion absorptions at velocities closer to the systemic velocity. These differences do not seem to be present for higher-ionization lines (e.g., see the C IV $\lambda$1550Å, panel), nor are there differences in the maximum blue-shifted velocities. From Steidel *et al.* (2010), ©AAS, reproduced with permission.

from the sample because it leads to large biases. Instead, another low-ion transition line should be investigated to allow this measurement (Prochaska and Wolfe, 2010).

Care has to be taken when comparing velocity shifts for elements having different ionization stages since they may not arise from the same location or ISM phase. For instance, the bulk of the gas flow, in a neutral phase $T < 10^4$ K, is traced by low ionization metals, such as Si I$\lambda$1260Å, O I, Si II$\lambda$1303Å, C II$\lambda$1334Å, Si II$\lambda$1526Å, Fe II$\lambda$1608Å, Si II$\lambda$1610Å, Al II$\lambda$1670Å, or Mg II$\lambda$2796, 2803Å. On the other hand the warm and hot gas, from the regions heated and shocked by the front of the flow, are probed by high ionization ions such as Si V$\lambda\lambda$1393, 1402Å, C IV$\lambda\lambda$1548, 1550Å, or N V$\lambda\lambda$1230, 1242Å (Steidel *et al.*, 2010).

**Example 3.14: *Large scale outflows in Lyman Break Galaxies***

A fraction of Lyman break galaxies have Ly$\alpha$ lines with P Cygni profiles, which result from the combination of an absorption line with a redshifted emission line, see panel (c) of Fig. 3.30. The commonly adopted model to explain the asymmetric shape of Ly$\alpha$ in distant galaxies involves a spherical symmetric outflow, see Fig. 3.32. The Ly$\alpha$ photons emitted towards the observer are absorbed by the neutral Hydrogen in the expanding shell, resulting in the observed absorption line. The red wings of the Ly$\alpha$ line are expected to be produced by backscatter photons from the receding side of the expanding shell. Part of the Ly$\alpha$ photons scattered by the outflowing material on the far side of the galaxy are re-emitted towards the observer's direction with a wavelength redshifted by several hundred km/s relative to the systematic velocity of the galaxy (see the description in Steidel *et al.* (2010)). For sufficiently large column densities ($N_H > 10^{20}$ cm$^{-2}$), the position of the main Ly$\alpha$ emission peak is generally redshifted by approximately twice the expansion velocity (Verhamme *et al.*, 2006). This model is consistent with the velocity shift of interstellar UV lines compared to the photospheric lines observed in the spectra of LBGs, as in Fig. 3.31. Both support that these galaxies host powerful outflows with velocities of hundreds of km s$^{-1}$.

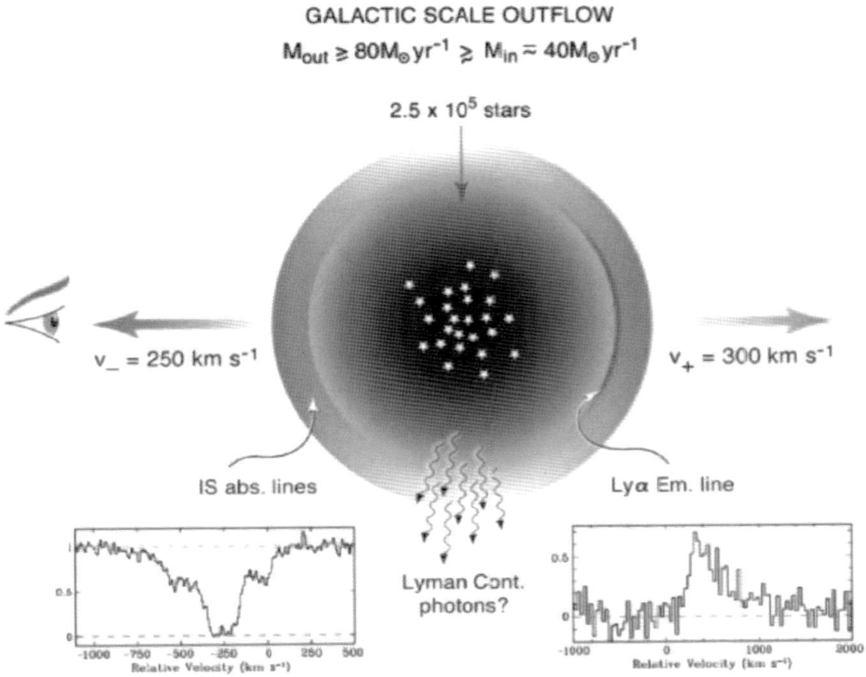


Fig. 3.32: Schematic representation of large-scale outflows in star-forming galaxies. The kinetic energy deposited by the starburst is coupled to the ambient interstellar medium and drives a galaxy-wide superwind whose existence is inferred from the blue-shift of interstellar absorption lines relative to the systemic velocity of the stars against which they are seen. The receding portion of the shell gives rise to a redshifted tail of Ly$\alpha$ emission. From Pettini *et al.* (2007) Fig. 8, reproduced with kind permission of Societa Italiana di Fisica.

## 3.8 Properties of stellar populations

The integrated spectrum of a galaxy is a fossil record of its history, because the age and metal content of young and old stars leave their imprints on the observed colors and absorption features. Scheiner (1899) compared the spectrum of Andromeda to the solar spectrum and found many similarities. He concluded that M31 was composed of solar-type stars and must therefore be a distant stellar system. Decades later, Tinsley (1968), Morgan and Osterbrock (1969), Spinrad and Taylor (1971) and many teams have subsequently tried to decompose the stellar component of galaxies into at simple set of

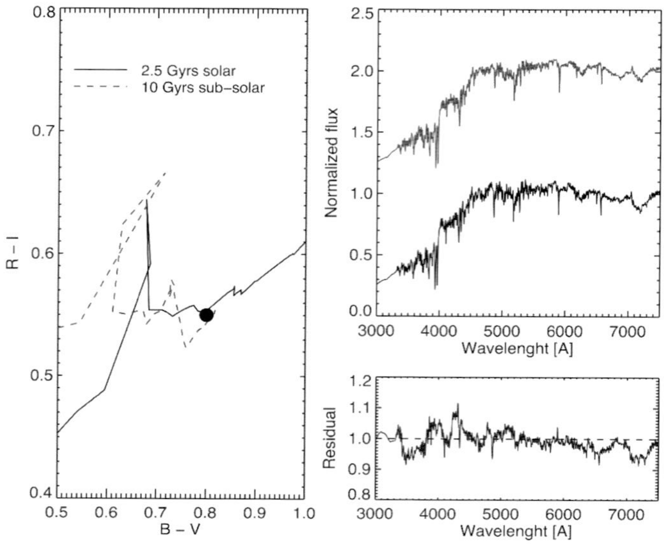


Fig. 3.33: Illustration of the age-metallicity degeneracy. Left panel: Evolution of the $R-I$ and $B-V$ colors for SSPs with solar (solid line) and sub-solar $0.4Z_{\odot}$ metallicities (dashed line), adopting a Chabrier universal IMF. The black circle corresponds to the position in the color diagram for a SSP of 10 Gyr with $Z = 0.4Z_{\odot}$ and for a 2.5 Gyrs with solar metallicity. It is not possible to disentangle between these two populations from their position in the color diagram due to the age-metallicity degeneracy. Upper right panel: visible spectrum of the previous stellar population (bottom spectrum: 2.5 Gyr old with solar metallicity; top spectrum: 10 Gyr with $Z = 0.4Z_{\odot}$). Bottom right panel: ratio between the two spectra. It shows that two populations can be differentiated using the strength of specific absorption lines, in particular those in the interval 3000–4500Å or ~7000Å.

single stellar populations (SSPs) of different ages and metallicities. The study of the stellar content of galaxies is now a commonly used tool to investigate the galaxy evolution.

However, there exists a main limitation when extracting stellar populations of distant galaxies from the analysis of their SED: the degeneracy between star ages and metallicity. Two stellar populations with the same value of age$\times Z^{3/2}$ have identical colors (Worthey, 1994 and see Fig. 3.33). The only way to overcome this difficulty is to use the information from optical absorption features arising from the stellar photospheres, which are selectively sensitive to the age or to the metallicity. Spectral analysis thus allows to better estimate the SFH and the amount of metals locked into stars, with far better accuracy than with SED fitting analysis.

The spectral analysis method to break degeneracies has its own limitations. First, it often focuses on a narrow wavelength region, in contrast with multi-band photometry, and are thus only sensitive to the population of stars dominating the studied spectral domain. For instance, UV is dominated by the continuum and photospheric lines of massive and young stars, while intermediate mass stars dominate the light budget in the NIR. A robust stellar population analysis demands multi-wavelength observations to avoid being biased towards specific populations. Secondly, SFHs derived from an integrated spectrum are also model-dependent. They are sensitive to the ingredients of stellar synthesis population models and to the way they are mixed, including to the assumed extinction model. Finally, the spectra of distant galaxies often suffer from low S/N ratios and only a qualitative estimation of the SFH (see Ex. *Estimating a coarse SFH of a distant galaxy*) can be obtained. In any case, the application of spectral analysis to distant galaxies should be restricted to spectra with S/N $> 20$ and when boundaries on metallicities and/or dust extinction are already constrained by other methods.

### 3.8.1 *Spectral Evolution of SSPs*

The integrated spectrum of distant galaxies exhibits spectral features of the type of stars that dominate the light budget. As a result, galaxies with recent star formation activity have spectra dominated by O- or B- type stars, while the spectra of quiescent galaxies have the characteristic features of late type stars. The presence of specific features can thus be directly associated to the age and metallicity of the stellar populations. Figure 3.34 shows the evolution of the spectrum of a SSP at different epochs after the burst. It assumes that all the stars were born at the same time, during an instantaneous burst, from a gas cloud having solar metallicity. We describe hereafter the main spectral features detected in the spectrum of such a SSP:

- $t < 100$ Myr. The spectrum is dominated by the light from O- and B-type stars: high fluxes at UV and blue wavelengths,

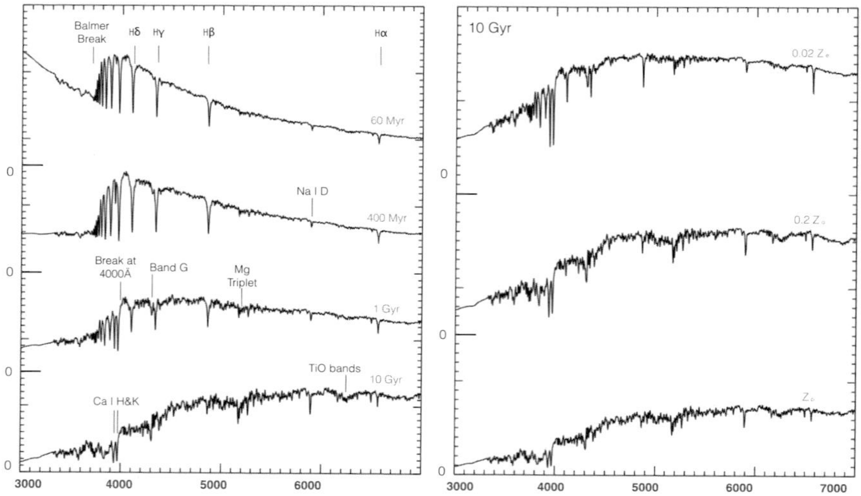


Fig. 3.34: Left-hand panel: solar metallicity SSP spectra of different ages (60 Myr, 400 Myr, 1 Gyr, and 10 Gyr), increasing from top to bottom, using the (Bruzual and Charlot, 2003) stellar library. Right-hand panel: 10 Gyr SSP spectra of different metallicity, increasing from top to bottom. For clarity, the spectra have been shifted by arbitrary amounts and smoothed to a resolution of 6Å.

strong absorption lines from helium and hydrogen at <5000Å, and absorption lines from highly ionized metals in hot stars such as Si V$\lambda$1400Å, and C IV$\lambda$1549Å. The presence of Wolf–Rayet stars (see Ex. *Identifying WR features*) and P-Cygni line profiles from stellar winds of hot O- or early B-type stars (Leitherer, 1995) are indicators of $\sim$ 10 Myr star formation bursts. At $t \sim 100$ Myr, the O- and B-type stars fade and the A-type stars are at the top of the main sequence.

- $t \sim 400$ Myr. The strengths of the Balmer absorptions are at their maximum. Metallic lines from Calcium (Ca I doublet 3969,3969Å, Ca II triplet 8498,8542,8662Å), Magnesium (Mg II 2800Å, Mg I triplet 5175Å), Sodium (Na I doublet 8183,8195Å), and Iron transitions start to appear and their strengths increases with time.
- $t \sim 1$ Gyr. The 4000Å, break, created by the accumulation of a large number of metallic lines, start to be visible and increases almost monotonically with time. It is also the case for the G

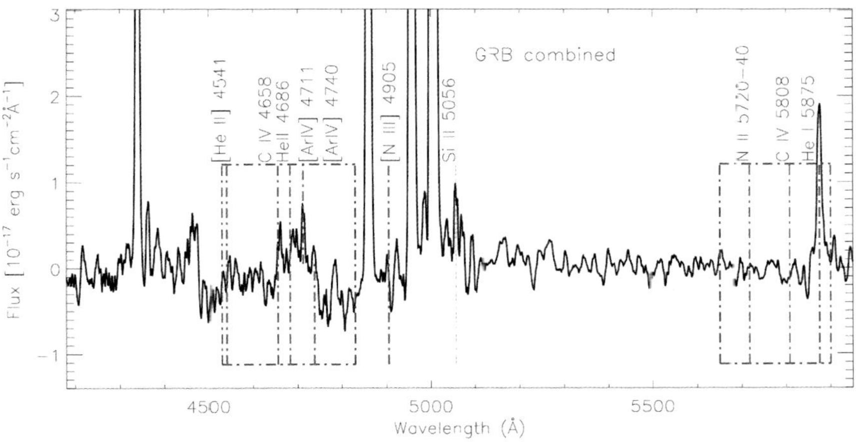


Fig. 3.35: Combined spectrum of 5 GRB host with S/N > 10 studied by Han *et al.* (2010). The WR bumps are delimited by blue dot-dashed lines. The WR lines are identified and marked with red dot-dashed lines.

band, at 4300Å, originated by the accumulation of cyanogen (CN) molecules and ionized iron transitions.

- $t > 4$ Gyr. Prominent 4000Å break and G band. The molecular bands from the titanium oxide molecules (TiO) produced in M-type stars are visible at 6180, 7150Å.

Metallicity affects the lifetime of stars, i.e., stars with higher metallicities evolve faster. In Table 3.5 we tabulated the common spectral features detected in the spectra of distant galaxies and their dependence on age and metal abundance.

**Example 3.15: *Identifying WR features***

Wolf–Rayet stars are massive ($M > 20$–$30\,M_\odot$), and are associated to recent star formation history <12 Myr. The main *Wolf–Rayet features* in galaxy optical spectra are two diagnostic cluster of blend broad emission lines, see Fig. 3.35: a *blue bump* around 4600–4680Å (Allen *et al.*, 1976; Kunth and Joubert, 1985) and a *red bump* around 5650–5800Å (Kunth and Schild, 1986; Dinerstein and Shields, 1986).

(*Continued*)

Table 3.5: Definition of Lick indices.

| Index | Blue bandpass | | Central bandpass | | Red bandpass | | Tracer | Source |
|---|---|---|---|---|---|---|---|---|
| CN1 | 4080.125 | 4117.625 | 4142.125 | 4177.125 | 4244.125 | 4284.125 | | Worthey *et al.* (1994) |
| HdA | 4041.600 | 4079.750 | 4083.500 | 4122.250 | 4128.500 | 4161.000 | Age | Worthey and Ottaviani (1997) |
| B4000 | 3750.000 | 3950.000 | — | — | 4050.000 | 4250.000 | Age and $Z$ | Gorgas *et al.* (1999) |
| H$\beta$ | 4827.875 | 4847.875 | 4847.875 | 4876.625 | 4876.625 | 4891.625 | Age | Worthey *et al.* (1994) |
| Ca4227 | 4211.000 | 4219.750 | 4222.250 | 4234.750 | 4241.000 | 4251.000 | $Z$ | Worthey *et al.* (1994) |
| G4300 | 4266.375 | 4282.625 | 4281.375 | 4316.375 | 4318.875 | 4335.125 | Age and $Z$ | Worthey *et al.* (1994) |
| Mg1 | 4895.125 | 4957.625 | 5069.125 | 5134.125 | 5301.125 | 5366.125 | $Z$ | Worthey *et al.* (1994) |
| Mg2 | 4895.125 | 4957.625 | 5154.125 | 5196.625 | 5301.125 | 5366.125 | $Z$ | Worthey *et al.* (1994) |
| Mgb | 5142.625 | 5161.375 | 5160.125 | 5192.625 | 5191.375 | 5206.375 | $Z$ | Worthey *et al.* (1994) |
| Fe5270 | 5233.150 | 5248.150 | 5245.650 | 5285.650 | 5285.650 | 5318.150 | $Z$ | Worthey *et al.* (1994) |
| Fe5335 | 5304.625 | 5315.875 | 5312.125 | 5352.125 | 5353.375 | 5363.375 | $Z$ | Worthey *et al.* (1994) |
| Fe5406 | 5376.250 | 5387.500 | 5387.500 | 5415.000 | 5415.000 | 5425.000 | $Z$ | Worthey *et al.* (1994) |
| Fe5709 | 5672.875 | 5696.625 | 5696.625 | 5720.375 | 5722.875 | 5736.625 | $Z$ | Worthey *et al.* (1994) |
| Fe5782 | 5765.375 | 5775.375 | 5776.625 | 5796.625 | 5797.875 | 5811.625 | $Z$ | Worthey *et al.* (1994) |
| NaD | 5860.625 | 5875.625 | 5876.875 | 5909.375 | 5922.125 | 5948.125 | Age and $Z$ | Worthey and Ottaviani (1997) |
| Ti$O$1 | 5816.625 | 5849.125 | 5936.625 | 5994.125 | 6038.625 | 6103.625 | Age and $Z$ | Worthey and Ottaviani (1997) |
| Ti$O$2 | 6066.625 | 6141.625 | 6189.625 | 6272.125 | 6372.625 | 6415.125 | Age and $Z$ | Worthey and Ottaviani (1997) |
| H$\delta_A$ | 4041.600 | 4079.750 | 4083.500 | 4122.250 | 4128.500 | 4161.000 | Age | Worthey and Ottaviani (1997) |
| H$\gamma_A$ | 4283.500 | 4319.750 | 4319.750 | 4363.500 | 4367.250 | 4419.750 | Age | Worthey and Ottaviani (1997) |
| H$\delta_F$ | 4057.250 | 4088.500 | 4091.000 | 4112.250 | 4114.750 | 4137.250 | Age | Worthey and Ottaviani (1997) |
| H$\gamma_F$ | 4283.500 | 4319.750 | 4331.250 | 4352.250 | 4354.750 | 4384.750 | Age | Worthey and Ottaviani (1997) |

**Example 3.15: *Continued***

Their intensity and shape depend on the emission line strength and velocity broadening of individual WR stars and thus can significantly vary from one galaxy to another.

The blue bump is the main WR feature in most star forming galaxies. It is composed of several WR lines, such as N V 4605, 4620Å,, N III 4634, 4640Å, C III/C IV4650, 4658Å, and He II 4686Å, and some nebular emission lines superposed on the bump, such as: [Fe III] 4658Å, He II 4686Å, He I + [Ar IV] 4711Å, and [Ar IV] 4740Å (Guseva *et al.*, 2000; Izotov and Thuan, 1998).

The red bump is usually much weaker. The features C IV 5808Å, and He I 5875Å, are commonly seen (Kunth and Schild, 1986; Dinerstein and Shields, 1986).

The WR bumps are generally weak compared to the strong nebular emission lines. A S/N > 10 in the continuum on both sides of the blue bump is required to detect them. The flux of the detected WR features can be used to estimate the number and type of WR stars in each galaxy. For instance, Schaerer and Vacca (1998) determined the mean absolute line luminosities of some specific WR subtypes.

In the UV, the main tracer of WR stars is the broad He I $\lambda$1640Å, emission line which traces the strong winds arising from these stars. This line has been frequently detected in the spectra of distant galaxies (Erb *et al.*, 2012; Shapley *et al.*, 2003; Dessauges-Zavadsky *et al.*, 2010). The He I line scales directly to the number, type and metallicity of the WR stars in the galaxy.

### 3.8.2 *Lick indices*

The first studies on stellar populations from visible spectra have focused on a set of specific absorption features, termed spectral indices, sensitive to stellar age, and metallicity [Fe/H] (Faber, 1973; Burstein *et al.*, 1984). The equivalent width of the observed indices can be directly compared to those measured in models in order to infer the ages and metallicities of the stellar populations. Models should have the same total broadening than the input spectrum.

The fine measurement of the equivalent width of spectral features is quite complex. Proper assessment of the continuum around an absorption feature is affected by the numerous faint and blended lines composing the spectral continuum itself. $W_\lambda$ depends on the wavelength window used to evaluate the continuum and the assumed boundaries of the absorption lines. Systems of spectral indices have been defined to enable the comparison of measurements of absorption lines. A spectral index is defined by three bands: two side bands (blue and red) for the estimation of the pseudo-continuum, and a central band containing the feature of interest (see Fig. 3.36 and example *Measure a Lick index*). The boundaries of each band is optimized in order to minimize contamination by faint absorption lines, and avoid the dependence of the indices on multiple parameters (see Table 3.5).

Standard spectral indices have been defined in the optical domain, e.g., Lick indices (see Table 3.5 and Worthey, 1994; Vazdekis *et al.*,

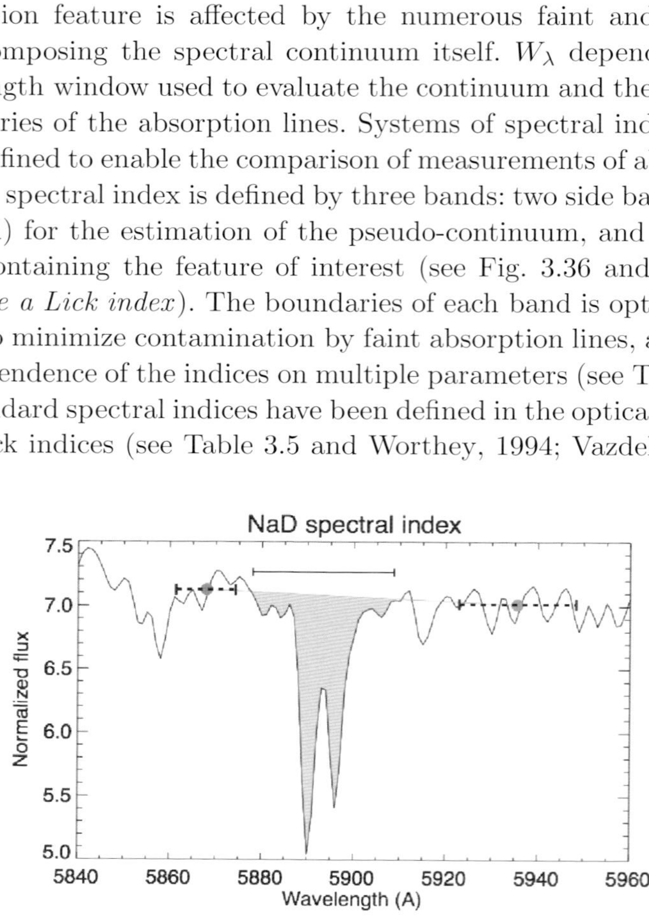


Fig. 3.36: Example of measurement of the NaD spectral index. The three bands defining the NaD index are indicated by three bars, including two with dashed lines for the continuum on the blue and red sides of the absorption feature, and a solid one for the central bandpass. The straight grey line represents the pseudo-continuum evaluated from the mean value of the two continuum windows. The index is measured by the integrated flux of the absorption line in the central bandpass under the pseudo-continuum (grey area).

2010, and in the UV and NIR range (Rix *et al.*, 2004; Keel, 2006; Maraston *et al.*, 2009; Cesetti *et al.*, 2013). Their analysis has the advantage of collapsing critical spectral information into few discrete values that can be analyzed using a parametrized SFH method as described in Chapter 2 (Kauffmann *et al.*, 2003c; Gallazzi *et al.*, 2005; Tremonti *et al.*, 2004b), or by decomposing the SFH into SSPs (Bica, 1988; Pelat, 1997; Boisson *et al.*, 2000; Trager *et al.*, 2008; Graves and Schiavon, 2008).

In many cases, observations need to be compared with studies plotting single index–index diagnostic diagrams, which implies using a standard resolution to measure line indices. Comparison of these indexes measured at different resolutions is affected by systematics, especially for indices such as Ca 4227 and Fe 4531 which are sensitive to the total broadening of the spectrum. For historical reasons, many extragalactic studies use the spectral indices from the Lick Observatory group (Worthey, 1994) defined for a resolution of 9Å, and the instrumental response of the Lick/IDS in the 1980s. The Lick system does not demand flux calibrated spectra, however it requires Lick standards for each observational setup. With the advent of new stellar libraries with better resolution and S/N, several authors have revisited the standard spectrophotometric system in which the indices can be measured. We encourage the reader to use the new line index system (LIS) from Vazdekis *et al.* (2010) that is based on the same wavelength definitions that the Lick/IDS system but can be used directly on flux calibrated spectra. Notice that the choice of the set of indices to fit is also critical. One should use a similar number of indices characterizing both the $\alpha$ and Fe elements.

**Example 3.16: *Measure a Lick index***

The definition of the Lick indexes is based on the classical expression of equivalent width, see Eq. 3.26. The equivalent width is sensitive to the assumed boundaries of the line and on the evaluation of the pseudo-continuum. Three regions are

(*Continued*)

**Example 3.16: *Continued***

defined for each index: two side bands (blue and red) for the estimation of the pseudo-continuum, and a central band containing the feature of interest.

Narrow lines, such as H$\beta$, H$\delta_a$, Mg$_b$, Fe I, G, Na, are expressed in angstroms (atomic index) and measured as follows:

$$W_\lambda = \int_{\lambda_1}^{\lambda_2} \left(1 - \frac{F_{\rm obs}(\lambda)}{F_{\rm cont}(\lambda)}\right) {\rm d}\lambda. \tag{3.32}$$

Multi-lines are expressed rather in magnitude, e.g., Mg, Mg$_2$, TiO, TiO$_2$:

$$\text{Mag} = -2.5 \log \left[ \left( \frac{1}{\lambda_2 - \lambda_1} \right) \int_{\lambda_1}^{\lambda_2} \frac{F_{\rm obs}(\lambda)}{F_{\rm cont}(\lambda)} {\rm d}\lambda \right], \tag{3.33}$$

where $F_{\rm obs}$ is the observed flux and $F_{\rm cont}$ is the pseudo-continuum interpolated at the midpoint of the central bandpass using the average flux of the continuum in the side bands $\langle F^{\rm red}_{\rm cont} \rangle$ and $\langle F^{\rm blue}_{\rm cont} \rangle$. The reader is referred to Trager *et al.* (1998) for a full description of the definition and measurement of spectral indexes. The computation of random uncertainties in the index measurements, together with an analytical estimate of uncertainties from S/N ratios, can be found in González (1993); Cardiel *et al.* (2003). Besides this, for the 4000Å, break, the $D4000$ index is defined as the ratio of the flux in the red continuum to that in the blue continuum (Bruzual A., 1983).

### 3.8.3 *Full spectra fitting*

Improvements of spectral libraries and models in terms of spectral resolution and wavelength coverage enables us not to restrict studies to portions of the spectrum, but instead make use of the full spectrum of a galaxy. *Full spectral fitting* methods use all the information contained within a spectrum: absorption features and continuum shape. The full spectral decomposition method assumes that the spectrum of a galaxy can be approximated by a linear combination of individual stellar populations. It relies on four ingredients:

- The observed spectrum with the associated uncertainties;
- A library made of individual templates $T^i_\lambda$: star clusters or synthesis model templates should spawn a grid spanning the full range of possible ages and metallicities expected in distant galaxies. Template spectra must have the same wavelength coverage and equal or higher spectral resolution compared to the observations;
- A merit function to be optimized, which includes the description of how the different components are mixed and the treatment of dust;
- An Optimization algorithm.

The spectrum of a galaxy is described as a grid of SSPs spanning a wide range of ages and metallicities, to which an attenuation function is applied and then convolved with the line-of-sight velocity dispersions of the stars. This multi-linear combination of nonlinear functions can be described by

$$F_{\text{obs}}(\lambda) = M_{\lambda 0} \sum_{j=1}^{N_\star} x_j T_j(\lambda)\, r_j(\lambda) \otimes G(v_\star, \sigma_\star), \tag{3.34}$$

where $M_{\lambda 0}$ is the scaling parameter for the normalization of the flux, $T_j$ are the SSP templates, and $x_j$ are the luminosity-weighted of the SSP. The reddening term $r_j(\lambda)$ is described as

$$r_j(\lambda) = 10^{-0.4(A^j_\lambda - A_{\lambda 0})}, \tag{3.35}$$

where $A_\lambda$ is the extinction at $\lambda$ (see Sec. 3.6.1). The line-of-sight velocity dispersion of stars $G(v_\star, \sigma_\star)$ is modeled as a Gaussian distribution centered at velocity $v_\star$ and with dispersion $\sigma_\star$. The aim of the full spectral fitting method is to extract the set of free parameters $[x_j, Av, v_\star, \sigma_\star]$ that best fit the observations, by minimizing the merit function expressed as

$$\chi^2 = \sum_\lambda \frac{O_\lambda - F(\lambda)^2}{\sigma^2_{O\lambda}}, \tag{3.36}$$

where $O_\lambda$ and $\sigma_{O\lambda}$ are respectively the flux and flux uncertainty on the observed spectrum at wavelength $\lambda$, and $F(\lambda)$ is the synthetic spectrum modeled by Eq. (3.34).

Equation (3.34) is the core of spectral synthesis analysis and it describes how the light from the different stellar populations is affected by dust and then mixed together. Given the large number of free parameters, Eq. (3.34) is often simplified by making assumptions on the dust, on the stellar velocity distribution, and/or by decreasing the number of templates $N_*$. We warn the reader against using models that are too simplistic such as those assuming that all the populations of stars are affected by a homogeneous screen of dust. This is likely not valid for star-forming galaxies, where young stars clusters are dust-enshrouded in dense environments (see Sec. 3.6.1).

There is now a large number of softwares dedicated to full spectra fitting: moped (Heavens *et al.*, 2000), ppxf (Cappellari and Emsellem, 2004), platefit (Tremonti *et al.*, 2004b; Lamareille *et al.*, 2006), STARLIGHT (Cid Fernandes *et al.*, 2005), stecmap (Ocvirk *et al.*, 2006), ULySS (Koleva *et al.*, 2009), vespa (Tojeiro *et al.*, 2007). They are based on different inversion methods and optimization algorithms, but according to the tests reported at IAU Symposium 241, they all give comparable results. Koleva *et al.* (2008) have validated the full spectrum fitting method on the integrated spectra of stellar clusters and have shown that this approach gives results consistent with Lick indices but three times more precise.

Full spectrum fitting analysis requires as input a flux calibrated spectrum and an associated mask indicating the regions to be excluded during the fitting process: emission lines from the ISM, spikes from cosmic rays and regions with low S/N ratio or affected by strong sky lines. Some full-spectra fitting softwares use the information of the slope of the stellar continuum. In this case the input is very sensitive to the accuracy of the flux-calibration of the observed spectrum (in a relative sense). A careful verification of the spectrophotometry should be done using photometric bands (see Sec. 3.4.2).

#### 3.8.3.1 *How many templates should be used?*

The elements of the library should be independent, i.e., one element should not be reproducible by the combination of other elements in the base, in order to avoid creating degeneracies. Using a diffusion map algorithm, Richards *et al.* (2009) investigated this issue and provided an optimization of the elements of the grid, age, and metallicity. Tojeiro *et al.* (2007) have shown that degeneracies can be reduced by being less demanding on the level of detail of the SFH. The number of elements in the base should be adjusted according to the quality of the observed spectrum and to the number of observable constraints (for spectral indices). Full spectrum fitting methods can recover, from spectra typical of the SDSS (S/N $\sim$ 20, optical wavelength range), between 2 and 4 stellar population ages when the metallicity is already bound by other observations. These values can be improved by a factor of 2 when going either from S/N = 20 to S/N = 50 or when enlarging the wavelength range to include not only the optical, but also the UV range. For distant galaxies, only a coarse SFH description, divided into three bins of stellar ages (young, intermediate, and old stellar populations) can be estimated with a good level of confidence (see Ex. *Estimating a coarse SFH of a distant galaxy*).

**Example 3.17: *Estimating a coarse SFH of a distant galaxy***

Figure 3.37 shows the spectrum of a compact LIRG galaxy at $z = 0.43$. The morphology and kinematics are consistent with the properties of a remnant, re-forming disk galaxy observed 0.5 Gyr after a gas-rich merger (Hammer *et al.*, 2009), implying strong star formation activity since at least $\sim$0.8 Gyr. As an independent test, the full spectral fitting technique was applied to the integrated spectrum in order to estimate a coarse SFH. It leads to a mass scaled stellar fraction of 20% with age <0.3 Gyr, 33% with age = 0.5–0.8 Gyr), the remainder being attributed to older ages (>2 Gyr). It is consistent with a recent formation of the disk in this galaxy.

#### 3.8.3.2 *Libraries of star cluster or of synthesis models*

Spectra of star clusters are usually favored over star spectral libraries (Bica and Alloin, 1986; Santos *et al.*, 1995; Piatti *et al.*, 2002;

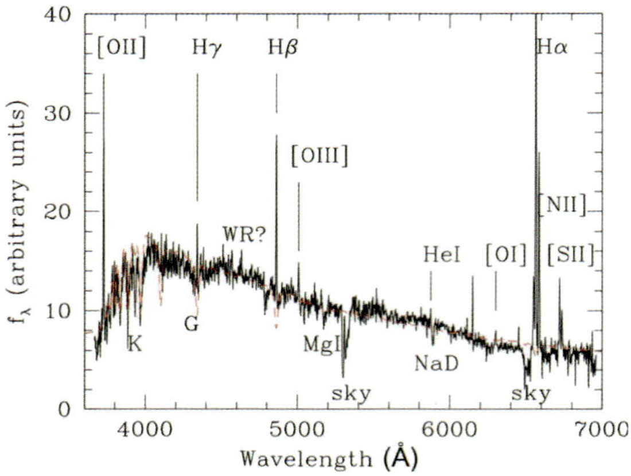


Fig. 3.37: This figure shows a full spectrum analysis of a distant galaxy. The rest-frame spectrum (in black) is fitted with a combination of single stellar population (in red), using the code STARLIGHT. Credit: Hammer *et al.* (2009), A&A, 496, 381, reproduced with permission ©ESO.

Schiavon *et al.*, 2005) because star clusters are mostly made of SSPs according to their formation scenario. Integrated spectra of star clusters are thus ideal templates for spectral decomposition because this approach is free of any assumption about the IMF and specificities in the stellar evolution.

Another possibility is to use SSP templates based on evolutionary population synthesis models (see their description in Sec. 2.6.3.1). These synthesis models allow generating homogeneous grids of SSP spanning a wide range of age and metallicity, and in which spectra are defined over a wide range of wavelengths. Since the templates are normalized to $1\,M_\odot$, estimates of stellar mass, SFR, $M/L$, and other quantities are straightforward. Bruzual A. and Charlot (1993) first introduced isochrone synthesis, followed by: Bruzual and Charlot (2003); Maraston (2005); Kotulla *et al.* (2009); Vazdekis *et al.* (2010). However, using them may induce uncertainties arising from stellar evolution models and star spectral libraries (empirical or theoretical, see Chapter 2). Quantities estimated with evolutionary population synthesis models can hence vary substantially from one model to another (Conroy *et al.*, 2009; Conroy and Gunn, 2010). To estimate the physical properties it is recommended to use several sets of evolutionary population synthesis codes in order to evaluate the dependence of the result on the models.

#### 3.8.3.3 *UV templates*

In very high-$z$ galaxies, only the rest-frame UV part of the spectrum can be observed. The rest-frame UV lines and UV libraries (e.g., Starburst99 (Leitherer *et al.*, 1999) and LavalSB (Robert *et al.*, 2003)) can be used to probe the properties of their young stellar population. The poor spectral type coverage offered by currently available empirical UV spectral libraries limits the age and metallicity diagnostic. New UV libraries however, such as the New Generation Stellar Library (Koleva and Vazdekis, 2012), may improve the evolutionary population synthesis models in this spectral domain.

### 3.8.4 *Velocity dispersion*

The absorption lines produced by the stellar atmosphere can also be used to estimate the internal motions of stars in distant galaxies. The width of the stellar features are widened by both the stellar rotation and velocity dispersion of the stars in the observed system. Velocity dispersion in the line-of-sight (LOSVD) can thus be measured from the width of stellar absorption lines, after accounting for the instrument spectral resolution (see Eq. (3.5)). In practice, the LOSVD can be estimated by fitting the continuum spectrum to the convolution of stellar population templates with a broadening function having the velocity and dispersion as free parameters. This can be done with software such as PPXF by Cappellari and Emsellem (2004).

In starburst and post-starburst galaxies one should avoid using absorbing Balmer lines because they are naturally broadened due to the fast rotation of massive intermediate-age stars.

## 3.9 Conclusion

The main difficulty in interpreting integrated spectra arises from the fact that different regions of a galaxy do not equally contribute

Table 3.6: Robust determination of physical quantities of distant galaxies.

| Quantity | 1st measurement | 2nd measurement | Final uncertainties |
|---|---|---|---|
| $A_V$ | Balmer line ratio (correct for absorption) | IR/visible ratio (aperture corrected) | <0.3 dex |
| SFR | IR | H$\alpha$ or H$\beta$ (extinction corrected) | <0.3 dex |
| Metallicities | $R_{23}$ (extinction correct) | [N II]/H$\alpha$ | ~0.1 dex |

Combination of measurements required for a robust determination of the physical quantities of distant galaxies. From Hammer (2005b).

to the integrated spectrum. The properties derived from integrated spectroscopy are not mean values representative of the entire galaxy, but are instead dominated by high surface brightness regions. For instance, abundances derived using $R_{23}$ can either sample all the ionized gas spread over the entire disk, or only one or few giant H II regions in which most of the star formation activity is concentrated.

To better control the numerous biases when deriving global properties, the physical parameters of faint and distant galaxies should therefore be assessed based on at least 2 independent measurements, as summarized in Table 3.6. Because each measurement will be affected by a different bias, such an approach will both enhance and refine the robustness of the results.

On the other hand, distant galaxies are far from homogeneous due to star formation occurring in clumps, the presence of dust, of various metallicity gradients, AGN, and overall the relatively large time scale of mixing during and after mergers. Radial gradients and spatial variations of the properties within galaxies may also create biases when comparing observations done with different apertures. The effect could be redshift dependent, i.e., it should be controlled when comparing galaxies at different epochs.

Some of these limitations can be tackled and solutions generally lead to the promising avenue that is offered by spatially-resolved spectroscopy which is described in Chapter 4.

## Bibliography


Ajiki M., Taniguchi Y., Fujita S. S. *et al.* (2003). A subaru search for Ly$\alpha$ emitters at redshift 5.7, *AJ* **126**, pp. 2091–2107.

Allen D. A., Wright A. E., Goss W. M. (1976). The dwarf emission galaxy He2-10, *MNRAS* **177**, pp. 91–97.

Alloin D., Collin-Souffrin S., Joly M. (1979). Line intensity data compilation for a sample of HII regions, *A&AS* **37**, pp. 361–366.

Asplund M., Grevesse N., Sauval A. J. *et al.* (2009). The chemical composition of the Sun, *ARA&A* **47**, pp. 481–522.

Atek H., Kunth D., Hayes M. *et al.* (2008). On the detectability of Ly$\alpha$ emission in star forming galaxies. The role of dust, *A&A* **488**, pp. 491–509.

Atek H., Kunth D., Schaerer D. *et al.* (2009). Empirical estimate of Ly$\alpha$ escape fraction in a statistical sample of Ly$\alpha$ emitters, *A&A* **506**, pp. L1–L4.

Baldwin J. A., Phillips M. M., Terlevich R. (1981). Classification parameters for the emission-line spectra of extragalactic objects, *PASP* **93**, pp. 5–19.

Becker R. H., Fan X., White R. L. *et al.* (2001). Evidence for Reionization at z˜6: Detection of a Gunn–Peterson trough in a $z = 6.28$ quasar, *AJ* **122**, pp. 2850–2857.

Benn C. R., Ellison S. L. (1998). Brightness of the night sky over La Palma, *NAR* **42**, pp. 503–507.

Bica E. (1988). Population synthesis in galactic nuclei using a library of star clusters, *A&A* **195**, pp. 76–92.

Bica E., Alloin D. (1986). A grid of star cluster properties for stellar population synthesis, *A&ASS* **66**, pp. 171–179.

Boisson C., Joly M., Moultaka J. *et al.* (2000) Stellar populations in active galactic nuclei. II. Population synthesis, *A&A* **357**, pp. 850–870.

Bruzual G., Charlot S. (2003). Stellar population synthesis at the resolution of 2003, *MNRAS* **344**, pp. 1000–1028.

Bruzual A. G. (1983). Spectral evolution of galaxies. I — Early-type systems, *ApJ* **273**, pp. 105–127.

Bruzual A. G., Charlot S. (1993). Spectral evolution of stellar populations using isochrone synthesis, *ApJ* **405**, pp. 538–553.

Buat V., Giovannoli E., Heinis S. *et al.* (2011). GOODS-Herschel: Evidence of a UV extinction bump in galaxies at z > 1, *A&A* **533**, A93.

Burstein D., Faber S. M., Gaskell C. M. *et al.* (1984). Old stellar populations. I — A spectroscopic comparison of galactic globular clusters, M31 globular clusters, and elliptical galaxies, *ApJ* **287**, pp. 586–609.

Calura F., Matteucci F. (2004). Cosmic metal production and the mean metallicity of the Universe, *MNRAS* **350**, pp. 351–364.

Calzetti D., Armus L., Bohlin R. C. *et al.* (2000). The dust content and opacity of actively star-forming galaxies, *ApJ* **533**, pp. 682–695.

Calzetti D., Kinney A. L., Storchi-Bergmann T. (1994). Dust extinction of the stellar continua in starburst galaxies: The ultraviolet and optical extinction law, *ApJ* **429**, pp. 582–601.

Cappellari M., Emsellem E. (2004). Parametric recovery of line-of-sight velocity distributions from absorption-line spectra of galaxies via penalized likelihood, *PASP* **116**, pp. 138–147.

Cardelli J. A., Clayton G. C., Mathis J. S. (1989). The relationship between infrared, optical, and ultraviolet extinction, *ApJ* **345**, pp. 245–256.

Cardiel N., Gorgas J., Sánchez-Blázquez P. *et al.* (2003). Using spectroscopic data to disentangle stellar population properties, *A&A* **409**, pp. 511–522.

Centurión M., Molaro P., Vladilo G. *et al.* (2003). Early stages of nitrogen enrichment in galaxies: Clues from measurements in damped Lyman alpha systems, *A&A* **403**, pp. 55–72.

Cesetti M., Pizzella A., Ivanov V. D. *et al.* (2013). The Infrared Telescope Facility (IRTF) spectral library: Spectral diagnostics for cool stars, *A&A* **549**, A129.

Charlot S., Fall S. M. (2000). A simple model for the absorption of starlight by dust in galaxies, *ApJ* **539**, pp. 718–731.

Cid Fernandes R., Mateus A., Sodré L. *et al.* (2005). Semi-empirical analysis of sloan digital sky survey galaxies — I. Spectral synthesis method, *MNRAS* **358**, pp. 363–378.

Conroy C., Gunn J. E. (2010). The propagation of uncertainties in stellar population synthesis modeling. III. Model calibration, comparison, and evaluation, *ApJ* **712**, pp. 833–857.

Conroy C., Gunn J. E., White M. (2009). The propagation of uncertainties in stellar population synthesis modeling. I. The relevance of uncertain aspects of stellar evolution and the initial mass function to the derived physical properties of galaxies, *ApJ* **699**, pp. 486–506.

Cuby J. G., Bottini D., Picat J. P. (1998). *Optical Astronomical Instrumentation, Society of Photo-Optical Instrumentation Engineers (SPIE) Conference Series*, Vol. 3355 (ed.), D'Odorico, S., pp. 36–47.

Cuillandre J.-C., Fort B. P., Picat J.-P. *et al.* (2003). *Society of Photo-Optical Instrumentation Engineers (SPIE) Conference Series, Society of Photo-Optical Instrumentation Engineers (SPIE) Conference Series*, Vol. 4841, (eds.) Iye, M. and Moorwood, A. F. M., pp. 1531–1538.

Dessauges-Zavadsky M., Calura F., Prochaska J. X. *et al.* (2007). A new comprehensive set of elemental abundances in DLAs. III. Star formation histories, *A&A* **470**, pp. 431–448.

Dessauges-Zavadsky M., D'Odorico S., Schaerer D. *et al.* (2010). Rest-frame ultraviolet spectrum of the gravitationally lensed galaxy "the 8 o'clock arc": Stellar and interstellar medium properties, *A&A* **510**, A26.

Dessauges-Zavadsky M., Péroux C., Kim T.-S. *et al.* (2003). A homogeneous sample of sub-damped Lyman $\alpha$ systems — I. Construction of the sample and chemical abundance measurements, *MNRAS* **345**, pp. 447–479.

Dinerstein H. L., Shields G. A. (1986). Reassessing the primordial helium abundance — New observations of NGC 4861 and CG 1116 + 51, *ApJ* **311**, pp. 45–57.

Dopita M. A., Evans I. N. (1986). Theoretical models for H II regions. II — The extragalactic H II region abundance sequence, *ApJ* **307**, pp. 431–440.

Dopita M. A., Kewley L. J., Heisler C. A. *et al.* (2000). A theoretical recalibration of the extragalactic H II region sequence, *ApJ* **542**, pp. 224–234.

Dutta R., Srianand R., Rahmani H. *et al.* (2014). A study of low-metallicity DLAs at high redshift and C II* as a probe of their physical conditions, *MNRAS* **440**, pp. 307–326.

Erb D. K., Quider A. M., Henry A. L. *et al.* (2012). Galactic outflows in absorption and emission: Near-ultraviolet spectroscopy of galaxies at $1 < z < 2$, *ApJ* **759**, 26.

Faber S. M. (1973). Variations in spectral-energy distributions and absorption-line strengths among elliptical galaxies, *ApJ* **179**, pp. 731–754.

Ferland G. J., Osterbrock D. E. (1986). IUE spectra and a resulting model of Seyfert 2 galaxies, *ApJ* **300**, pp. 658–668.

Filippenko A. V. (1982). The importance of atmospheric differential refraction in spectrophotometry, *PASP* **94**, pp. 715–721.

Fitzpatrick E. L. (1999). Correcting for the effects of interstellar extinction, *PASP* **111**, pp. 63–75.

Flores H., Hammer F., Thuan T. X. *et al.* (1999). 15 Micron infrared space observatory observations of the 1415 + 52 Canada–France redshift survey field: The cosmic star formation rate as derived from deep ultraviolet, optical, mid-infrared, and radio photometry, *ApJ* **517**, pp. 148–167.

Fontana A., Ballester P. (1995). A midas tool for the analysis of absorption spectra, *The Messenger* **80**, pp. 37–41, available at: http://cdsads.u-strasbg.fr/abs/1995, Provided by the SAO/NASA Astrophysics Data System.

Fox A. J., Savage B. D., Wakker B. P. (2005). Measurement of noisy absorption lines using the apparent optical depth technique, *AJ* **130**, pp. 2418–2423.

Gallazzi A., Charlot S., Brinchmann J. *et al.* (2005). The ages and metallicities of galaxies in the local universe, *MNRAS* **362**, pp. 41–58.

Garnett D. R. (1992). Electron temperature variations and the measurement of nebular abundances, *AJ* **103**, pp. 1330–1337.

Glazebrook K., Bland-Hawthorn J. (2001). Microslit nod-shuffle spectroscopy: A technique for achieving very high densities of spectra, *PASP* **113**, pp. 197–214.

González J. J. (1993). *Line Strength Gradients and Kinematic Profiles in Elliptical Galaxies*, Ph.D. thesis, University of California, Santa Cruz. *Source*: Dissertation Abstracts International, Vol. 54-05, Sec. B, p. 2551.

Gordon K. D., Clayton G. C., Misselt K. A. *et al.* (2003). A quantitative comparison of the small magellanic cloud, large magellanic cloud, and milky way ultraviolet to near-infrared extinction curves, *ApJ* **594**, pp. 279–293.

Graves G. J., Schiavon R. P. (2008). Measuring ages and elemental abundances from unresolved stellar populations: Fe, Mg, C, N, and Ca, *ApJS* **177**, pp. 446–464

Guseva N. G., Izotov Y. I., Thuan T. X. (2000). A spectroscopic study of a large sample of wolf-rayet galaxies, *ApJ* **531**, pp. 776–803.

Hammer F. (2005a). Growth of spirals: Secular or driven by mergers? arXiv: astroph/0509907.

Hammer F., Flores H., Elbaz D. *et al.* (2005b). Did most present-day spirals form during the last 8 Gyr?. A formation history with violent episodes revealed by panchromatic observations, *A&A* **430**, pp. 115–128.

Hammer F., Flores H., Lilly S. J. *et al.* (1997). Canada–France redshift survey. XIV. Spectral properties of field galaxies up to $z = 1$, *ApJ* **481**, pp. 49–82.

Hammer F., Flores H., Schaerer D. *et al.* (2006). Detection of Wolf-Rayet stars in host galaxies of gamma-ray bursts (GRBs): are GRBs produced by runaway massive stars ejected from high stellar density regions? *A&A* **454**, pp. 103–111.

Hammer F., Flores H., Yang Y. B. *et al.* (2009). A forming, dust-enshrouded disk at z = 0.43: the first example of a massive, late-type spiral rebuilt after a major merger? *A&A* **496**, pp. 381–387.

Han X. H., Hammer F., Liang Y. C. *et al.* (2010). The Wolf-Rayet features and mass-metallicity relation of long-duration gamma-ray burst host galaxies, *A&A* **514**, A24.

Hanuschik R. W. (2003). A flux-calibrated, high-resolution atlas of optical sky emission from uves, *A&A* **407**, pp. 1157–1164.

Hayes M., Scarlata C., Siana B. (2011). Central powering of the largest Lyman-$\alpha$ nebula is revealed by polarized radiation, *Nature* **476**, pp. 304–307.

Heavens A. F., Jimenez R., Lahav O. (2000). Massive lossless data compression and multiple parameter estimation from galaxy spectra, *MNRAS* **317**, pp. 965–972.

Izotov Y. I., Stasińska G., Meynet G. *et al.* (2006). The chemical composition of metal-poor emission-line galaxies in the data release 3 of the sloan digital sky survey, *A&A* **448**, pp. 955–970.

Izotov Y. I., Thuan T. X. (1998). Reexamining the helium Abundance of I ZW 18, *ApJ* **497**, pp. 227–237.

Izotov Y. I., Thuan T. X., Lipovetsky V. A. (1994). The primordial helium abundance from a new sample of metal-deficient blue compact galaxies, *ApJ* **435**, pp. 647–667.

Jansen R. A., Franx M., Fabricant D. (2001). [O II] As a tracer of current star formation, *ApJ* **551**, pp. 825–832.

Juneau S., Dickinson M., Alexander D. M. *et al.* (2011). A new diagnostic of active galactic nuclei: revealing highly absorbed systems at redshift >0.3, *ApJ* **736**, p. 104.

Kauffmann G., Heckman T. M., Tremonti C. *et al.* (2003). The host galaxies of active galactic nuclei, *MNRAS* **346**, pp. 1055–1077.

Kauffmann G., Heckman T. M., White S. D. M. *et al.* (2003b). Stellar masses and star formation histories for 10 5 galaxies from the sloan digital sky survey, *MNRAS* **341**, pp. 33–53.

Keel W. C. (2006). Stellar abundances and molecular hydrogen in high-redshift galaxies: The far-ultraviolet view, *AJ* **131**, pp. 2755–2762.

Kennicutt R. C., Evans N. J. (2012). Star formation in the milky way and nearby galaxies, *ARA&A* **50**, pp. 531–608.

Kennicutt, Jr. R. C. (1992). The integrated spectra of nearby galaxies — General properties and emission-line spectra, *ApJ* **388**, pp. 310–327.

Kennicutt, Jr. R. C. (1998). Star formation in galaxies along the hubble sequence, *ARA&A* **36**, pp. 189–232.

Kenworthy M. A., Hanson M. M. (2004). Minimizing strong telluric absorption in near-infrared stellar spectra, *PASP* **116**, pp. 97–102.

Kewley L. J., Dopita M. A. (2002). Using strong lines to estimate abundances in extragalactic H II regions and starburst galaxies, *ApJS* **142**, pp. 35–52.

Kewley L. J., Ellison S. L. (2008). Metallicity calibrations and the mass-metallicity relation for star-forming galaxies, *ApJ* **681**, pp. 1183–1204.

Kewley L. J., Groves B., Kauffmann G. *et al.* (2006). The host galaxies and classification of active galactic nuclei, *MNRAS* **372**, pp. 961–976.

Kewley L. J., Heisler C. A., Dopita M. A. *et al.* (2001). Optical classification of southern warm infrared galaxies, *ApJS* **132**, pp. 37–71.

Kewley L. J., Maier C., Yabe K. *et al.* (2013). The cosmic BPT diagram: Confronting theory with observations, *ApJ* **774**, L10.

Kobulnicky H. A., Kewley L. J. (2004). Metallicities of $0.3 < z < 1.0$ Galaxies in the GOODS-North field, *ApJ* **617**, pp. 240–261.

Koleva M., Prugniel P., Bouchard A. *et al.* (2009). ULySS: A full spectrum fitting package, *A&A* **501**, pp. 1269–1279.

Koleva M., Prugniel P., Ocvirk P. *et al.* (2008). Spectroscopic ages and metallicities of stellar populations: Validation of full spectrum fitting, *MNRAS* **385**, pp. 1998–2010.

Koleva M., Vazdekis A. (2012). Stellar population models in the UV. I. Characterisation of the new generation stellar library, *A&A* **538**, A143.

Kornei K. A., Shapley A. E., Erb D. K. *et al.* (2010). The relationship between stellar populations and Ly$\alpha$ emission in Lyman break galaxies, *ApJ* **711**, pp. 693–710.

Kotulla R., Fritze U., Weilbacher P. *et al.* (2009). GALEV evolutionary synthesis models — I. Code, input physics and web interface, *MNRAS* **396**, pp. 462–484.

Kroupa P., Weidner C., Pflamm-Altenburg J *et al.* (2013). *The Stellar and Sub-Stellar Initial Mass Function of Simple and Composite Populations*, Planets, Stars and Stellar Systems Vol. 5, (eds.) Oswalt, T. D. and Gilmore, G., Springer Science, Business Media Dordrecht, p. 115, ISBN 978-94-007-5611-3.

Kunth D., Joubert M. (1985). Wolf-Rayet stars in 'lazy' galaxies — A statistical approach, *A&A* **142**, pp. 411–420.

Kunth D., Mas-Hesse J. M., Terlevich E. *et al.* (1998). HST study of Lyman-alpha emission in star-forming galaxies: The effect of neutral gas flows, *A&A* **334**, pp. 11–20.

Kunth D., Schild H. (1986). Further galaxies with Wolf–Rayet emission features, *A&A* **169**, pp. 71–77.

Kwok S. (2007). *Physics and Chemistry of the Interstellar Medium*. University Science Books.

Lamareille F., Contini T., Le Borgne J. *et al.* (2006). Spectrophotometric properties of galaxies at intermediate redshifts ($z \sim 0.2$–$1.0$). I. Sample description, photometric properties and spectral measurements, *A&A* **448**, pp. 893–906.

Leitherer C., Schaerer D., Goldader J. D. *et al.* (1999). Starburst99: Synthesis models for galaxies with active star formation, *ApJS* **123**, pp. 3–40.

Liang Y. C., Hammer F., Flores H. (2006a). Significant evolution of the stellar mass–metallicity relation since $z \sim 0.65$, *A&A* **447**, pp. 113–119.

Liang Y. C., Hammer F., Flores H. *et al.* (2004). Misleading results from low-resolution spectroscopy: From galaxy interstellar medium chemistry to cosmic star formation density, *A&A* **417**, pp. 905–918.

Liang Y. C., Hammer F., Yin S. Y. *et al.* (2007). The direct oxygen abundances of metal-rich galaxies derived from electron temperature, *A&A* **473**, pp. 411–421.

Liang Y. C., Yin S. Y., Hammer F. *et al.* (2006b). The oxygen abundance calibrations and N/O abundance ratios of ~40,000 SDSS star-forming galaxies, *ApJ* **652**, pp. 257–269.

Lilly S. J., Le Fevre O., Crampton D. *et al.* (1995). The Canada–France redshift survey. I. Introduction to the survey, photometric catalogs, and surface brightness selection effects, *ApJ* **455**, p. 50.

López-Sánchez Á. R., Dopita M. A., Kewley L. J. *et al.* (2012). Eliminating error in the chemical abundance scale for extragalactic H II regions, *MNRAS* **426**, pp. 2630–2651.

Maraston C. (2005). Evolutionary population synthesis: Models, analysis of the ingredients and application to high-z galaxies, *MNRAS* **362**, pp. 799–825.

Maraston C., Nieves Colmenárez L., Bender R. *et al.* (2009). Absorption line indices in the UV. I. Empirical and theoretical stellar population models, *A&A* **493**, pp. 425–444.

Massey P., Hanson M. M. (2010). Astronomical Spectroscopy, arXiv-1010.5270.

McCall M. L., Rybski P. M., Shields G. A. (1985). The chemistry of galaxies. I — The nature of giant extragalactic H II regions, *ApJS* **57**, pp. 1–62.

McGaugh S. S. (1991). H II region abundances — Model oxygen line ratios, *ApJ* **380**, pp. 140–150.

Milutinovic N., Ellison S. L., Prochaska J. X. *et al.* (2010). Ionization corrections in a multiphase interstellar medium: lessons from a $z_{abs} \sim 2$ sub-DLA, *MNRAS* **408**, pp. 2071–2082.

Morgan W. W., Osterbrock D. E. (1969). On the classification of the forms and the stellar content of galaxies, *AJ* **74**, p. 515.

Moustakas J., Kennicutt, Jr. R. C., Tremonti C. A. (2006). Optical star formation rate indicators, *ApJ* **642**, pp. 775–796.

Moustakas J., Zaritsky D., Brown M. *et al.* (2011). Evolution of the stellar mass-metallicity relation since $z = 0.75$, arXiv-1112.3300.

Noll S., Burgarella D., Giovannoli E. *et al.* (2009). Analysis of galaxy spectral energy distributions from far-UV to far-IR with CIGALE: Studying a SINGS test sample, *A&A* **507**, pp. 1793–1813.

Noll S., Kausch W., Barden M. *et al.* (2012). An atmospheric radiation model for Cerro Paranal. I. The optical spectral range, *A&A* **543**, A92.

Noterdaeme P. (2008). High redshift Damped Lyman-alpha systems: Study of molecular hydrogen, available at: https://tel.archives-ouvertes.fr/ tel-00414784, Les chapitres 2, 4 et 5 sont publiés dans A&A et reproduits inté gralement en anglais dans la thèse., Université Pierre et Marie Curie - Paris VI, Ph.D. Theses, available at: https://tel.archives-ouvertes.fr/ tel-00414784/file/phd_noterdaeme.pdf.

Noterdaeme P., Ledoux C., Petitjean P. *et al.* (2007). Excitation mechanisms in newly discovered H_2-bearing damped Lyman-$\alpha$ clouds: Systems with low molecular fractions, *A&A* **474**, pp. 393–407.

Ocvirk P., Pichon C., Lançon A. *et al.* (2006). STECMAP: STEllar content from high-resolution galactic spectra via maximum *a posteriori*, *MNRAS* **365**, pp. 46–73.

Osterbrock D. E. (1989). *Astrophysics of Gaseous Nebulae and Active Galactic Nuclei.* Astrophysics of gaseous nebulae and active galactic nuclei, 2nd. ed. by D.E. Osterbrock and G.J. Ferland. Sausalito, CA: University Science Books.

Pagel B. E. J., Simonson E. A., Terlevich R. J. *et al.* (1992). The primordial helium abundance from observations of extragalactic H II regions, *MNRAS* **255**, pp. 325–345.

Partridge R. B., Peebles P. J. E. (1967). Are young galaxies visible? II. The integrated background, *ApJ* **148**, p. 377.

Pasquali A., Pirzkal N., Larsen S. *et al.* (2006). Slitless grism spectroscopy with the Hubble space telescope advanced camera for surveys, *PASP* **118**, pp. 270–287.

Patat F. (2008). The dancing sky: 6 years of night-sky observations at Cerro Paranal, *A&A* **481**, pp. 575–591.

Pelat D. (1997). A new method to solve stellar population synthesis problems with the use of a data base, *MNRAS* **284**, pp. 365–375.

Péroux C., Dessauges-Zavadsky M., D'Odorico S. *et al.* (2003). A homogeneous sample of sub-damped Lyman $\alpha$ systems — II. Statistical, kinematic and chemical properties, *MNRAS* **345**, pp 480–496.

Pettini M. (2006). *The Fabulous Destiny of Galaxies: Bridging Past and Present*, (eds.) Le Brun, V. and Mazure, A. and Arnouts, S. and Burgarella, D., p. 319.

Pettini M., Rix S. A., Steidel C. C. *et al.* (2002). New observations of the interstellar medium in the Lyman break galaxy MS 1512-cB58, *ApJ* **569**, pp. 742–757.

Pettini M., Shapley A. E., Steidel C. C. *et al.* (2001). The rest-frame optical spectra of Lyman break galaxies: Star formation, extinction, abundances, and kinematics, *ApJ* **554**, pp. 981–1000.

Pettini M., Steidel C. C., Adelberger K. L. *et al.* (2007). Lyman break galaxies: A ten-year perspective, *NCimB* **122**, pp. 1043–1055.

Piatti A. E., Bica E., Clariá J. J. *et al.* (2002). Integrated spectral evolution of Galactic open clusters, *MNRAS* **335**, pp. 233–240.

Pilyugin L. S. (2000). On the oxygen abundance determination in H bt II regions. The problem of the line intensities — oxygen abundance calibration, *A&A* **362**, pp. 325–332.

Pilyugin L. S., Thuan T. X. (2005). Oxygen abundance determination in H II regions: The strong line intensities-abundance calibration revisited, *ApJ* **631**, pp. 231–243.

Prochaska J. X. (2006). On the perils of curve-of-growth analysis: Systematic abundance underestimates for the gas in gamma-ray burst host galaxies, *ApJ* **650**, pp. 272–280.

Prochaska J. X., Howk J. C., Wolfe A. M. (2003). The elemental abundance pattern in a galaxy at z = 2.626, *Nature* **423**, pp. 57–59.

Prochaska J. X., Wolfe A. M. (2010). On the Analysis of DLA Kinematics, *ArXiv e-prints*.

Puech M., Flores H., Yang Y. B. *et al.* (2012). *Society of Photo-Optical Instrumentation Engineers (SPIE) Conference Series*, *Society of Photo-Optical Instrumentation Engineers (SPIE) Conference Series*, Vol. 8446, 84467L.

Rauch M. (1998). The Lyman Alpha Forest in the Spectra of QSOs, *ARA&A* **36**, pp. 267–316.

Richards J. W., Freeman P. E., Lee A. B. *et al.* (2009). Accurate parameter estimation for star formation history in galaxies using SDSS spectra, *MNRAS* **399**, pp. 1044–1057.

Rix S. A., Pettini M., Leitherer C. *et al.* (2004). Spectral modeling of star-forming regions in the ultraviolet: Stellar metallicity diagnostics for high-redshift galaxies, *ApJ* **615**, pp. 98–117.

Robert C., Pellerin A., Aloisi A. *et al.* (2003). Synthetic high-resolution line spectra of star-forming galaxies below 1200 Å, *ApJS* **144**, pp. 21–45.

Rodrigues M., Cirasuolo M., Hammer F. *et al.* (2012a). *Society of Photo-Optical Instrumentation Engineers (SPIE) Conference Series*, *Society of Photo-Optical Instrumentation Engineers (SPIE) Conference Series*, Vol. 8450, 84503H.

Rodrigues M., Flores H., Puech M. *et al.* (2010). *Society of Photo-Optical Instrumentation Engineers (SPIE) Conference Series*, *Society of Photo-Optical Instrumentation Engineers (SPIE) Conference Series*, Vol. 7735, 77356L.

Rodrigues M., Foster C., Taylor E. *et al.* (2016). Galaxy and Mass Assembly (GAMA): Improved emission lines measurements in four representative samples at 0.07< z <0.3, *A&A*, **590**, A18, p. 16.

Rodrigues M., Hammer F., Flores H. *et al.* (2008). IMAGES IV: Strong evolution of the oxygen abundance in gaseous phases of intermediate mass galaxies from $z \sim 0.8$, *A&A* **492**, pp. 371–388.

Rodrigues M., Puech M., Hammer F. *et al.* (2012b). A decrease of the gas exchanges between galaxies and the intergalactic medium, from 12 to 6 billion years ago, *MNRAS* **421**, pp. 2888–2903.

Rola C., Pelat D. (1994). On the estimation of intensity for low S/N ratio narrow emission lines, *A&A* **287**, pp. 676–684.

Rousselot P., Lidman C., Cuby J.-G. *et al.* (2000). Night-sky spectral atlas of OH emission lines in the near-infrared, *A&A* **354**, pp. 1134–1150.

Salpeter E. E. (1955). The luminosity function and stellar evolution. *ApJ* **121**, p. 161.

Santos, Jr. J. F. C., Bica E., Dottori H. *et al.* (1995). Synthesis of the integrated spectrum of metal-rich globular clusters using the HR diagram and a stellar library. *A&A* **303**, p. 753.

Schaerer D. (2003). The transition from population III to normal galaxies: Lyalpha and He II emission and the ionising properties of high redshift starburst galaxies, *A&A* **397**, pp. 527–538.

Schaerer D., Vacca W. D. (1998). New models for Wolf–Rayet and O star populations in young starbursts, *ApJ* **497**, pp. 618–644.

Schaerer D., Verhamme A. (2008). 3D Ly$\alpha$ radiation transfer. II. Fitting the Lyman break galaxy MS 1512-cB58 and implications for Ly$\alpha$ emission in high-z starbursts, *A&A* **480**, pp. 369–377.

Scheiner J. (1899). On the spectrum of the great nebula in Andromeda. *ApJ* **9**, pp. 149–150.

Schiavon R. P., Rose J. A., Courteau S. *et al.* (2005). A Library of integrated spectra of galactic globular clusters, *ApJS* **160**, pp. 163–175.

Shapley A. E., Reddy N. A., Kriek M. *et al.* (2014). The MOSDEF survey: Excitation properties of $z \sim 2.3$ Star-forming Galaxies, now published in 2015 *ApJ*, **801**, p. 88.

Shapley A. E., Steidel C. C., Pettini M. *et al.* (2003). Rest-frame ultraviolet spectra of $z \sim 3$ Lyman Break Galaxies, *ApJ* **588**, pp. 65–89.

Smette A., Sana H., Noll S. *et al.* (2015). Molecfit: A general tool for telluric absorption correction. I. Method and application to ESO instruments, *A&A* **576**, A77.

Spinrad H., Taylor B. J. (1971). The stellar content of the nuclei of nearby galaxies. I. M31, M32, and M81, *ApJS* **22**, p. 445.

Starck J.-L., Siebenmorgen R., Gredel R. (1997). Spectral analysis using the wavelet transform, *ApJ* **482**, p. 1011.

Stasińska G. (2007). What can emission lines tell us? arXiv:0704.0348.

Stasińska, G., Prantzos, N., Meynet, G., Simón-Díaz, S., Chiappini, C., Dessauges-Zavadsky, M., Charbonnel, C., Ludwig, H.-G., Mendoza, C., Grevesse, N., Arnould, M., Barbuy, B., Lebreton, Y., Decourchelle, A., Hill, V., Ferrando, P., Hébrard, G., Durret, F., Katsuma, M. and Zeippen, C. J. (eds.) (2012). *Oxygen in the Universe*, *EAS Publications Series*, Vol. 54.

Stasińska G., Vale Asari N., Cid Fernandes R. *et al.* (2008). Can retired galaxies mimic active galaxies? Clues from the sloan digital sky survey, *MNRAS* **391**, pp. L29–L33.

Steidel C. C., Erb D. K., Shapley A. E. *et al.* (2010). The structure and kinematics of the circumgalactic medium from far-ultraviolet spectra of $z \sim= 2-3$ Galaxies, *ApJ* **717**, pp. 289–322.

Steidel C. C., Giavalisco M., Dickinson M. *et al.* (1996). Spectroscopy of Lyman break galaxies in the Hubble deep field, *AJ* **112**, p. 352.

Steidel C. C., Hunt M. P., Shapley A. E. *et al.* (2002). The population of faint optically selected active galactic nuclei at $z \sim 3$, *ApJ* **576**, pp. 653–659.

Steidel C. C., Rudie G. C., Strom A. L. *et al.* (2014). Strong nebular line ratios in the spectra of $z \sim 2-3$ Star-forming Galaxies: First results from KBSS-MOSFIRE, now published in 2014, *ApJ*, **795**, p. 165.

Storchi-Bergmann T., Calzetti D., Kinney A. L. (1994). Ultraviolet to near-infrared spectral distributions of star-forming galaxies: Metallicity and age effects, *ApJ* **429**, pp. 572–581.

Talia M., Mignoli M., Cimatti A. *et al.* (2012). GMASS ultradeep spectroscopy of galaxies at $z \sim 2$. VI. Star formation, extinction, and gas outflows from UV spectra, *A&A* **539**, A61.

Tinsley B. M. (1968). Evolution of the stars and gas in galaxies, *ApJ* **151**, p. 547.

Tojeiro R., Heavens A. F., Jimenez R. *et al.* (2007). Recovering galaxy star formation and metallicity histories from spectra using VESPA, *MNRAS* **381**, pp. 1252–1266.

Trager S. C., Faber S. M., Dressler A. (2008). The stellar population histories of early-type galaxies — III. The Coma cluster, *MNRAS* **386**, pp. 715–747.

Trager S. C., Worthey G., Faber S. M. *et al.* (1998). Old stellar populations. VI. Absorption-line spectra of galaxy nuclei and globular clusters, *ApJS* **116**, pp. 1–28.

Tremonti C. A., Heckman T. M., Kauffmann G. *et al.* (2004). The origin of the mass-metallicity relation: Insights from 53,000 star-forming galaxies in the sloan digital sky survey, *ApJ* **613**, pp. 898–913.

Tresse L., Rola C., Hammer F. *et al.* (1996). The Canada–France redshift survey — XII. Nature of emission-line field galaxy population up to $z = 0.3$, *MNRAS* **281**, pp. 847–870.

Vacca W. D., Cushing M. C., Rayner J. T. (2003). A method of correcting near-infrared spectra for telluric absorption, *PASP* **115**, pp. 389–409.

Vazdekis A., Sánchez-Blázquez P., Falcón-Barroso J. *et al.* (2010). Evolutionary stellar population synthesis with MILES — I. The base models and a new line index system, *MNRAS* **404**, pp. 1639–1671.

Verhamme A., Schaerer D., Atek H. *et al.* (2008). 3D Ly$\alpha$ radiation transfer. III. Constraints on gas and stellar properties of $z \sim 3$ Lyman break galaxies (LBG) and implications for high-z LBGs and Ly$\alpha$ emitters, *A&A* **491**, pp. 89–111.

Verhamme A., Schaerer D., Maselli A. (2006). 3D Ly$\alpha$ radiation transfer. I. Understanding Ly$\alpha$ line profile morphologies, *A&A* **460**, pp. 397–413.

Villar-Martín M., Cerviño M., González Delgado R. M. (2004). Nebular and stellar properties of a metal-poor HII galaxy at $z = 3.36$, *MNRAS* **355**, pp. 1132–1142.

Vladilo G., Centurión M., Bonifacio P. *et al.* (2001). Ionization properties and elemental abundances in damped Ly$\alpha$ systems, *ApJ* **557**, pp. 1007–1020.

Vladilo G., Centurión M., Levshakov S. A. *et al.* (2006). Extinction and metal column density of HI regions up to redshift z = 2, *A&A* **454**, pp. 151–164.

Wild V., Charlot S., Brinchmann J. *et al.* (2011). Empirical determination of the shape of dust attenuation curves in star-forming galaxies, *MNRAS* **417**, pp. 1760–1786.

Wolfe A. M., Gawiser E., Prochaska J. X. (2005). Damped Ly$\alpha$ Systems, *ARA&A* **43**, pp. 861–918.

Worthey G. (1994). Comprehensive stellar population models and the disentanglement of age and metallicity effects, *ApJs* **95**, pp. 107–149.

Yan R., Ho L. C., Newman J. A. *et al.* (2011). AEGIS: Demographics of x-ray and optically selected active galactic nuclei, *ApJ* **728**, 38.

Yang Y., Flores H., Hammer F. *et al.* (2008). IMAGES. I. Strong evolution of galaxy kinematics since $z = 1$, *A&A* **477**, pp. 789–805.

Yang Y., Rodrigues M., Puech M. *et al.* (2013). Accurate sky continuum subtraction with fibre-fed spectrographs, *Messenger* **151**, pp. 10–13.

Yin S. Y., Liang Y. C., Hammer F. *et al.* (2007). Empirical strong-line oxygen abundance calibrations from galaxies with electron-temperature measurements, *A&A* **462**, pp. 535–546.

Zafar T., Watson D., Elíasdóttir Á. *et al.* (2012). The properties of the 2175 Å extinction feature discovered in GRB afterglows, *ApJ* **753**, p. 82.

Zheng X. Z., Hammer F., Flores H. *et al.* (2004). HST/WFPC2 morphologies and color maps of distant luminous infrared galaxies, *A&A* **421**, pp. 847–862.

## Chapter 4

# Integral Field Spectroscopy

## 4.1 Introduction

The morphology of distant galaxies has largely been investigated using the (Hubble Space Telescope) (HST), revealing the strong evolution of disk fraction as well as peculiar galaxies (see, e.g., Guzmán *et al.*, 1997; Brinchmann *et al.*, 1998; Zheng *et al.*, 2004, and Sec. 2.4). However, star formation may affect the appearance of distant galaxies and it is essential to know whether or not their internal motions are also perturbed. This is achieved exclusively by integral field spectroscopy, which disperses the light from each pixel sampling a galaxy (see Fig. 4.1). If this can be done for all pixels, the result is a three-dimensional (3D) data cube rather than a simple two-dimensional (2D) image.

3D spectroscopy circumvents the major weaknesses of 1D spectroscopy (see Sec. 3.9) including those related to aperture correction and sky subtraction accuracy, because the distant galaxies are generally completely sampled by the (Integral Field Unit[1]) (IFU), as well as their surrounding sky background. However, 3D spectroscopy of distant galaxies is obviously limited by surface-brightness, i.e., the already faint emissions from these extended objects is further divided into several pixels. For distant galaxies, only bright emission lines can be sampled with sufficient S/N. Fortunately, the fraction of emission line galaxies increases by a factor 3 between z = 0 and

[1]IFU instruments are a subset of 3D-imaging techniques, also known as 3D spectroscopy; other techniques rely on generating a path difference between interfering beams using electromechanical scanning techniques, e.g., Fabry–Perot interferometers.

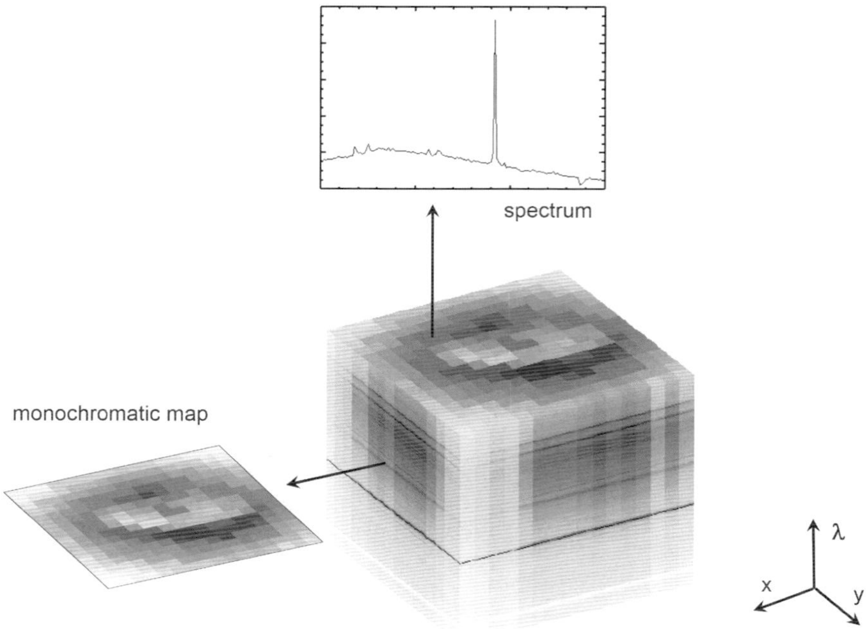


Fig. 4.1: 3D dataset (data cube) obtained as the result of 3D spectroscopy: it includes one spectral ($\lambda$) and two spatial (x,y) coordinates. The data cube can be visualized as a stack of quasi-monochromatic images, or, alternatively, as an assembly of n×m spectra (Credit: M.M. Roth).

z = 0.6, and beyond that, the fraction reaches 80% of L* galaxies (Hammer *et al.*, 1997). Additionally, numerous distant galaxies are starbursts and can be observed with 3D spectroscopy, which is naturally related to the increasing gas abundance at higher redshifts (Rodrigues *et al.*, 2012).

In these instruments, the spatial sampling of the targets in the focal plane is provided by an IFU, which decomposes the light using either optical fibers, lens-let arrays, or slicing mirrors (see Fig. 4.2). The first application in astronomy was demonstrated by Vanderriest (1980), who built a 27× 43 $arcsecond^2$ IFU consisting in 169 hexagonal fibers and 36 additional fibers for sky subtraction. Fibers collect the light from specified locations within the focal plane and reformat it into a single entrance spectrograph slit (see Fig. 4.2).

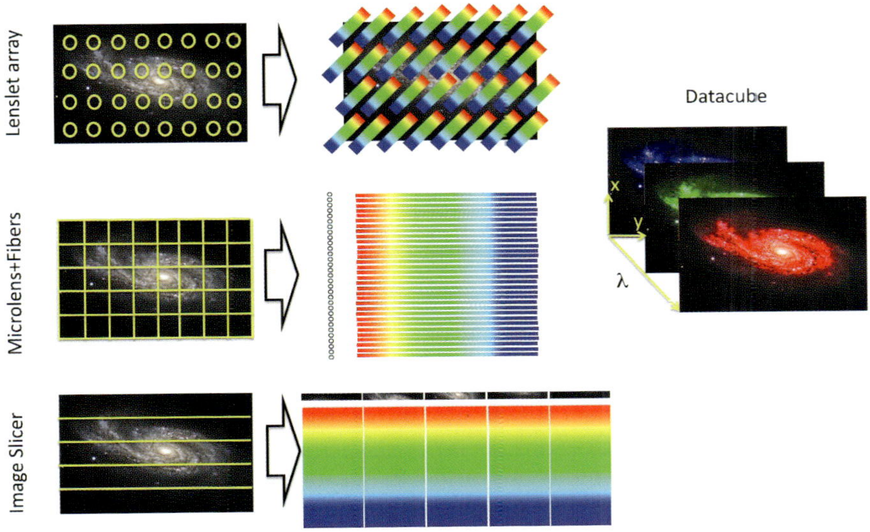


Fig. 4.2: The three main IFU techniques used to divide the light from the focal plane onto 3D spectrographs (inspired from Allington-Smith, 2006).

Another concept proposed by Courtes (1982) uses a lens-let array, which samples the light using an array of microlenses and sends the resulting beams directly into the spectrograph. This was first used on-sky with the TIGER instrument (Bacon *et al.*, 1995). Although conceptually simpler, it results in overlapping spectra on the detector. Both techniques were first combined by Afanasiev (1990) to increase the spectra density on the detector with minimal crosstalk between them. The last technique uses an optical component called "image slicer" (first designed by Bowen, 1938), which simultaneously divides the light of the objects into slices and reformats them onto a single entrance spectrograph slit (Weitzel *et al.*, 1996).

The first measurements of galaxy rotation were made by Wolf (1914) on M81 and Slipher (1914) on M31, even before the community agreement on the extragalactic nature of galactic nebulae. The need for 3D spectroscopy had since been increasing, for example, van der Kruit (1976) used long slit spectra with different P.A.s to sample the overall galaxy emission. To study distant galaxies, IFU spectrographs have been built at several 8–10 m telescopes in the optical

Table 4.1: Main characteristics of the integral field units instaled on 8m class telescopes.

| Telescope/Instrument | Throughput [%] | Crosstalk[2] [%] | Multiplex | FoV [arcsec$^2$] | Spaxels[3] [arcsec] | $\lambda$ range [$\mu$m] | R |
|---|---|---|---|---|---|---|---|
| GEMINI/GMOS | 30 | 30 | 1 | $5 \times 7$ | 0.2 | 0.4–1.1 | 3,000 |
| KECK/OSIRIS | 25 | 10 | 1 | $0.6 \times 1.3$ | 0.02 | z–J–H–K | 3,800 |
| | 25 | 10 | 1 | $3.2 \times 6.4$ | 0.1 | z–J–H–K | 3,800 |
| MAGELLAN/IMACS | 20 | 30 | 1 | $5 \times 7$ | 0.2 | 0.35–0.9 | 1,800–1,0000 |
| VLT/FLAMES-GIRAFFE[1] | 8 | <10 | 1 | $11 \times 7$ | 0.52 | 0.35–0.9 | 9,000 |
| | 8 | <10 | 1 | $6.6 \times 4.2$ | 0.30 | 0.35–0.9 | 9,000 |
| | 8 | <10 | 15 | $3 \times 2$ | 0.52 | 0.35–0.9 | 9,000 |
| VLT/KMOS | 30 | < 3 | 24 | $2 \times 2$ | 0.2 | 0.8–2.2 | 3,000 |
| VLT/MUSE | 25 | < 3 | 1 | $60 \times 60$ | 0.2 | 0.35–0.90 | 1,500–4,000 |
| VLT/SINFONI | 25 | 8 | 1 | $0.8 \times 0.8$ | $0.0125 \times 0.025$ | 1.10–2.45 | 1,500–4,000 |
| | 25 | 8 | 1 | $3 \times 3$ | $0.050 \times 0.10$ | 1.10–2.45 | 1,500–4,000 |
| | 25 | 8 | 1 | $8 \times 8$ | $0.125 \times 0.250$ | 1.10–2.45 | 1,500–4,000 |
| VLT/VIMOS | 10 | 50 | 1 | $27 \times 27$ | 0.33 | 0.37–1.15 | 200–3,000 |
| | 10 | 50 | 1 | $54 \times 54$ | 0.67 | 0.37–1.15 | 200–3,000 |
| VLT/X–SHOOTER | 30 | 7 | 1 | $1.8 \times 4.0$ | $0.15 \times 0.6$ | 0.3–2.45 | 6,000–12,000 |

[1]Fiber-fed instruments are labeled with. [2]Crosstalks on the detector between adjacent slicer or fibers, given in %. [3]IFU spaxel size, given as the width of the on-sky square aperture or full rectangular dimensions when relevant.

and the near infrared. Table 4.1 lists some of the present facilities that are used for such observations. These instruments have been especially designed to recover the maximum flux per spaxel to detect their selected targets. They are unique tools to study the evolution of the galaxy kinematics, for example, the fraction of disk-like galaxies detected at different redshifts.

In Sec. 4.2, we give advice for selecting an IFU instrument and the optimized observational procedures depending on the scientific program. Section 4.3 presents different steps for reducing the data, including an optimized sky subtraction method. Section 4.4 deals with the measurements of the emission lines and their reformatting into a data cube for mapping physical properties. Section 4.5 describes the analysis of distant galaxy kinematics, showing their limitations and then proposing a simple methodological morpho-kinematics classification.

## 4.2 Basics of IFU instruments

Given the way the light of the object is sampled, the goal of IFU instruments is to get enough spaxels with sufficient S/N in their associated spectra to study the spatial distributions and profiles of emission or absorption lines. This implies a subtle trade-off between spatial and spectral resolutions. The spectral resolution also helps to tackle contamination from skylines, whose peak intensities may at times be hundreds of times greater than that of the observed galaxy emission lines (see Sec. 3.3.3).

The strategy for selecting an instrument such as one listed in Table 4.1 depends on the wavelength of interest, galaxy luminosity (or that of their emission lines) and spatial/spectral resolutions, but also on the need for accurate flux calibration, sky emission subtraction, and measurements of the different line moments.

In the data cube, spatial and spectral voxels[2] are contaminated by their neighbors due to the 3D point spread function (PSF) of

[2]3D spectroscopy generates $(X, Y, \lambda)$ cubes, populated by 3D pixels (also called "Voxels"). A spaxel is the sum along the wavelength dimension of all pixels at a given (X,Y) location: a spectrum can be associated to each of them.

the instrument. In principle, the spectral and spatial PSFs are independent of the position in the field of view, as often assumed in image processing. Disentangling the spatial from the line spread function (LSF or spectral PSF) is a subject of debate, including in other scientific domains. In practice, it is generally assumed that both PSFs can be separated, though this implies *ad minima* to avoid further contaminations such as those from spectral crosstalk (see Sec. 4.2.5). The latter should be minimized for proper measurements of second-order moments (e.g., velocity dispersion) or for observing objects with strong luminosity variations from one spaxel to another. Conversely, if the IFU is used as a simple wide light collector for detecting ultra-faint sources, high instrument throughput should be the favored criterion.

### 4.2.1 *Fiber instruments*

Fiber arrays were originally used because of their versatility. Fibers can readily reformat a 2D field into a 1D slit to make optimal use of the detector area without compromising the length of the spectrum (Fig. 4.3), i.e., spectral bandwidth. Because the light is reformatted in perfect alignment on the slit spectrograph, fiber spectrographs are known to be mechanically very stable, and they are not affected by higher order contamination. Hence, fiber spectrographs are better suited for high spectral resolutions than image slicers.

The focal-ratio degradation (FRD) is due to the fact that fibers can reduce the incident f-ratio, implying transmission losses (see Figs. 1 and 6 of Ramsey 1988). To minimize these losses, fiber arrays are generally used in association with a microlenslet array which focuses the light into an array of spots that are then captured by the fibers. Figure 4.3 illustrates this for FLAMES/GIRAFFE, which was installed in 2002 on the VLT.

Recent advances have allowed fiber-fed instruments to reach a total throughput similar to those of image slicers, thanks to the introduction of microlenses at each extremity of the fiber and then use hexagonal microlens arrays, which are 20% more efficient than the square arrays to collect the light.

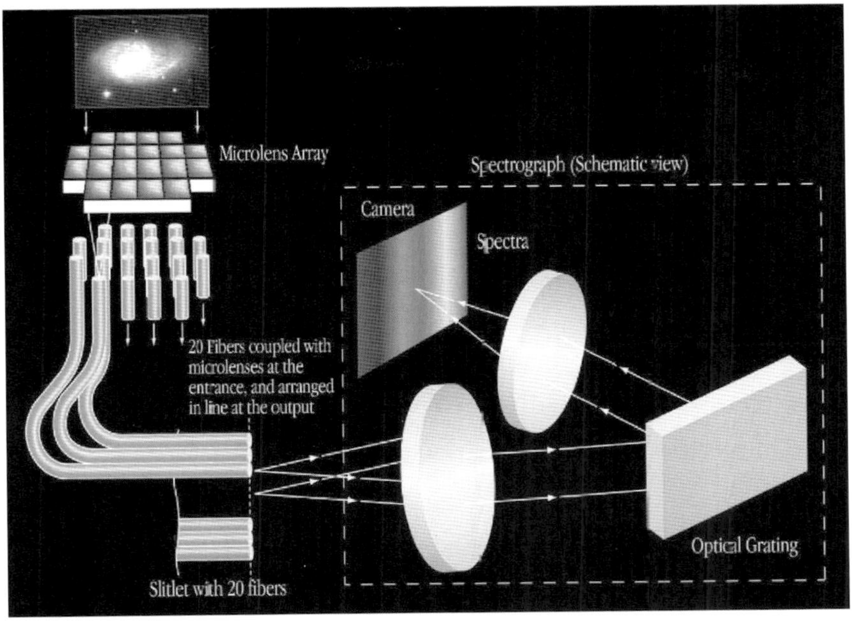


Fig. 4.3: Fiber array arrangement of the FLAMES/GIRAFFE IFU. Credit: ESO.

### 4.2.2 *Image slicer instruments*

The image slicer is now the most widely used technique in IFU instruments, in particular in the near infrared, because of the high throughput of the optical system (however, see Sec. 4.2.1 for drawbacks compared to fiber instruments) VLT/SINFONI, VLT/KMOS, or Keck/OSIRIS are examples of image slicer instruments (see Table 4.1).

At first glance, an image slicer can be seen as a multi-slit system. The field of view is divided into N strips on all of which the light is redirected by a set of mirrors along N different light paths towards the common entrance slit of the grating instrument (see Fig. 4.4). The use of tilted mirrors gives some freedom to optimize the coupling of the telescope and spectrograph leading to an efficient system with minimal dead space on the detector.

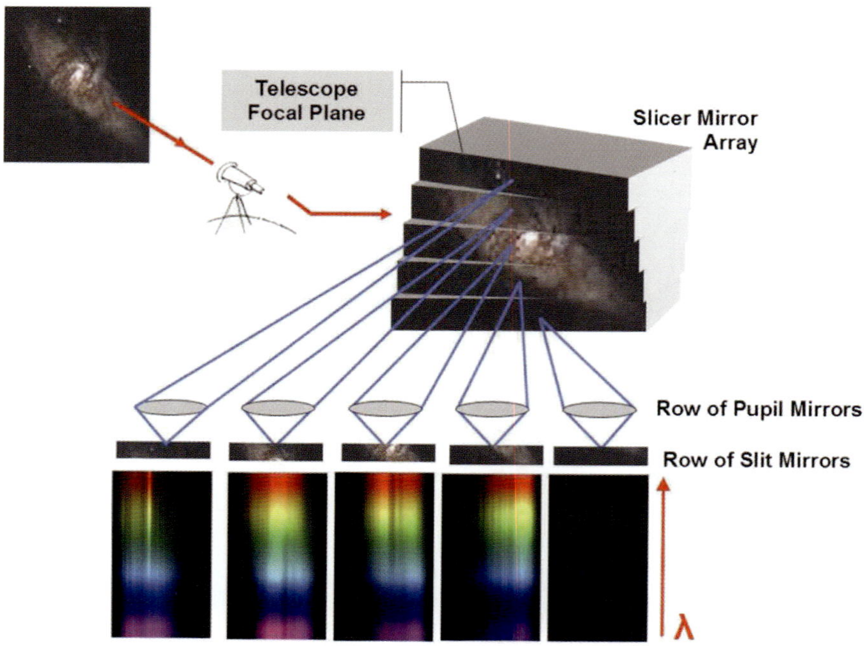


Fig. 4.4: Principle of the image slicer. The image from the telescope is redirected by mirrors tilted at different angles, and then aligned on the detector. Reproduced from Ealet *et al.* (2008) with kind permission.

### 4.2.3 *Compromise between resolutions and S/N*

Since small resolution elements strongly limit the S/N in a background limited regime (i.e., when sky is brighter than the target), it follows:

$$S/N \sim \sqrt{(\Delta x \times \Delta y \times \Delta\lambda)} \times \mathit{Eff} \times t, \tag{4.1}$$

where $\Delta x$, $\Delta y$ are the spatial elements of resolution, $\Delta\lambda$ is the spectral element of resolution, $t$ is the exposure time and *Eff* is the total throughput of the instrument. A trade-off between the spatial and spectral resolution must be considered, in particular, for faint objects. Increasing the spectral and/or the spatial resolution unavoidably leads to loose the detection of the faintest regions, generally at the galaxy outskirts (see e.g., Fig. 4.5). Calculations of performances (through an ETC) are thus recommended and should be done for different areas (spaxels) of the IFU.

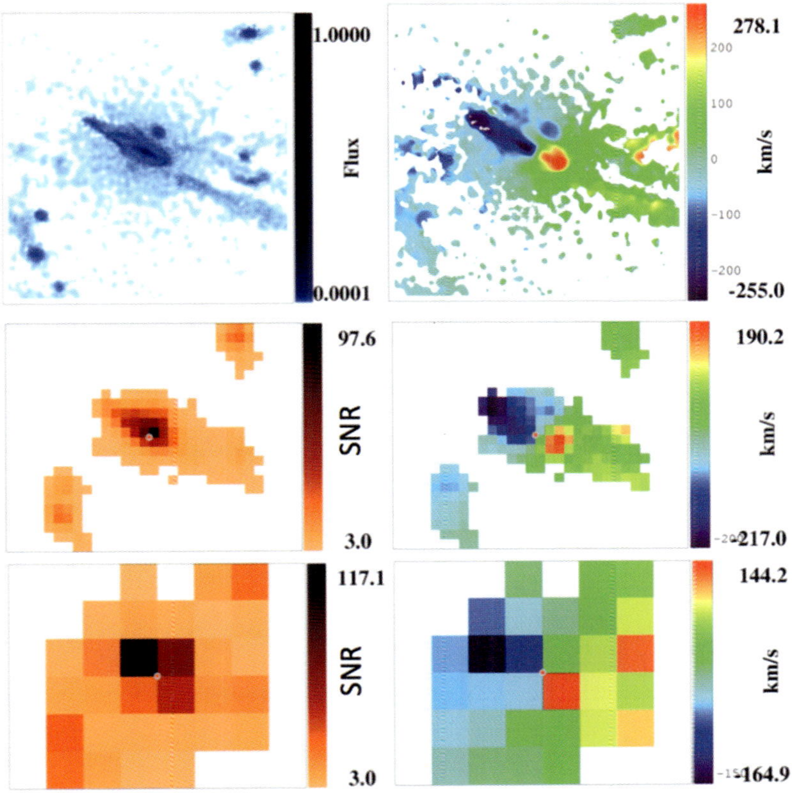


Fig. 4.5: Simulated observations of a galaxy merger illustrating how the spatial resolution affects the detection of faint structures. The simulation was carried out in the frame of an early Phase A study for an E-ELT/MOS assuming $m_{AB} = 24$ at $z = 3$ and a 10-h exposure. The upper panels show the simulated galaxy (stellar image and velocity field), while the central and lower panels show the recovered stellar image coded in S/N (left) and the velocity field (right). The lower panels have a spatial resolution degraded by a factor of 4 compared to the central ones, and allow the recovery of large scale motions in the outskirts.

### 4.2.4 *Spatial resolution*

Spatial resolution defines the size of the smallest spatial feature that can be detected. It is determined by the observational seeing and spaxel size (see Sec. 2.2.1). Oversampling the seeing leads to a decrease of the S/N per spaxel, while undersampling causes a loss in resolution. Figure 4.6 shows how the PSF size is the key element

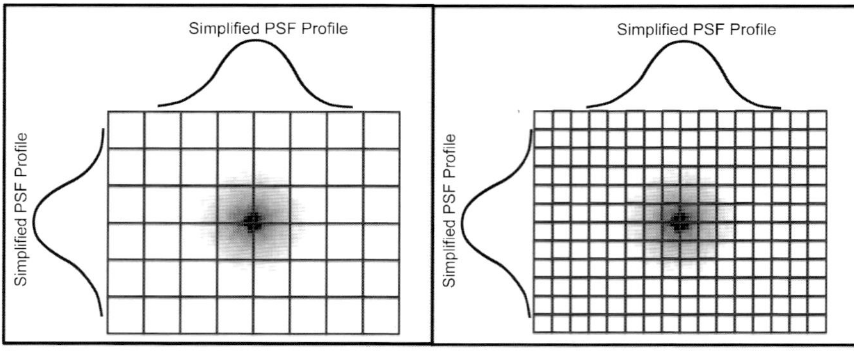


Fig. 4.6: Illustration of how a star image contaminates adjacent spaxels at two different samplings. The upper-left curves show how the different spaxels are illuminated. *Left*: for an optimized sampling (0.3 arcsec spaxel for a 0.8 arcsec PSF FWHM), most of the stellar light is included within >2 spaxels per axis as required by the Nyquist–Shannon sampling theorem. *Right*: decreasing the spaxel size by a factor of 2 does not yield higher spatial resolution but leads to lower S/N per spaxel.

in resolving structures: the information originating within one spatial element of resolution (the ~4 central spaxels on the left panel) spreads out, contaminating adjacent spaxels.

Figure 4.5 shows the S/N and galaxy structures detected at two resolutions differing by a factor of 4 and under the same observational conditions. At high resolution, one can detect the central regions of the galaxy in more detail (see Ex. *High spatial resolution kinematics of high-z galaxies*), but the outer regions are not detected due to the associated low S/N per spaxel. At low resolution, the large-scale motions in the galaxy halo can be detected, but the central details are lost. This is also illustrated with FLAMES/GIRAFFE observations where the large spaxel size merely samples a 1 arcsec seeing. Nevertheless, these remain well suited for sampling large-scale motions (~ 7 kpc) in moderately distant (z~ 0.6) galaxies with angular sizes in the 2–3 arcsec range.

**Example 4.1: *High spatial resolution kinematics of high-z galaxies***

The AO laser guide star of the SINFONI and OSIRIS spectrographs can improve the spatial resolution by a factor

(*Continued*)

**Example 4.1: *Continued***

of 4 when compared to seeing-limited observations. It provides exquisite details on massive distant galaxies that allows testing the presence of an AGN in the galaxy center (see Fig. 4.7). Such observations require from 10 h to 24 h of exposure time with 8 m class telescopes, and are limited by the spaxel size and the surface brightness of the objects. Novel techniques with MOAO (e.g., with MOSAIC on the E-ELT) will enable the simultaneous observation of much fainter targets, and are expected to provide yet-inaccessible resolved rotation curves for samples of distant galaxies.

### 4.2.5 *Spectral resolution and crosstalk*

For wavelengths above 700nm, spectroscopy is strongly affected by OH skylines and telluric absorption lines. Many scientific cases require working in-between these skylines, which entails a spectral resolution higher than 3000 (although 5000–6000 would be best, see Sec. 3.3.3). In the visible, when observing z = 0.5–1.5 galaxies, the [*O II*]3726–3729Å doublet is an ideal line to observe, but it requires a resolution of at least R = 6000 to be properly resolved.

Resolutions in excess of 5000 may also significantly help to remove unresolved skylines when they coincide with the source emissions. As shown in Fig. 4.8 (see also Sec. 4.3), the noise from emission skylines can significantly alter the capacity of the data reduction to recover, e.g., complex line profiles. Resolutions $\sim$ 3000 are available for most 3D instruments (for example, VLT/VIMOS, SINFONI, KMOS and MUSE). It allows centroid determination of emission lines such as H$\alpha$, but not the measurement of its line profile or velocity dispersion when smaller than 75 km/s (see Fig. 4.8).

Spectral crosstalk occurs as a result of the imperfect projection of the spectra onto the detector pixels. It is defined as the fraction of light contaminating a spectrum due to adjacent spectra (see Fig. 4.9). It is driven by compromises made during the optical design phase of a spectrograph (see Table 4.1) and may have a considerable impact

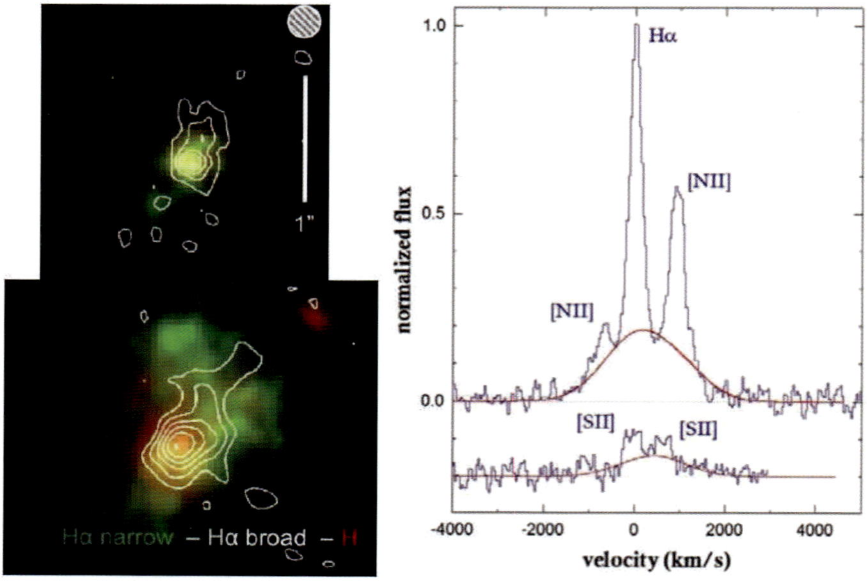


Fig. 4.7: *Left*: Distribution of the broad-line emission (white contours) overlaid on maps of narrow Hα emission (green tones) from deep high-resolution SINFONI+AO observations, and on H-band emission (red tones) from HST NICMOS/NIC2 and WFC3/IR. Targets are two $M_\star \geq 10^{11}\ M_\odot$ galaxies observed at z∼ 2.4 by Förster Schreiber *et al.* (2014), ©AAS, reproduced with permission. H-band emission samples intermediate age stars at rest (0.47 μm) and coincides with the overlaid broad-line emission. The white vertical bar and the filled circle indicate the angular scale and the size of the FWHM, respectively. *Right*: Zoom-in on the Hα+[N II] complex (top) and the [S II] doublet (bottom) of the nuclear spectrum, which is the sum of the six massive galaxies observed by Förster Schreiber *et al.* (2014). The FWHM of the broad nuclear emission (red lines) is 1500 ± 240 km/s, and the [N II]6584/Hα ratio is 0.60 ± 0.01 for this average spectrum.

on the ability to reach a given scientific goal (see Ex. *Small crosstalk or large field of view?*).

**Example 4.2: *Small crosstalk or large field of view?***

VLT/VIMOS is an example of an instrument maximizing the FoV (see Table 4.1) but which leads to very large amounts of crosstalk, i.e., 50% assuming uniform spaxel illumination. Large crosstalks strongly limit the capability of an IFU to map physical properties because of the flux contamination

(*Continued*)

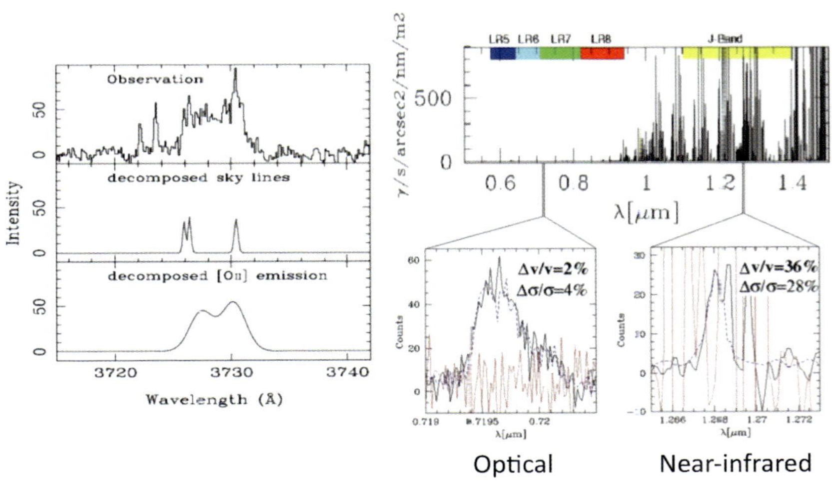


Fig. 4.8: Impact of spectral resolution on the sky subtraction in IFU instruments. *Left*: the top panel shows the observed [O II] doublet emission, for which three night skylines are easily identified with FLAMES/GIRAFFE at R = 9000. The central panel shows the isolated components of the night skylines from simultaneous line fitting, and the bottom panel shows the reconstructed [O II] emission line doublet (see, e.g., Yang *et al.*, 2008). *Right*: A comparison between performances of two IFU instruments observing the same object either in the visible (FLAMES/GIRAFFE, using the [O II] doublet) or in the NIR (SINFONI, using the H$\alpha$ line). The sky spectrum is shown in the top panel, and the uncorrected lines are shown in the bottom panel. The flux of the [O II] doublet is assumed to be half that of H$\alpha$ (Gallagher *et al.*, 1989). Also provided is the accuracy in recovering velocity and dispersion after sky subtraction. The skylines (red) are so intense in the NIR that the moderate resolution offered by SINFONI is hardly sufficient to remove them. The lower throughput of the FLAMES/GIRAFFE instrument is counterbalanced by (1) the advantage of observing the [O II] line at wavelengths free of skylines, and (2) the higher spectral resolution (9000 instead of 3000), which collectively allow very efficient removal of sky emission lines.

**Example 4.2: *Continued***

from neighboring fiber projections onto the detector. It is recommended to restrict the use of large crosstalk instruments to mapping low-order moments of emission lines such as intensities or position/velocities. Conversely, the design of FLAMES/GIRAFFE has been chosen to minimize crosstalk (less than 10%). This instrument can be safely used to measure line widths (see Figs. 4.8 and 4.9), at the cost of offering a small FoV per IFU (i.e., $3 \times 2$ arcsec$^2$).

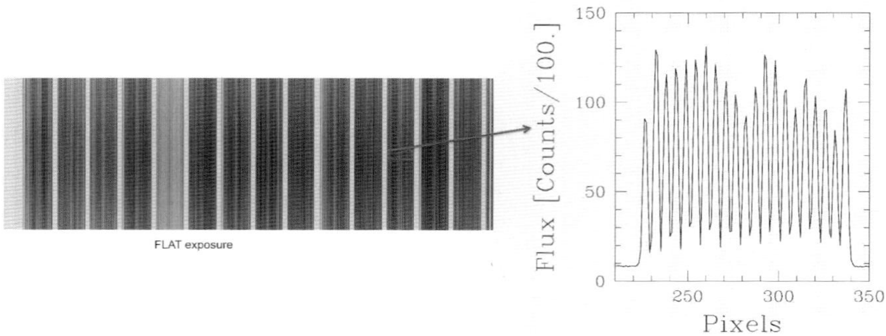


Fig. 4.9: Fiber-to-fiber spectral crosstalk contamination when minimized at the optical definition stage of the spectrograph. *Left*: a flat exposure of FLAMES/GIRAFFE in the IFU mode. *Right*: extracted image profile displaying limited overlap between adjacent spectra (of only 10% in this case).

### 4.2.6 *Observing with IFU instruments*

#### 4.2.6.1 *Astrometry*

IFUs at 8m telescopes aim at observing faint targets, which are generally small (a few arcsec in diameter). The astrometric accuracy of an observed field can be 0.25 arcsec even for high Galactic latitude cosmological fields, using, e.g., the USNO-B or the UCAC2 catalogs. This allows most IFUs to target faint objects using the "blind offset" technique. The idea is to use a bright star from the catalog and near the object position, point the telescope to that star and then move "blindly" the telescope to the object.

#### 4.2.6.2 *Observational strategy*

This important step is mainly described in Sec. 3.3. The goal of IFU observations is to reach "Background-Limited Performance".[3] Here, we concentrate on specific aspects of IFU observations. Some

[3]This detection limit is approximately the sky background noise and is quite difficult to achieve because it requires getting a sufficiently high S/N on the background. This is even more difficult at infrared wavelengths because the exposure time is limited by the rapid saturation and variability of strong skylines.

IFUs have a FoV significantly larger than the galaxy, allowing the simultaneous sampling of both object and sky. Observations can then be conducted in starring or nodding mode (i.e., using small IFU offsets around the object so as to sample it within two different IFU quadrants). Alternatively, if the IFU FoV is too small for accurate sky sampling, one should follow an ON/OFF strategy where object and sky are observed sequentially (see Sec. 3.3.3 and Ex. *Strategy to reduce the sky signal with IFUs*).

**Example 4.3: *Strategy for reducing the sky signal with IFUs***

The SINFONI IFU FoV is up to four times larger than the majority of distant galaxies at $z = 1–2$ (1–2 arcsec). Object and sky can thus be observed simultaneously at no loss of exposure time when nodding the source within the SINFONI FoV. Conversely, with instruments such as VLT/X-SHOOTER, KMOS or Keck/OSIRIS, it becomes necessary to use a beam-switching sequence following an ON (or "A") and OFF (or "B") exposure which sequentially samples the object then the sky (see a detailed description in Sec. 3.3.3).

## 4.3 Data reduction of IFU observations

Both image-slicer and fiber-fed instruments have quite complex data reduction procedures. Raw data from image slicers resemble those from multi-slit spectrographs (see Fig. 4.10, left panel), which is useful for calibration purposes. However, the sky subtraction scheme is far more challenging because the spectrum within each subslit is a complex mixture of galaxy and sky emissions (see Fig. 4.4). Figure 4.10 (right panel) shows that fiber-fed IFU spectra are better organized and aligned along a single, broad slit (see also Fig. 4.2). The sky subtraction scheme remains complex however because the individual fiber response may depend on mechanical torsions.

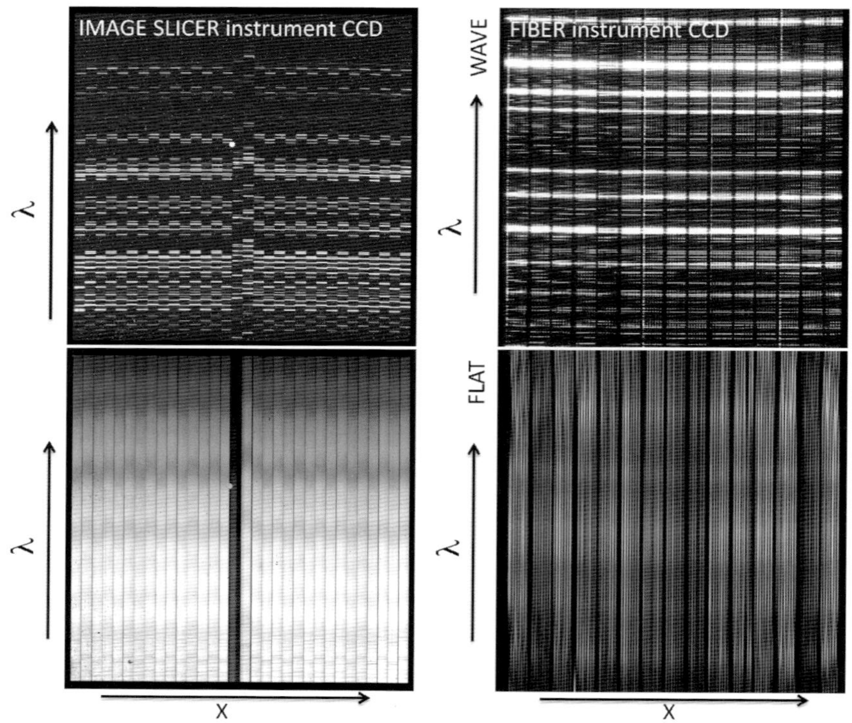


Fig. 4.10: Spectral format of the data on the detector for an image slicer (left) and fiber IFU (right) with raw data of two calibration observations (top: wavelength lamp, bottom: flat field). *Left*: subslit traces are shown in the flat field frame, which are not aligned in the wavelength frame. *Right*: The flat field shows the trace of each fiber, which are organized in groups of fibers (1 per IFU) resulting in the same number of aligned spectra in the top panel.

### 4.3.1 *Reduction of fiber data*

The first step is similar to standard fiber spectrograph data reduction: bias and flat field correction, wavelength calibration with lamps exposures and sky subtraction (lines and continuum) using sky exposures. Whenever possible, calibration exposures need to be carried out using the same orientation and fiber configuration than the one used for observing scientific targets. We summarize below specific aspects of the fiber-fed IFU data reduction.

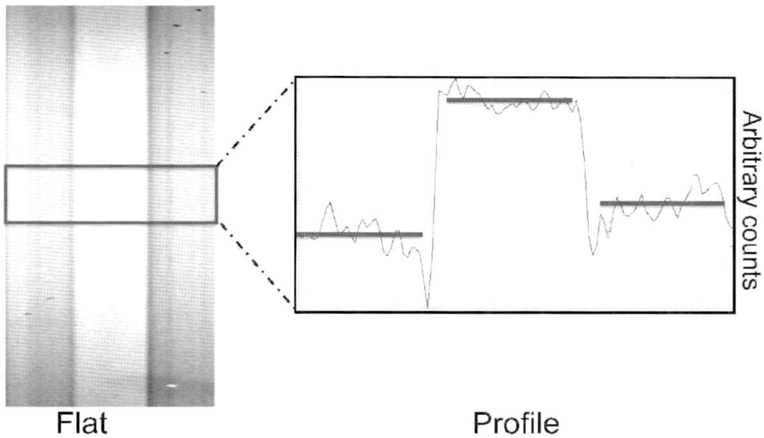


Fig. 4.11: Profile for three slices in a flat exposure extracted within a given $\lambda$ range (see rectangle on the left panel). They show different responses that can be normalized using the median intensity (see horizontal thick lines) along the profile from the right frame. Fibers also provide different responses when they are assembled to form an IFU. However, fiber-to-fiber variations can be due to mechanical torsions and require a more complex treatment to be corrected (see Sec. 4.3.1).

#### 4.3.1.1 *Fiber-to-fiber response and sky subtraction*

The major limitation arises from the difference in fiber throughput, an issue shared with slicer IFUs (see Fig. 4.11). Here, the throughput can change from fiber to fiber, likely due to differential torsion between fibers.

The standard approach to correct this is to acquire a sky flat[4] which is assumed to be observed in the same conditions as the scientific target. Ideally, the flat should be observed just before or after the target (and is then called an "attached flat"). It is used to generate a normalized response which accounts for the fiber-to-fiber chromatic response. Its construction requires rejecting sky lines and cosmic rays by setting masked-out spectral regions which depend on the resolution of the spectrograph.[5] This fiber response correction

[4] A dome flat may be used instead if a moving screen has been purposely implemented on the telescope dome.

[5] Note that at resolutions smaller than 300,000, the skylines are not resolved.

technique can also be applied to observations of spectrophotometric stars, allowing accurate flux calibration of the spectra.

The above procedure allows one to reach an accuracy of 1% on the total sky background. Sky subtraction has been often considered a major issue for the fiber-fed instruments (see however the discussion in Sec. 3.3 and Sharp and Parkinson 2010). In principle, IFUs allow reaching the theoretical limit on the sky subtraction, which corresponds to a few tenths of percents (Yang *et al.*, 2013).

#### 4.3.1.2 *Correction of crosstalk*

Crosstalk occurs when fibers projected onto the detector are too close, and this affects the few instruments favoring a large FoV (see Section 4.2.5). An unwelcome effect arises when bright emissions (including those from sky lines) contaminate neighboring elements on the detector. To (partially) compensate these effects, the pipeline can use multi-Gaussian functions to separate the different elements or simply reject some fibers from the reduction. In case of significant crosstalks, this technique can recover the centroid of the emission, but not the line profile or its dispersion, despite the numerous efforts which have been made to correct this effect (e.g., VIMOS IFU observations).

### 4.3.2 *Reduction of image slicer data*

The procedure to subtract the bias, dark and to apply dome-flat corrections are equivalent to those used in multi-slit instruments. Sky subtraction is done after using dedicated sky exposures except if there are enough IFU spaxels free of galaxy emissions which can then be used to sample directly the sky signal (see Ex. *Strategy for reducing the sky signal with IFUs*). In the latter case, one has first to account for the difference of response of each subslit and Fig. 4.11 illustrates how this can be handled in the case of a flat image.

#### 4.3.2.1 *Alignment of slices and PSF*

Image slicer instruments may have problems related to the misalignment of the subslits and the elliptical shape of the PSF. These two problems can be corrected using a detector distortion map produced using dedicated observations with a pinhole mask.

#### 4.3.2.2 *Possible problems with image slicers*

The contamination between subslits can be a major problem in a few cases, e.g., with SINFONI for which even after distortion correction, there is substantial overlap (or crosstalk) at the edges of the subslits (see Fig. 4.12). This can be problematic when strong emission skylines of the sky contaminate an adjacent subslit at different wavelengths. This can be corrected by rejecting the contaminated fraction of each subslits (see Fig. 4.12).

### 4.3.3 *Correcting for atmospheric dispersion*

IFU observations can be affected by atmospheric dispersion, especially at the blue end of the visible spectrum and in absence of an ADC. For example, the IFU mode of X-SHOOTER is made up of three subslits of $0.6 \times 4$ $arcsec^2$ which are affected by atmospheric dispersion because of the large wavelength coverage (see Fig. 4.13). In such a case, the trace of faint galaxy spectra with undetected continuum can be easily lost. To correct this effect, one can use the trace of telluric star spectra which need to be observed under similar conditions (air mass) as the target, and bracketing the science

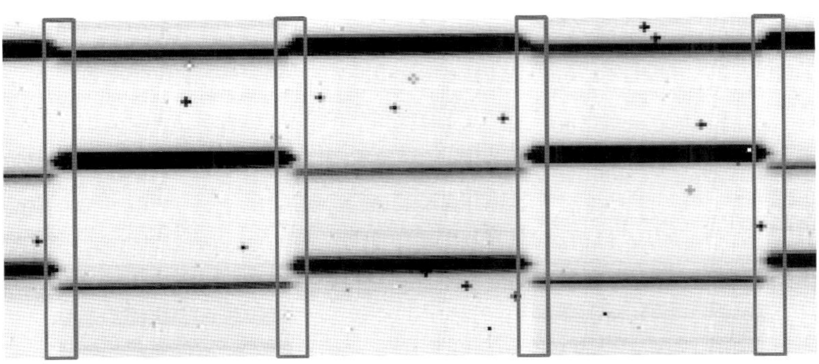

Fig. 4.12: Zoom on the raw data of a science frame from the VLT/SINFONI instrument. The small vertical rectangles indicate the regions of crosstalk between adjacent slits. Losses due to contamination may represent 5–8% of the subslit sizes.

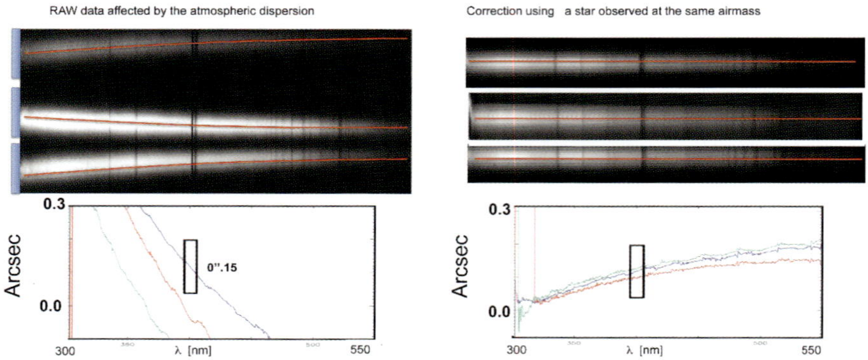


Fig. 4.13: Traces of a star observed at airmass 1.1 within the three slices of the X-SHOOTER/IFU UBV arm. *Left*: On top is shown the trace of the same star that illuminates the three slits along the wavelength direction, evidencing the effect of atmospheric dispersion. The latter affects the top and bottom slits in an opposite direction compared to the central one because of the optical design. The bottom-left panel shows the normalized effect in each subslit. The blue, green, and red traces are respectively: the central and two lateral slit-lets, and the small rectangle represents the pixel scale of the final data-cube (0.15 arcsec). *Right*: On top is shown the same object after atmospheric dispersion correction, using a star observation at similar airmass and during the same night. The result is shown in the bottom-right where the previous correction has been applied on the X-SHOOTER three slit-lets. The resulting precision on the correction is less than 0.1 arcsec.

exposure. This strategy ensures that both observations are affected in the same way, so that the residual distortion between both edges of the spectra remains smaller than the spaxel size (0.15 arcsec, see rectangle box). The galaxy spectra can then be accurately realigned in the subslits.

### 4.3.4 *Spatial, spectral smoothing, and binning*

#### 4.3.4.1 *Spatial smoothing*

Spatial smoothing improves the S/N, but can reduce the spatial resolution. If the PSF FWHM is sampled by much more than 2 spaxels, smoothing is always useful, otherwise any smoothing or rebinning strongly degrades the resolution. A balance must therefore be found between improving the S/N while maintaining a sufficient resolution (see Ex. *Optimizing emission line S/N in the near-IR at moderate resolution*). The simplest way to smooth is just to apply a spatial $3\times3$

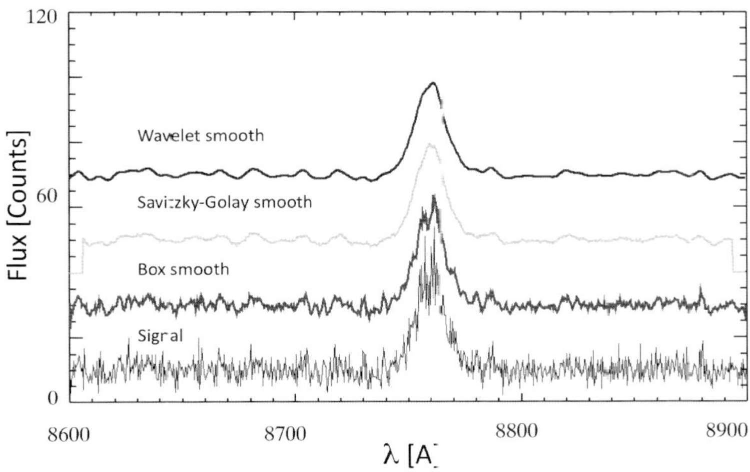


Fig. 4.14: Illustration of the different line shapes that result from different smoothing algorithms, all using the same box size.

point filter (see Clare 1997). However, such a smoothing is possible only for galaxies having a large number of resolution elements, i.e., it is generally not applicable for seeing-limited observations of (small) distant galaxies.

Nevertheless, it may become useful to increase locally the S/N within specific regions such as the galaxy halo, by degrading the spatial resolution to detect the faintest emissions (see, e.g., Fig. 4.5). Cappellari and Copin (2003) developed a numerical method that applies an adaptive spatial binning (or Voronoi binning) for 2D data to reach a chosen constant S/N per bin, defining new spaxels of different sizes (see also Sec. 2.2.2.2). This solution, when applicable, is certainly the best method to balance S/N and resolution.

#### 4.3.4.2 *Spectral smoothing*

Spectral smoothing can change the shape of the emission line, and affect its different moments (velocity and dispersion). Among the spectral smoothing methods, we can mention the moving average filter or box filter, the Savitzky–Golay filtering (Vetterling *et al.*, 1992) and the wavelet decompositions (Vegte, 2001). Figure 4.14 shows the effect of these three methods at low and high S/N. The moving average filter smoothes data by replacing each data point with the average of the neighboring data points defined within the window, and is equivalent to low-pass filtering.

The Savitzky–Golay filtering can be considered a generalized moving average filter by performing an unweighted linear least-squares fit using a polynomial of a given degree. This method is more efficient at preserving higher moments of the peak such as the line width (velocity dispersion), and should be preferred e.g., for kinematic studies.

To increase the detection sensitivity, multiple Gaussian profiles may be fitted simultaneously to several emission lines. The line profile fitting can be conducted using a $\chi^2$ minimization procedure that takes into account the noise level, including that close to atmospheric OH sky emissions (see Ex. *Fitting three lines together*).

### 4.3.5 *Optimizing sky subtraction with NIR IFUs*

In the NIR range, residuals from the subtracted skylines are so intense that they can still affect the detection of target emission lines (see Sec. 3.3.3). The target has to be carefully chosen in redshift so as to avoid contamination of the galaxy emission lines by strong skylines, otherwise the emission line moments cannot be measured. Figure 4.15 demonstrates how spatial and temporal variations of the OH skylines can alter the data and the validity of its reduction and analysis. It is thus mandatory to scrutinize the raw data cube to investigate the presence of undesirable OH skylines prior to data reduction and analysis.

The following Ex. *Optimizing emission line S/N in the near-IR at moderate resolution* presents KMOS/VLT J band observations of a $z = 0.649$ galaxy. Figure 4.16 illustrates the considerable gain when properly removing the skyline residuals. Combined with Fig. 4.15, it demonstrates that an R of 3000 is hardly sufficient to properly remove OH skylines. This is because such a spectral resolution translates into a velocity resolution of 100 km$s^{-1}$, which corresponds to a significant fraction of the velocity gradient observed in galaxies, especially those observed at moderate inclinations. This can be avoided using resolutions larger than 9000 (see Fig. 4.8).

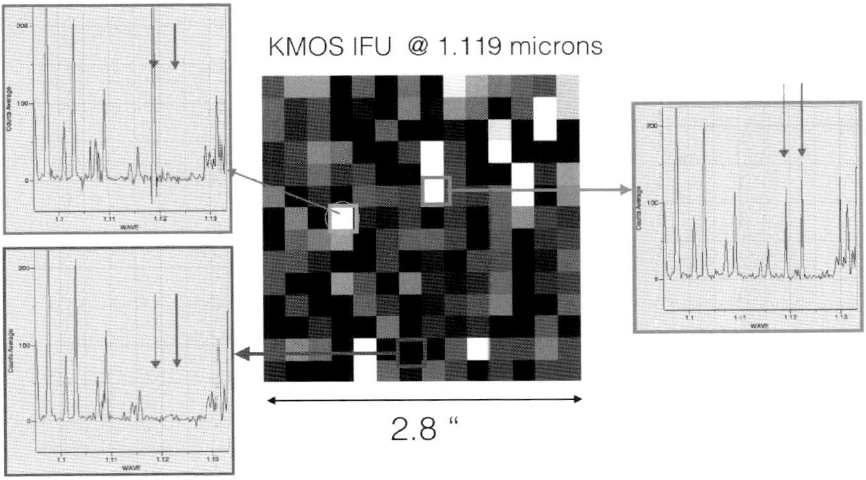


Fig. 4.15: Spatial (and sometimes temporal) variations of the OH skylines within a KMOS data cube. Although the target has been chosen to warrant H$\alpha$ in a wavelength range devoid of OH skylines, their variations at different positions in the KMOS IFU jeopardizes the data reduction and analysis if not accounted for. It also evidences the considerable temporal and spatial variations of OH skylines, which justify favoring instruments with sufficiently high spectral resolution.

**Example 4.4: *Optimizing emission line S/N in the near-IR at moderate resolution* (*KMOS*)**

Figure 4.16a shows the spectrum of a spaxel after data reduction: the H$\alpha$ line at 1.082 $\mu$m (in blue) is hardly recognizable within the numerous spikes of sky subtraction residuals. In many spaxels, the S/N of the line is well below the detection threshold if the sky residuals are taken into account in the noise determination. Figure 4.16b shows the resulting velocity field for such a nonoptimized data reduction. The S/N has to be estimated in a region free of skylines, because sky subtraction residuals artificially enhance the noise which then result in underestimated S/Ns. The three-steps procedure to properly estimate the S/N is as follows:

(1) Extracting the mean spectrum of the data cube and create a mask of the wavelength ranges affected by sky residuals.

(Continued)

**Example 4.4: *Continued***

Spikes (both positive and negative) are easy to detect with a peak detection algorithm because they always exhibit the same signature restricted to a single resolution element (see Fig. 4.16a): here, one negative spike followed by a positive spike. Flag the regions to be masked $\sim \pm$ 2 pixels around the spikes.[6]

(2) The masked data cube can then be used to properly estimate the S/N in each spaxel. The data cube can be smoothed spatially using a Gaussian filter of width similar to the PSF of the observations. Spatial smoothing can significantly increase the S/N of the lines at the cost of reducing the spatial resolution of the observation. A Gaussian smoothing should always be applied in the cube after the spikes have been masked out, otherwise the noise would propagate over many pixels in the spectra. Here a Gaussian filter of 1.5 spaxel (0.3 arcsec) in a $3 \times 3$ spaxel box has been used for an observational seeing of 0.5 arcsec FWHM.

(3) Using the masked and smoothed data cube, the emission line of the object can be fitted with a Gaussian profile to measure the position, width, and flux. The noise is computed from the unmasked pixels on both sides of the line. Figure 4.16c shows the final velocity field map obtained by using the spatial smoothing and spike rejection algorithms.

## 4.4 Mapping galaxy properties from a data cube

The re-alignment of the slices or fibers can be easily handled by the instrument pipeline allowing one to extract the spectra and reconstruct a data cube (see, e.g., Fig. 4.1). Data cubes can be analyzed in order to construct maps of several galaxy properties, which are derived from measurements of their emission lines. This

[6]This assumes that the spectral resolution is sampled by two pixels; the exact $\lambda$ range to be flagged depends on the spectral resolution, i.e., the higher is the better.

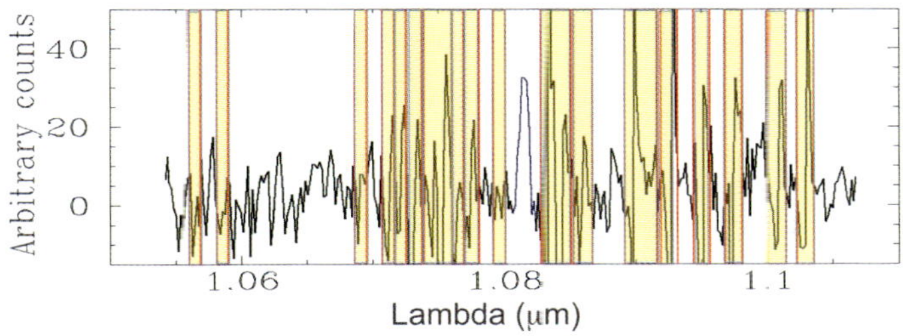


(a)

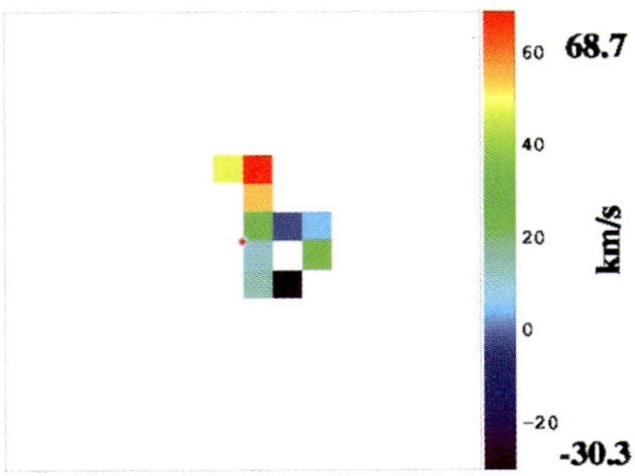


(b)

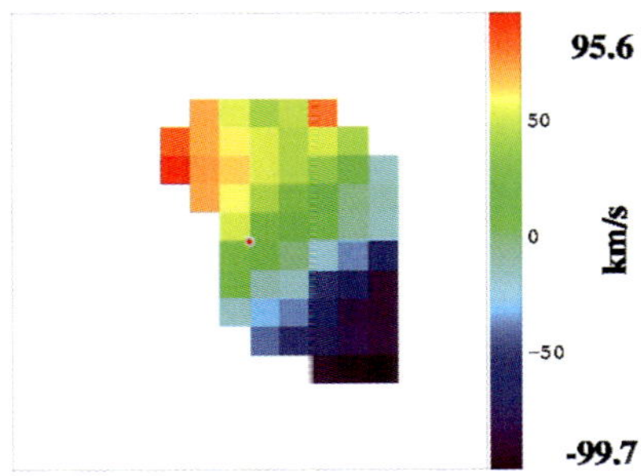


(c)

Fig. 4.16: (a) Zoom on a spectrum provided by the sum of sky-subtracted spectra over the spaxels covering a z = 0.649 galaxy. The S/N in the continuum is estimated on a much larger spectral window as provided by KMOS. The numerous spikes from sky-residuals are masked (yellow and red vertical lines) within $\pm 2$ pixels. (b) Velocity field without masking residual spikes. (c) Same as (b) after masking the skyline residuals and applying a $3 \times 3$ Gaussian filter of 1.5 spaxel (0.5 arcsec).

includes kinematics and properties of the gaseous phase such as SFR, extinction, or metal abundances. Stellar population studies based on absorption lines are to-date limited to z = 0.3 for 8m telescopes, underlying the current difficulty to detect the continuum with sufficiently high S/N.

### 4.4.1 *Measuring emission lines: methods and error budget*

The physical and chemical properties of the ionized gas phase in the galaxies may be studied after measuring fluxes, positions, and profile shapes of the most prominent emission lines, such as [O II]$\lambda\lambda$3726,3729Å, H$\beta$, [O III]$\lambda\lambda$4959,5007Å, H$\alpha$, [N II] $\lambda\lambda$6549,6584Å, and [S II]$\lambda\lambda$6716,6731Å. For a preliminary analysis of a data cube, it is recommended to visualize by eye each individual spectrum to detect possible artifacts or contamination due to sky correction residuals or unexpected faint skylines.

The following example illustrates how errors can be estimated after a Gaussian fit, how they can be optimized in some configurations, and finally how difficult it can be to derive accurate maps of distant galaxy properties. In this case, the presence of gravitational lensing allows robust constraints to be established on the metallicity radial profile for a z = 2.49 source (see Fig. 4.17). In general, this kind of study can only be achieved for the brightest distant galaxies and only along an average radial profile because of the low S/N in the external regions of the galaxy, and after using Voronoi binning (see Sec. 4.3.4.1).

**Example 4.5: *Fitting three lines together***

Analyses of Keck/OSIRIS 3D data have used Gaussian profiles to fit three emission lines ([N II] $\lambda\lambda$6548-6583 and H$\alpha$), simultaneously (Yuan *et al.*, 2011). The search for emission structures with a multiple fit permits the detection of even weak emissions (see Fig. 4.17). In the NIR, about one out of three lines unavoidably lies near a strong OH sky line and the procedure needs to take into account the correspondingly higher noise level. The line profile fitting was conducted using a $\chi^2$ minimization procedure, and the result can be checked using the fiducial ratio of the [N II] lines ([N II]6583/[N II]6548 = 3). Notice that in this particular example, the S/N is further enhanced by gravitational lensing.

Measurements of emission line fluxes are detailed in Sec. 3.5.2. Here, we focus on how to measure the centroid and width of the emission lines which are tracing the velocity and velocity dispersion of the gas, respectively.

Measuring these properties requires adequate S/N over a large enough number of resolution elements. The line profile S/N can be defined as the mean value of the flux at a given emission as follows

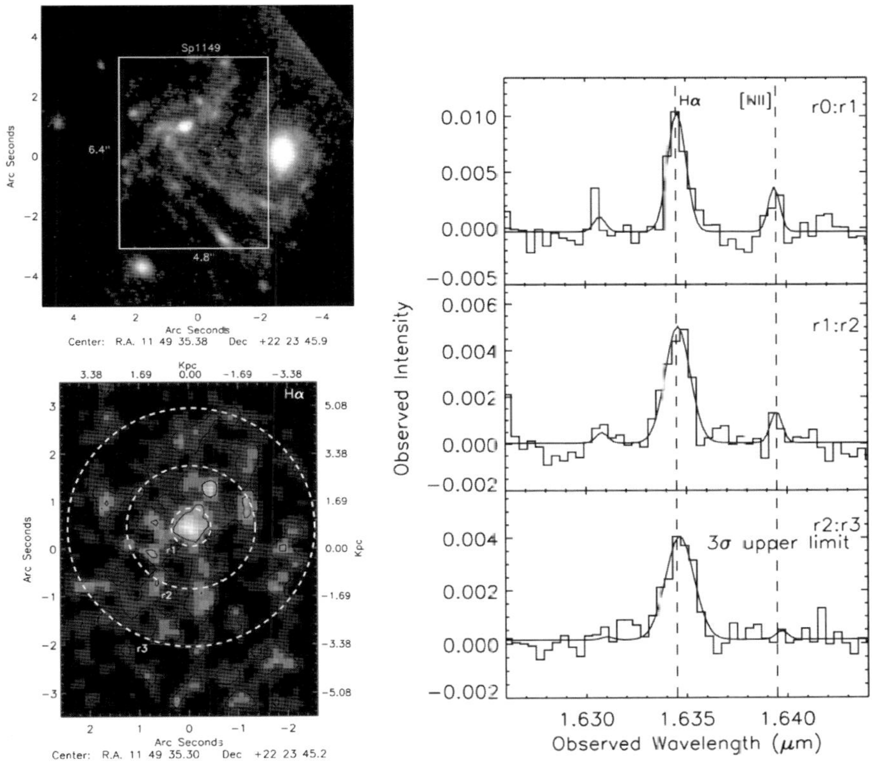


Fig. 4.17: *Top-left*: HST F814W and F555W band color image of a galaxy at $z = 1.49$, which is gravitationally lensed by the MACS J1149.5+2223 cluster (from Yuan *et al.* (2011), ©AAS, reproduced with permission). The white rectangular box shows the $4.8 \times 6.4$ FoV of the Keck/OSIRIS IFU. *Bottom-left*: the detected Hα emission from the OSIRIS datacube showing three rings in which the lines are searched. *Right*: detection in each ring of Hα and [N II]6548-6584 emission lines, allowing one to constrain the metallicity, which decreases from the center to the outskirts.

(Flores *et al.*, 2006):

$$\mathrm{S/N} = \Sigma_i^n (S_i/(\sqrt{n} \times \sigma_{cont})), \tag{4.2}$$

where $n$ is the number of spectral resolution elements in which the emission is detected, $S_i$ is the emission line flux at a given voxel $i$, and $\sigma_{cont}$ is the noise measured on the continuum. Only spaxels with sufficient S/N should be considered for further analysis. For a single emission line (e.g., Hα), a S/N $\geq 5$ should be considered (see

Sec. 3.5.2), while for a doublet such as [O II], the lower threshold S/N $\geq 3$ can be considered acceptable (Flores *et al.*, 2006).

At infrared wavelengths, the S/N varies strongly as a function of wavelength due to the numerous OH skylines (see Fig. 4.8). The total noise is not only due to the sky background continuum noise but also to the spatial residues of the correction. These originate from spatial and temporal variations of the sky emission lines (see Fig. 4.15) and further complicate uncertainty estimates. Figure 4.8 further illustrates the need for high spectral resolution in removing faint and narrow skylines that populate spectra at $\lambda > 0.72\ \mu$m.

### 4.4.2 *Mapping physical properties*

Physical properties can be derived after combining measurements from several emission lines, e.g., their flux ratios. As shown in Sec. 3.5.2, in this case the S/N of the individual lines has to be properly taken into account to estimate the uncertainty of the ratio. For example, $R_{23}$ depends on the [O II] 3727/H$\beta$ ratio (see Sec. 3.6.5), which needs to be corrected for extinction. Then the uncertainty on $R_{23}$ depend also on the Balmer line ratio uncertainty. This is particularly problematic for 3D spectroscopy because the associated S/N is generally quite low.

3D spectroscopy allows one to construct maps of physical and chemical properties, and then determine how they vary across different regions of the same galaxy. If the S/N is large enough, extinction (through the H$\alpha$/H$\beta$ ratio) can be measured for each element of resolution, but also: SFR (through extinction-corrected H$\alpha$ or H$\beta$), metallicity (through $R_{23}$), electron density (through e.g., the [O II]3226/[O II]3729 ratio), and diagnostics for identifying the presence of AGN (see 3.5.1). Recall that this often requires a flux calibrated datacube, except if the lines used are at very similar wavelengths, e.g., the [O III]/H$\beta$, [O II]3226/[O II]3729, or [N II]/H$\alpha$ flux ratios. The latter is often used for assessing the metallicity (see Sec. 3.6.5) as illustrated in Fig. 4.17).

Within each spaxel one can measure the emission lines with an associated flux and its corresponding S/N. We recommend applying

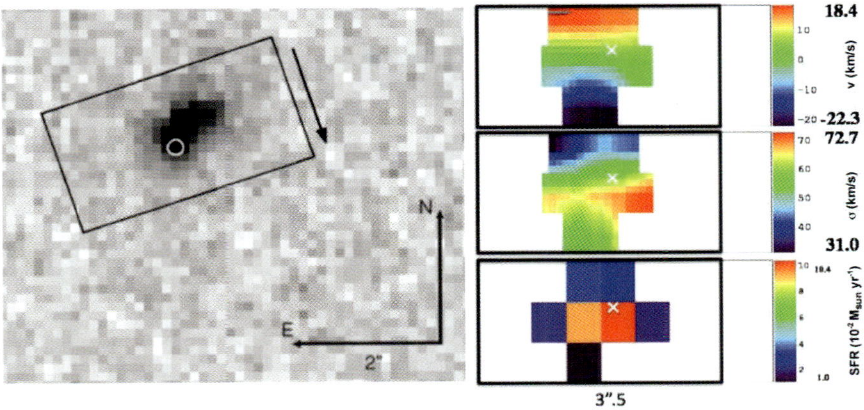


Fig. 4.18: *Left*: HST H-band image of a GRB host galaxy observed using the VLT/X-SHOOTER IFU (black rectangle). The black arrow represents the direction of the IFU maps. *Left, from top to bottom*: Velocity field, dispersion, and star formation maps estimated from H$\alpha$. The velocity and velocity dispersion maps are interpolated to spaxels of $0.1 \times 0.1$ arcsec for clarity. The SFR map is reconstructed from the H$\alpha$ emission line with a spatial resolution of $0.5 \times 0.6$ arcsec (there is no extinction deduced from the observations, see Vergani *et al.* (2011)). The afterglow (white circle and crosses in the right and left panels, respectively) occurs near the SFR peak. Credit: Vergani *et al.* (2011), A&A, 535, 127, reproduced with permission ©ESO.

the recipes described in Sec. 3.5.2 to each spaxel of the map. This includes dealing with low S/N spaxels and testing the presence of an AGN through diagnostic diagrams. Figure 4.18 shows the example of a star formation map constructed using the X-SHOOTER instrument, which allows linking SFR properties with kinematics, while Fig. 4.19 presents three electron density maps derived from the [O II] doublet ratio. The latter reveals the ionized gas density variations, which allows to test for gas compression or ionization by, e.g., mergers or SNs.

### 4.4.3 *Mapping the kinematics*

The velocity of the gas within each spaxel can be derived from the measurement of the line centroid (and its associated error, see Liang 2007). The systemic velocity (or redshift) of the object can be measured by the $\sigma$-clipped mean of the spatially-resolved velocities. A velocity field can be constructed from the line centroids of each

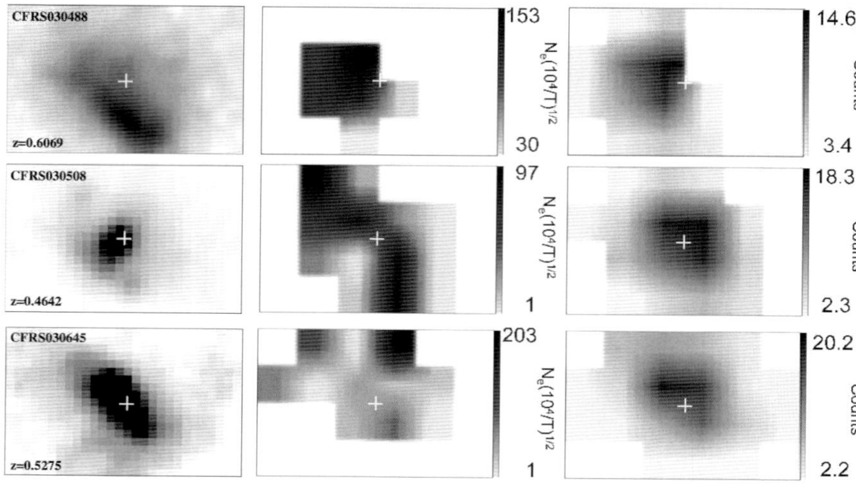


Fig. 4.19: Electron density maps derived from the [O II] doublet ratio maps of three distant galaxies. All maps have a scale of 0.1 arcsec/pix and show the same FoV (3 × 2 arcsec). From left to right: I band HST imaging, electron density map (5 × 5 interpolation), and [O II] counts distribution map (5 × 5 interpolation). Credit: Puech *et al.* (2006), A&A, 455, 131, reproduced with permission ©ESO.

spaxel, after subtraction of the systemic velocity. A sigma map can be constructed from the measured dispersion in each spaxel after correcting for instrumental dispersion, which is preferentially measured on a skyline (Davies *et al.*, 2011).

A Gaussian fit is the best choice to determine the centroid of the emission lines for spectral resolutions below 10000. A simple Gaussian profile can be used because the detected emission corresponds to the sum of all the flux illuminating one spaxel. Higher spectral resolutions may reveal a difference between the fitted Gaussian and the real emission, which results in a systematic uncertainty. Figure 4.8 shows an example of how even faint skylines may affect the fit. We recommend the user to develop customized line simulations. For example, the simulations developed by Yang *et al.* (2008) are illustrated in the Ex. *Simulations to derive the S/N on velocities and dispersion* and have been used to validate the S/N limit of 3 for the [O II] doublet. The simulations must account for the instrument and its setup, and this allows verifying the validity and limits of the observations.

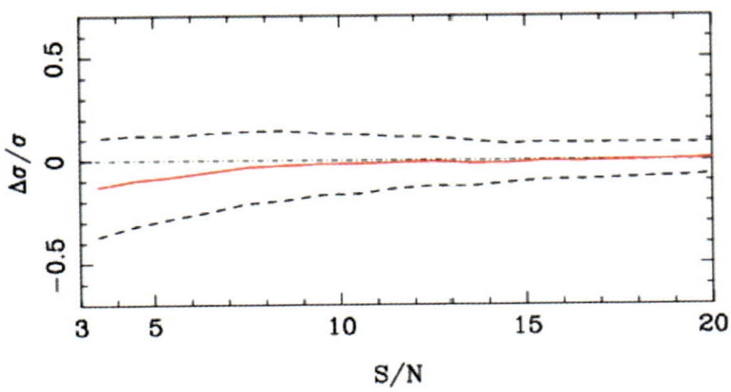


Fig. 4.20: Monte Carlo simulations showing how the error on $\sigma$ (red line, 68% probability in dashed lines) depends on the line S/N. Credit: Yang *et al.* (2008), A&A, 477, 789, reproduced with permission ©ESO.

**Example 4.6: *Simulations to set the S/N threshold of the velocity dispersion***

Yang *et al.* (2008) performed 50,000 Monte Carlo simulations of artificial [O II] emission line doublets with S/N ranging from 3 to 40, after assuming that they have been observed with FLAMES/GIRAFFE. An example of skyline removal is presented in Fig. 4.8 (left panel, as well as the performance on measuring velocity and dispersion, bottom-right panel). Figure 4.20 shows how the error on velocity dispersion depends on the S/N ($\Delta\sigma/\sigma$ = 20%, 15%, and 5% at S/N = 3, 10, and 15, respectively).

## 4.5 Kinematics of distant galaxies

### 4.5.1 *A fundamental reference: kinematics of nearby galaxies*

From 1950 to 1990, astronomers established kinematic maps from radio ($H_I$) and optical ($H\alpha$) observations which demonstrated that a majority of local spiral galaxies can be described as rotating disks[7]

[7]Rotation can be clearly identified in a velocity field thanks to isovelocity curves which describe a so-called spider diagram (see e.g., Rogstad *et al.*, 1974; Epinat *et al.*, 2010).

(Plummer, 1911), with the main kinematic gradient well aligned with respect to the principal morphological axis. For a large fraction of them, the rotation curve (i.e., the circular velocity as a function of radius to the galaxy center) follows a universal law described, e.g., by Salucci *et al.* (2007). The broad plateau observed at large radii in the rotation curves of local galaxies is still considered today as one of the most robust evidence for the existence of dark matter. Non-circular motions triggered by instabilities (including bars or warps) are present even in simple regular galaxies such as lenticular to late type spirals. 3D spectroscopy thus becomes necessary to provide a full description of these galaxies as it provides spatially-resolved kinematics, which in turn constitutes the most accurate method for deriving rotation curves, disk-to-bulge dynamical decompositions, or dynamical masses. Note that observations of emission and absorption lines at optical, millimetric, and radio wavelengths do not sample identical regions of the galaxies, although optical and radio observations generally provide similar results within the optical disk.

The fraction of peculiar galaxies increases sharply at higher redshift (see e.g., Guzmán *et al.*, 1997; Brinchmann *et al.*, 1998; Delgado-Serrano *et al.*, 2010). Since most of them are found to be out of equilibrium, it is impossible to assess whether bulk motions in these galaxies are dominated by rotation. However, local rotating disks remain a fundamental reference that should be used to characterize their kinematics and dynamics, since they represent the equilibrium state towards which such galaxies are expected to relax.

#### 4.5.1.1 *Modeling the velocity fields of rotating disks*

Analyzing a velocity field consists in finding the main kinematic parameters of the object. These are: the kinematic principal axis and inclination, the dynamical center, and the circular velocity as a function of radius, i.e., the rotation curve. In an ideal rotating thin disk, the bulk motions depict circular orbits in the galaxy plane, projected on-sky as ellipses which orientation depends on the disk inclination and kinematic P.A. The observed line-of-sight radial velocity $v_{obs}$ extracted along such ellipses can be described as (see, e.g., Krajnović

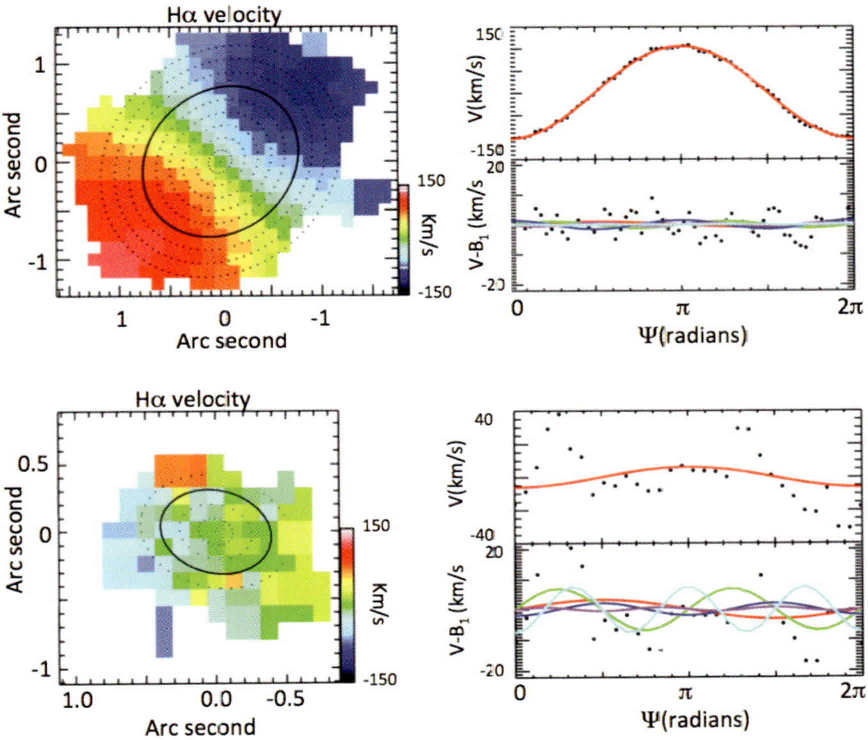


Fig. 4.21: *Top-left*: Velocity field based on a toy disk model with "kinemetric" ellipses overlaid (with $v_r = v_z = 0$ in Eq. 4.3). *Bottom-left*: Same but for an observed merger (IRAS 15206+3342) redshifted at $z \sim 2$. *Left panels*: Kinemetric expansion as a function of angle along the solid ellipse. On top is shown the measured velocities (black dots) and the B1 coefficient (red line) which fits the disk but not the merger; on the bottom are shown the residuals from this fit (black dots) and the higher order coefficients measured as a function of: (A1, red; A2, green; B2, blue; A3, magenta; and B3, cyan (from Shapiro *et al.* (2008), ©AAS, reproduced with permission).

*et al.*, 2006, and top-left panel of Fig. 4.21):

$$v_{obs}(r, \psi) = v_{sys} + v_{rot}(r, \psi) \cos\psi \sin i + v_r(r, \psi) \sin\psi \sin i + v_z(r, \psi) \cos i, \qquad (4.3)$$

where $v_{sys}$ is the systemic velocity of the galaxy (i.e., corresponds to the galaxy redshift),[8] $v_{rot}$ is the circular velocity in the azimuthal

[8]the two are related through $z = \sqrt{\frac{1+v_{sys}/c}{1-v_{sys}/c}} - 1$, which is often approximated by $z = v_{sys}/c$ at small distances.

$\psi$ direction (i.e., the rotation curve), $v_r$ is the radial (or streaming) velocity in the disk plane, and $v_z$ traces vertical motions. The azimuthal angle $\psi$ is measured from the projected major axis in the plane of the galaxy, $r$ is the radius of a circular ring in that same plane (or the semi-major axis length of the ellipse once projected on sky), and $i$ is the disk inclination ($i = 0$ for a disk seen face-on).

In practice, observed velocity fields are fitted assuming an ideal rotation, i.e., with no radial ($v_r$) or out-of-the plane ($v_z$) motion. In the local Universe, Eq. 4.3 is widely used for fitting the ellipses, keeping the kinematic parameters free for each of them. This corresponds to the so-called tilted-ring model (Rogstad *et al.*, 1974). With limited spatial resolution and S/N, such an approach cannot be used and Eq. 4.3 is then fitted assuming no variation of the kinematic parameters with $r$. The velocity field resulting from the best rotation model is then subtracted from the observed one. The corresponding residual map is a powerful tool which can be used to interpret the contribution of each kinematic parameter being fitted in Eq. 4.3. In turn, these allow constraining the best rotation model (see Fig. 4.22 and Warner *et al.*, 1973).

### 4.5.1.2 *Kinematic analysis of galaxies: "kinemetry"*

Numerous methods and softwares have been developed to perform automated fitting of kinematic parameters in nearby galaxies, such as the tilted-ring model (see above). Kinemetry is a generalization of surface photometry decomposition applied to the

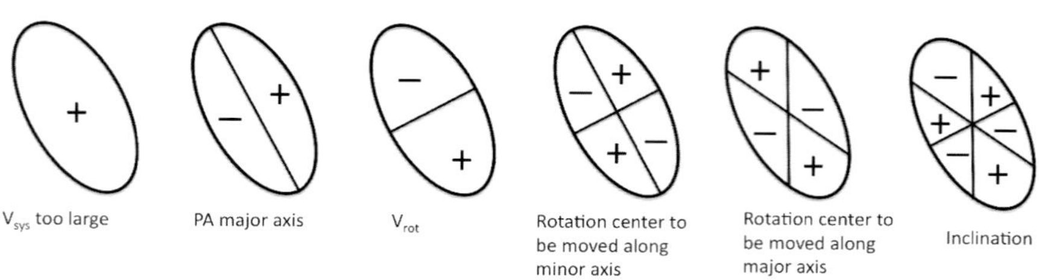


Fig. 4.22: Residuals after fitting the velocity field with an ideal rotation model (Eq. 4.3 with $v_r = v_z = 0$). Each panel indicates which parameter should be modified in order to improve the fit accuracy (inspired from Fig. 8 of Warner *et al.*, 1973).

high-order moments of the velocity distribution. It is derived from IFU observations (Krajnović *et al.*, 2006). This method uses maps of the different emission line moments (corresponding to the velocity field, velocity dispersion, and high-order moment maps) which it decomposes into harmonic expansions using a Fourier decomposition along ellipses. As illustrated in Fig. 4.21, the velocity field of a pure rotational disk presents many symmetries which are accurately represented by such a decomposition. The latter is therefore well-suited for distinguishing possible asymmetries in regular galactic rotations (see below). As such, it has also been tentatively used to analyze 3D observations of a few high redshift galaxies (see Sec. 4.5.3)

Equation 4.3 decomposes the observed velocity field into motions along elliptical rings (see top-left panel of Fig. 4.21). In a purely rotating disk, circular motions along such ellipses describe periodic variations (see top-right panel of Fig. 4.21). Variations of observed radial line-of-sight velocities along a given ellipse are described by a $\cos\psi$ term (see Eq. 4.3), with the velocity peaking along the galaxy major axis ($\psi = 0$, $\pi$, $2\pi$) and decreasing to zero towards the minor axis ($\psi = \pi/2$, $3\pi/2$, see also Shapiro *et al.*, 2008). Fourier analysis is particularly well-suited to characterize such periodic signals. Kinemetry identifies the ellipses best sampling the circular motions by minimizing the Fourier harmonic terms $A_n(r)\sin(n\psi)$ and $B_n(r)\cos(n\psi)$ which are sensitive to the ellipse geometry (see Eq. (6) of Krajnović *et al.*, 2006). The observed velocity field $v_{obs}(r, \psi)$ is then divided into a number of optimal elliptical rings yielding velocity profiles which can be described by a finite (and small) number of Fourier coefficients $A_n$ and $B_n$. In this decomposition, the $B_1(r)$ coefficient represents the circular velocity in each ring (i.e., the rotation curve, see Krajnović *et al.*, 2006), while the other coefficients (normalized to $B_1$) represent deviations from pure circular motions.

In extended local galaxies with very high S/N observations, these coefficients have been shown to accurately identify asymmetric components such as bars/radial inflow, lopsidedness/warps, multiple components, and spiral substructure (Schoenmakers *et al.*, 1997; Wong *et al.*, 2004; Krajnović *et al.*, 2006). However, for distant galaxies the kinemetry analysis needs to be limited to the first order coefficients (see Shapiro *et al.*, 2008), because of the low S/N and the small number of resolved spatial elements (see Sec. 4.5.2).

### 4.5.2 *Kinematics of distant galaxies and beam smearing*

The Nyquist–Shannon sampling theorem indicates that the spatial PSF FWHM should be sampled by at least 2 spaxels. Even massive distant galaxies have small apparent sizes, with a half-light radius in the continuum decreasing from 1 to 0.35 arcsec from z = 0.6 to z = 2 (in the case of galaxies with stellar masses larger than $\log(M_{star}/M_{\odot}) = 10$, see Dahlen *et al.*, 2007). Fortunately, ionized gas is found to be more extended than the continuum (see Ex. *Size of galaxies vs. sampling*), which considerably helps 3D spectroscopy of distant galaxies. Beam smearing remains a considerable limitation due to the relatively large size of the seeing compared to that of galaxies.

#### 4.5.2.1 *Conditions to establish a rotation curve*

Determining the rotation curves of nearby spiral galaxies is an easy task because they are often sampled with thousands of resolution elements for seeing-limited observations. The beam smearing effect can be defined as the ratio between the optical (Holmberg) radius of a galaxy to the half-power beam width, which in optical can be approximated to the observed PSF FWHM. A rotation curve may be accurately determined from a velocity field if this ratio is greater than or equal to 7 (Bosma 1978; see also Fig. 4.24). It is thus generally impossible to generate a rotation curve for most distant and faint galaxies which observed sizes range from 1 to 3 arcsec, as illustrated in Fig. 4.23. Adaptive optics can improve the situation for bright distant galaxies, but at the cost of missing out on the outer faint regions which have a too low S/N due to the small spaxel size. Clearly, telescopes with diameters larger than 8 m and equipped with adaptive optics will be necessary to provide rotation curves of representative samples of distant galaxies (Puech *et al.*, 2010b).

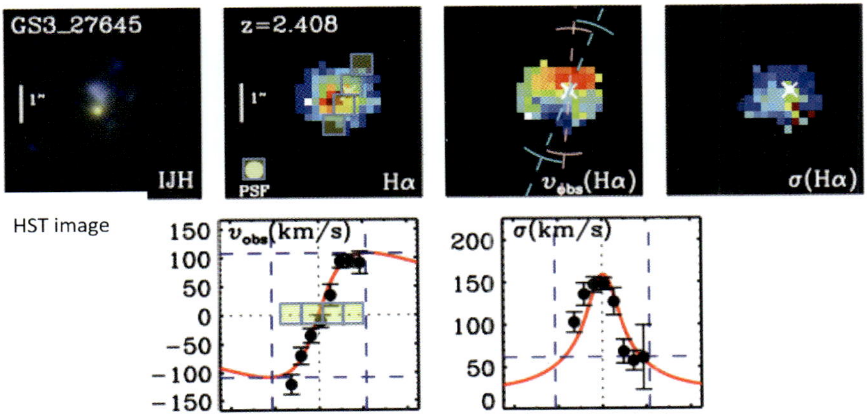


Fig. 4.23: Example of a distant galaxy at z = 2.4 observed with VLT/KMOS (from Wisnioski *et al.* (2015), ©AAS, reproduced by permission). In the upper panels from left to right, the HST color image, the Hα flux, velocity field, and dispersion maps. The ionized gas is sampled by less than four resolution elements along the kinematical P.A. (see the superimposed squares in the middle-left panel), though it would require 14 to sample a rotation curve (Bosma, 1978). Bottom panels show $v_{rot}$ and $\sigma$ profile along the P.A.

#### 4.5.2.2 *Impact of beam smearing on the rotation curve of distant galaxies*

Verifying the number of resolved elements is strongly recommended before performing any velocity field analysis. Estimates of maximal or flat velocities[9] in distant galaxies are necessary to assess their mass or the evolution of the Tully Fisher relation (see 5.3.1). However, beam smearing triggers a smoothing of the peak region in the velocity field as shown in Fig. 4.24, leading to an underestimation of the maximal velocity. The latter can be estimated after correcting for beam smearing. First, it requires ensuring that the galaxy is indeed a regular disk without significant residuals of past or on-going merging (see Sec. 4.5.3). Second, high resolution (space-based) imagery at rest-frame wavelengths is required to characterize old stars (i.e., at $\lambda \gg 0.4\ \mu$m) as the latter are assumed to follow the disk rotation and

[9] Flat velocities are assumed to be reached at large radii, beyond the optical radius, while maximal velocities are often measured at 2.2 × the disk scale length and can be affected by bulge motions (see definitions in Pizagno *et al.*, 2007 and in Epinat *et al.*, 2010).

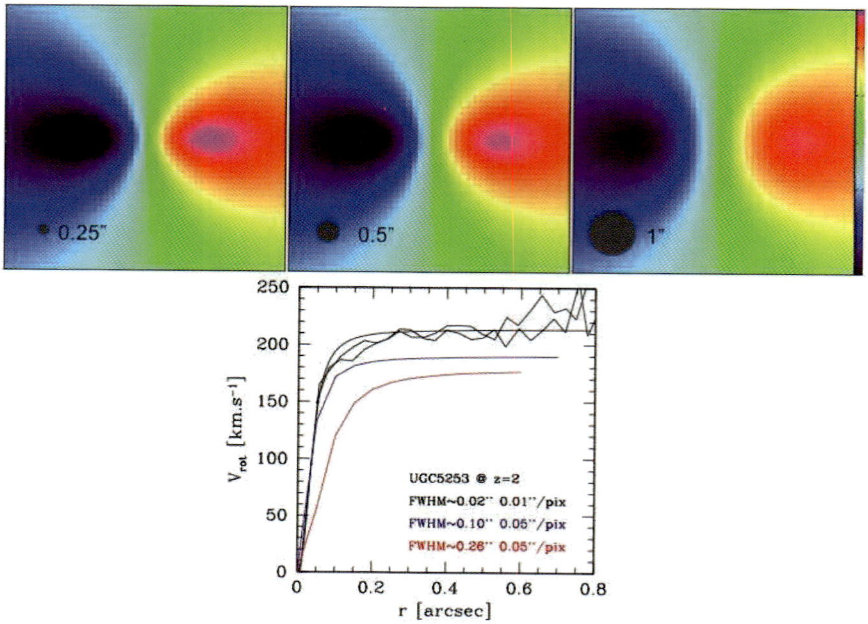


Fig. 4.24: *Top*: Effects of the beam smearing on a simulated velocity field and rotation curve. From left to right, the seeing (represented by a dark disk on the three images) increases from 0.25 to 1 arcsec. The pixel size is 0.125 arcsec. The disk scale length is set to 5 kpc (observed at $z = 1.7$), the inclination is 45° and the maximum velocity in the plane of the disk is 200 km s$^{-1}$ (from Fig. 4 of Epinat *et al.* (2010), MNRAS, 401, 2113, reproduced with permission of Oxford University Press on behalf of the Royal Astronomical Society). *Bottom*: Rotation curve of UGC5253 redshifted at $z = 2$. The original rotation curve on each side of the galaxy is shown in black, in the case of a perfect sampling, along with an arctan fit. The other curves show how $V_{\max}$ (here equals to $V_{flat}$) is underestimated when degrading the resolution to $\sim 0.1$ arcsec (MOAO FWHM case, blue curve) and then to 0.26 arcsec (GLAO FWHM case, red curve), respectively (adapted from Fig. 12 of Puech *et al.*, 2010b).

provide estimates of the mass center, P.A., inclination as well as disk scale length. Third, simulations are mandatory to evaluate the effect of blurring of the velocity peak due to the beam smearing effect (see Fig. 4.24, bottom panel). The correct velocity peak ($V_{max}$) can then be recovered (see Ex. *How to estimate* $V_{max}$ *and* $V_{flat}$ *for distant galaxies?*). To note however that this method can be detrimentally affected by the presence of a bulge (see, e.g., Fig. 1 of Puech *et al.*, 2008).

**Example 4.7: *How to estimate* $V_{\max}$ *and* $V_{\text{flat}}$ *for distant galaxies?***

In this example, one has obtained IFU observations of a distant rotating disk galaxy. Given the low number of resolution elements available for distant galaxies observed with IFUs, it is not possible to accurately determine their rotation curves. These elements are however sufficient to determine the maximal velocity (Bosma, 1978). First, the mass center, P.A. and the inclination (i) have to be determined from imagery. Second, one can estimate $\Delta V_{obs} = (V_M - V_m)/(2\sin(i))$, where $V_M$ and $V_m$ are respectively the maximal and minimal velocities determined from the IFU observation along the P.A. However, $\Delta V_{obs}$ may not account for all of the velocity amplitude because of beam smearing (see Fig. 4.24). Notice that the latter may also be affected by low S/N spaxels over the galaxy outskirts. To recover a better approximation of $V_{\max}$ one needs to simulate an artificial data cube assuming the observation of an ideal rotating disk with V $= V_0 \times (2/\pi) \times arctan(r/rt)$ (e.g., Courteau (1997)) at the observed P.A. and i The model should account for the observational seeing, the instrument properties, and the apparent size of the galaxy. This apparent size can be defined as the half-light radius, assumed to be 1.678 $\times$ the disk scale-length, or the radius at which the velocity is maximal ($V_{\max}$ at 2.2 $\times$ the disk scale-length). A grid of $V_{\max}$ is then constructed and optimized so as to match the $\Delta V_{\text{obs}}$ value and if possible, the dispersion map peak. This method was first used by Flores *et al.* (2006) then refined by Yang *et al.* (2008) and Puech *et al.* (2008), leading to $V_{\max}$ values $\sim$ 25% larger than $\Delta V_{obs}$ for z $\sim$ 0.65 IMAGES galaxies. In case of late type spirals, the bulge is too small to have a significant contribution to the maximal value of the dispersion map, and hence $V_{flat} = V_{\max}$. While preparing such a model, it is recommended to verify the galaxy type (e.g., on the basis of the B/T ratio) from space-based imagery at red wavelength rest-frame.

#### 4.5.2.3 *Impact of beam smearing on the σ maps of distant galaxies*

The significant distortions induced by beam smearing on distant galaxy kinematics complicates the interpretation of their kinematics. In contrast, when investigating $\sigma$ maps, beam smearing actually produces a remarkable signature which is very helpful to identify a pure rotation (see Sec. 4.5.3). Figure 4.25 shows a disk rotation curve which exhibits a considerable gradient near the galaxy center which is mostly included within a single resolving element. This produces a strong peak in the central region of the dispersion map. The peak shape and intensity depends on the inclination, velocity gradient slope, and velocity gradient of the rotation curve. It provides a unique and essential signature to test the rotational nature of a distant galaxy velocity field (see Sec. 4.5.3).

**Example 4.8: *Size of galaxies vs sampling***

Table 1 of Puech *et al.* (2010a) shows that for $z = 0.65$ galaxies, the average [O II] radius is 1.5 ±0.4 arcsec, i.e., larger than their optical half-light radii. Their diameters are then sampled by less than four spatial resolution elements[10] with a 0.8 arcsec seeing in the visible. At $z = 2$, H$\alpha$ emissions are, on average, resolved by the same number of elements (see Fig. 4.23): the much smaller galaxy sizes are compensated by the better image quality (0.4–0.5 arcsec FWHM) in the NIR.

For this feature to be practical in identifying pure rotation in distant galaxies, it is thus essential to accurately establish the location of the dispersion peak (see Ex. *Precision on the σ peak location*) as well as that of the dynamical center which is defined as the center between the two velocity extrema. These locations are then compared to the morphological center, after accounting for the relative astrometry (see Sec. 4.2.6).

[10]Note that in the case of the FLAMES/GIRAFFE IFU, these calculations produce even worse results because the spaxel size of 0.52 arcsec further increases the FWHM to 1.04 arcsec.

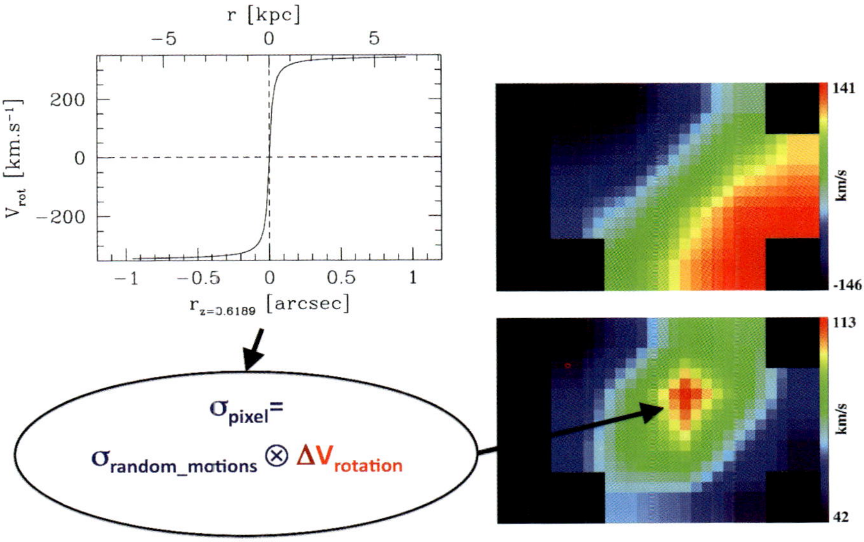


Fig. 4.25: Illustration of how the large spaxel size of an IFU (here that of FLAMES/GIRAFFE, shown as red vertical lines in the upper-left panel) captures the large scale motions (here a rotation) at the galaxy center where the velocity abruptly varies from the negative to the positive maximum velocity plateau. Exact formulation of the convolution with random motions can be found in Epinat *et al.* (2010). Because the velocity gradient easily dominates the actual $\sigma$ random motions (except if the galaxy is seen face-on), this naturally produces a peak at the center of the dispersion map.

### Example 4.9: *Precision on the $\sigma$ peak location*

Numerical simulations are necessary to estimate the precision on the $\sigma$ peak location. Let us consider a local rotating disk galaxy (UGC 5253) and redshift it to $z = 0.7$, assuming that it will be observed by the FLAMES/GIRAFFE IFU (spaxel = 0.52 arcsec). Figure 4.26 shows that the simulated peak is well recovered at the mass center (see crosses on the maps), even after moving or rotating the IFU grid. It leads to a precision in the detection of the peak equal to a half spaxel (0.26 arcsec) for S/N > 4. Less than half of this value may be expected when using NIR 3D spectroscopy, given the expected increase of FWHM seeing.

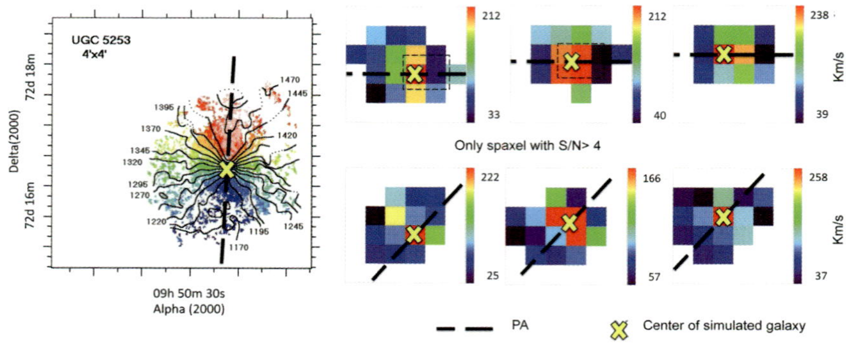


Fig. 4.26: Performances of the FLAMES/GIRAFFE IFU for determining the peak position of the dispersion map. The galaxy UGC 5253 (see the original velocity field in the left panel from Garrido *et al.*, 2002a) has been redshifted and rescaled to $z = 0.7$. In the three top-right panels, the simulated dispersion maps are shown, as well as the mass or velocity centers (see cross). From left to right, the simulated observation of the redshifted galaxy, same but after a shift of the IFU by half a spaxel, and then by one spaxel. The bottom row shows the same, but after a rotation of the IFU by 45°.

### 4.5.3 *Classifications based on pure kinematics*

Kinematics of distant galaxies may appear complex to interpret because of the limitations due to the current spatial resolution and S/N obtained from ground-based, 8m class telescopes.

For distant galaxies, 3D spectroscopy can only identify large scale and large amplitude deviations that occur in disturbed, unvirialized systems, such as mergers, while small perturbations caused by lopsidedness, warps, and bars are hardly or not detected (see, e.g., Shapiro *et al.*, 2008). A simple way to study kinematics is to assume that a galaxy is a rotating disk, and then investigate the model deviations to verify the validity of this assumption.

There are currently two methods that have been proposed to classify the spatially-resolved kinematics of distant galaxies: one based on the "kinemetry" (Shapiro *et al.*, 2008), and the other one using the "$\sigma$ centering" method (Flores *et al.*, 2006).

#### 4.5.3.1 *Classification based on kinemetry*

Figure 4.21 summarizes the principle of such a classification. From the kinemetry decomposition, the amplitude of each Fourier harmonic order is given by $k_n = \sqrt{A_n^2 + B_n^2}$. The information about kinematic asymmetries is contained in higher order terms than $k_1$ (see Sec. 4.5.1.2). In distant galaxies, the number of spatially resolved elements and their associated S/N is limited so that in practice the average amplitude of velocity asymmetries has been defined using a limited number of terms:

$$k_{avg} = (k_2 + k_3 + k_4 + k_5)/4. \tag{4.4}$$

If necessary, this can be applied to both the velocity field and dispersion map but with an even smaller number of terms. Then, for both maps, one can derive the ratio of the average velocity asymmetries amplitude to the circular motions as measured by $B_1$ (Shapiro *et al.*, 2008):

$$v_{asym} = k_{avg,v}/B_1, \tag{4.5}$$

$$\sigma_{asym} = k_{avg,\sigma}/B_1. \tag{4.6}$$

This method solely relies on velocity asymmetries and can be used to distinguish unvirialized systems or mergers from any type of spiral galaxy, including those having a prominent bulge. This is justified because both disk and bulge have symmetric signatures in a regular galaxy.

Figure 4.27 presents the results of such a classification. Calibrating the limit between asymmetric (supposedly mergers-related) and rotating disks requires representative simulations based on artificially redshifted nearby disks, mergers as well as ideal disks. One can then define the parameter $K_{asym} = \sqrt{v_{asym}^2 + \sigma_{asym}^2}$, for which Shapiro *et al.* (2008) established $K_{asym} = 0.5$ as an optimal empirical limit (see Fig. 4.27).

The classification based on kinemetry has the advantage of relying on a robust mathematical model which successfully distinguishes and classifies perturbations in local and regular galaxies (e.g., Krajnović *et al.* (2006) for early type galaxies). However, it requires a finer adjustment of the limit distinguishing mergers from disks. The latter may strongly depend on the model and associated template library used to calibrate the statistical limit between mergers and rotating disks. For example, Fig. 4.27 (right panels) evidences that major mergers cannot be distinguished from turbulent disks (Perret, 2014). Epinat *et al.* (2009, 2010) also noticed a degeneracy between velocity

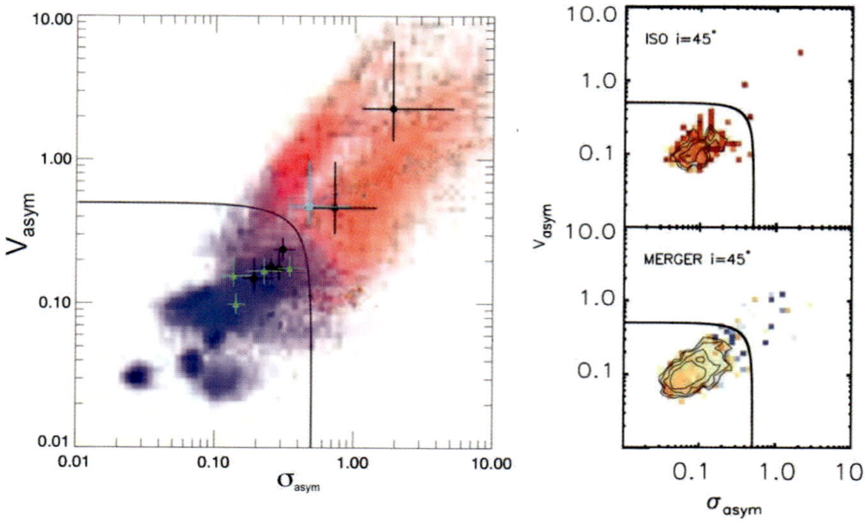


Fig. 4.27: *Left panel*: $V_{asym}$ vs. $\sigma_{asym}$ diagnostic diagram (see Shapiro *et al.* (2008) for details). A large fraction of disk templates (blue points) are in a region close to (0,0), while the merger templates (red points) have $V_{asym}$ and $\sigma_{asym}$ larger than 0.5. The line shows $K_{asym} = 0.5$ found by Shapiro *et al.* (2008) to distinguish disks from merger systems. *Right Panels*: The same figure constructed by Perret (2014) reproduced with kind permission, for both turbulent disks (ISO, top) and mergers (bottom) using RAMSES simulations and $i = 45°$. It evidences that gas-rich mergers cannot be easily distinguished from turbulent disks (see, e.g.,Bournaud *et al.*, 2007).

field and dispersion map asymmetries (see Fig. 4.27). Perhaps the main limitation of the "kinemetry" method, when applied to galaxies at high redshift, has been revealed by Förster Schreiber *et al.* (2009), since they have been able to apply "kinemetry" on only 15 out of 62 SINS galaxies, corresponding to those having enough S/N and spatial resolution.[11]

#### 4.5.3.2 *Classification based on "σ centering"*

Figure 4.25 evidences that at low spatial resolution, rotating disks have a strong signature in their dispersion map, expressed as a central peak at the kinematical center. This has led Flores *et al.* (2006) to

[11] This may change when using IFU instruments on extremely large telescopes equipped with adaptive optics.

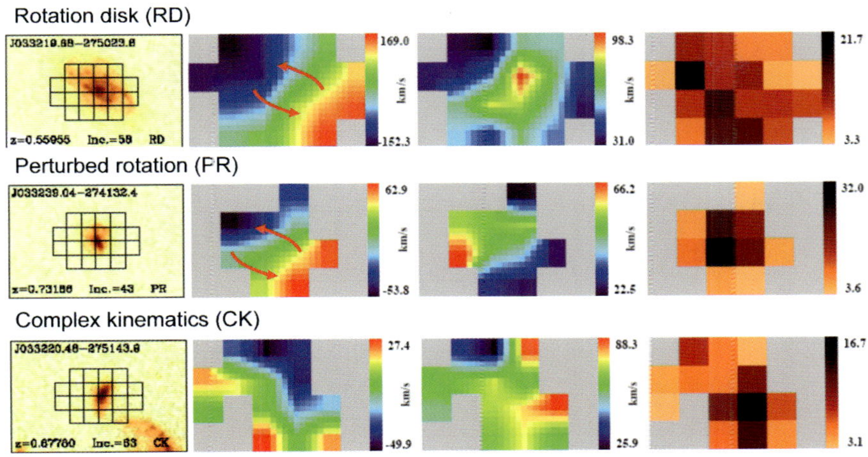


Fig. 4.28: Example of velocity fields for three kinds of distant galaxy kinematics (from top to bottom: RD, PR and CK). From left to right: the HST image with superposed IFU grid, the velocity field and the $\sigma$ map after a $5 \times 5$ lineal binning, and the S/N map. The red arrows point to the evidence for rotation in the velocity field panels. Credit: Yang *et al.* (2008), A&A, 477, 789, reproduced with permission ©ESO.

define a two-parameter classification, for which the first one is:

$$\Delta r = |c_\sigma - c_{VF}|, \tag{4.7}$$

which is the distance between the $\sigma$ peak ($c_\sigma$) and dynamical center ($c_{VF}$). Flores *et al.* (2006) and Yang *et al.* (2008) have been able to visually classify the kinematics of IMAGES galaxies using this parameter together with the alignment of both kinematic and optical P.A. It has lead to three classes of kinematics (see also Hung *et al.*, 2015) which are illustrated in Fig. 4.28: rotating disks (RC, with $\Delta$r $\sim 0$), perturbed rotations (PR, with $\Delta$r $> 0$ and signature of rotation in the velocity field) and finally complex kinematics for other, near-chaotic velocity fields (CK). For a perfect disk, the precision on $\Delta$r is that found on the peak of the emission line (see Section 4.5.2.3 and Fig. 4.26). The second parameter $\epsilon$ depends on the $\sigma$ peak intensity which results from the convolution of the velocity gradient (at the assumed rotating disk center) to the dispersion (due to random motions) (see Fig. 4.25). It leads to:

$$\epsilon = |\sigma_{obs} - \sigma_{mod}|/\sigma_{mod}, \tag{4.8}$$

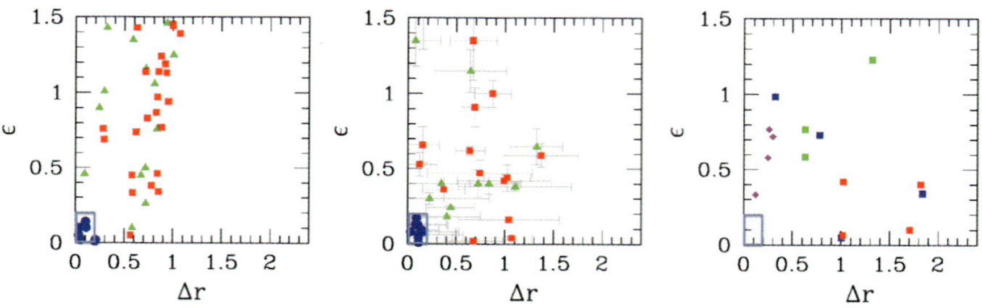


Fig. 4.29: Diagnostic diagrams of galaxy kinematics based on the "$\sigma$ centering" method and applied to three different samples for which the blue dots, green squares and red triangles represent the rotation disk (RD), perturbed rotation (PR), and complex kinematics (CK) galaxies, respectively. Notice that rotating disks lie in a quite compact region, defined by the errors on the kinematics of an ideal disk (see Yang *et al.* (2008) and the small box $\Delta$ r$\sim \epsilon \sim 0$). *Left*: Simulated rotating disks and mergers, for which orbits and pericenters have been defined from the grid of merger templates defined in Hammer *et al.* (2009). Here galaxies have been assumed to be seen at different view angles to test the robustness of the method. *Middle*: Same for galaxies at $z = 0.4{-}0.8$ observed by IMAGES (from Yang *et al.*, 2008). *Right*: Same applied to the SINFONI (Förster Schreiber *et al.*, 2006) and OSIRIS (Law *et al.*, 2009) observations of $z = 2{-}3$ galaxies. Blue dots and red squares represent respectively rotating disks and dispersion-dominated or mergers, according to the classification of Förster Schreiber *et al.* (2006), while magenta diamonds correspond to the single category of perturbed galaxies defined by Law *et al.* (2009).

where $\sigma_{obs}$ and $\sigma_{mod}$ are the observed and modeled peaks, respectively. Random motions can be modeled from the $\sigma$ map after the proper removal of the peak center (see e.g., Davies *et al.*, 2011). Modeling the velocity field is required to derive $\sigma_{mod}$ as done in Fig. 4.25. Error estimates on this parameter can be derived from Monte-Carlo simulations (see e.g., Yang *et al.*, 2008).

Figure 4.29 shows the results of the "$\sigma$ centering" method, indicating its ability to distinguish simulated disks from mergers. The latter also has the advantage of simplicity since the spatial centering is easy to determine by visual examination of both velocity field and dispersion maps. The centering is also valid for turbulent disks as most of them show a peak in the center of their $\sigma$ maps (Perret, 2014).

The "$\sigma$ centering" method may however fail on nearly face-on disks ($i < 10$–$15°$) for which the $\sigma$ component due to beam smearing is diluted (see, e.g., Epinat *et al.*, 2010). Galaxies with prominent, massive bulge may also be associated with nonnull values of $\epsilon$ due to the gravitational impact of the bulge on the rotation curve (see, e.g., the M31 rotation curve from Chemin *et al.*, 2009). It is recommended to verify the inclination and B/T ratio from high spatial resolution imagery prior to classifying the galaxy kinematics and to account for them when modeling the velocity field and its associated $\sigma$ map (see Fig. 1 of Puech *et al.*, 2008). Perhaps the main limitation of the "$\sigma$ centering" method is caused by its reduced accuracy for classifying galaxies dominated by their velocity dispersion (Epinat *et al.*, 2009 and see Bender, 1990 and Krajnović *et al.*, 2006 for studies of such kind of kinematics).

Several other methods have been proposed in the literature, generally based on variations of the above "kinemetry" and "$\sigma$ centering" methods (see e.g., Bellocchi *et al.*, 2012; Epinat *et al.*, 2009; and Hung *et al.*, 2015). The next section evidences how the classification based on pure kinematics may lead to important misinterpretations.

### 4.5.4 *The kinematic classification*

Kinematics provides the only way to infer whether a galaxy is dominated by a relaxed rotation or by strong gravitational perturbations often linked with major merger processes. Kinematics are also required to estimate the mass and establish the fundamental relationship between mass and velocity (see Sec. 5.3.1).

Even in the distant Universe, a rotating disk is assumed to be virialized and it is then easy to recognize because it is a simple dynamical system with a large number of remarkable geometrical signatures (see Sec. 4.5.1). When it is old enough, such a system should show many symmetries between its old star content and kinematics.[12] Phenomena such as mergers or galaxy interactions at various stages may produce strong misalignments between the gas and

[12] If a disk galaxy forms from a uniform sphere of gas in solid body rotation, the resulting disk shows a profile close to an exponential, although it is slightly more concentrated (Mestel, 1963; Gunn, 1982; Dutton, 2009). If it is formed through hierarchical scenario, then dissipationless stars are likely aggregating in the center, while viscous, cold gas is more affected by tides (Hopkins *et al.*, 2008). Thus, stars virialize within a dynamical time and are expected to better trace the potential of a galaxy at equilibrium. Up to our

old stars. This evidences the limitation of classifications based on pure kinematics. To avoid confusion and misinterpretation, a kinematic classification has to account for all the expected properties of a rotation-dominated virialized system.

In fact, early 3D spectroscopic surveys may have lacked accompanying high spatial resolution[13] imagery. It is methodologically incorrect to recover P.A. and inclination from the kinemetry by minimizing the asymmetry in the velocity field: such attempts are too limited by the small number of spatial resolution elements, and forcing the kinematics to be symmetric is a nondemonstrated assumption on the galaxy nature that may lead to wrong conclusions (Krajnović *et al.*, 2006).

It is then mandatory to know how the old stars are distributed in order to measure the inclination, the P.A., and the mass center of the galaxy (Hammer *et al.*, 2009; Hung *et al.*, 2015). This requires high spatial resolution imagery (see Sec. 2.4.2). This is because (1) IFUs only sample the ionized gas and, (2) 3D spectroscopy of distant galaxies is limited by S/N and the number of spatial resolution elements (see Sec. 4.5.2). Lacking this, properties derived exclusively from kinematics ("kinemetry" or "$\sigma$ centering" methods) can generally only yield an upper limit to the number of rotating disks in a sample (Hung *et al.*, 2015).

Figure 4.30 proposes a refinement of this, with a decision tree for distinguishing rotating disks from nonvirialized systems. The latter are supposed to be mostly major mergers, because conversely to supplementary minor processes, they impact the gas motions on large spatial scales (5–7 kpc). Mergers are however not necessarily responsible for all observed unvirialized kinematics, e.g., relatively low mass galaxies may have also experienced efficient outflows leading to a large value of $\Delta$PA (generally close to $\sim 90°$). Note that using the

knowledge, there is no counterexample of a virialized galaxy showing no stars in their center and with stellar population only in the outskirts.

[13]The center and P.A. can be retrieved from ground-based imagery, at the expense of much larger uncertainties (typically 0.3 compared to 0.05 arcsec), while the inclination is almost not recoverable, especially for well inclined objects.

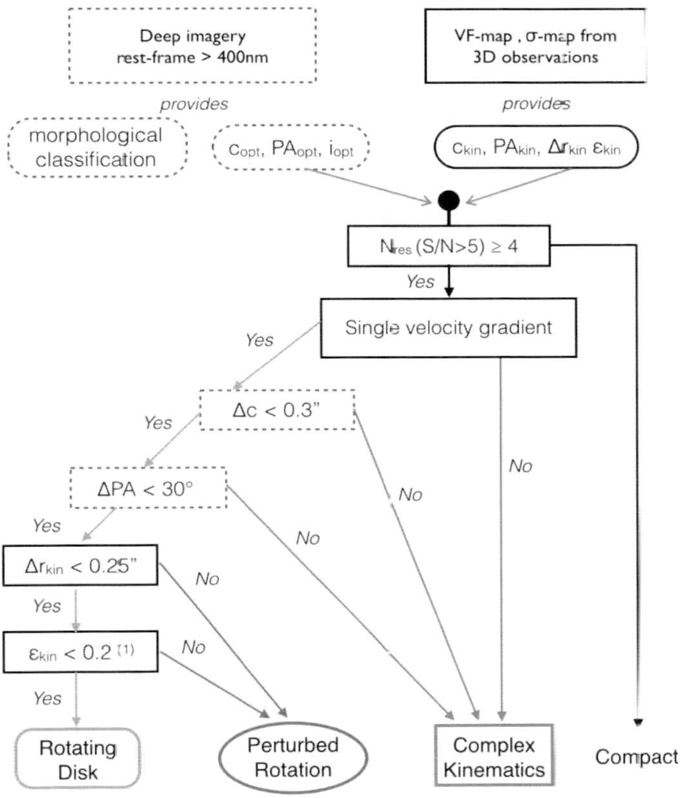


Fig. 4.30: Full kinematics decision tree for classifying the nature of distant galaxies. It refers directly to the morphological decision tree (Delgado-Serrano *et al.*, 2010, and see Fig. 2.10) and thoroughly compares morphological parameters associated to old stars and then to mass, to those derived from kinematics of the ionized gas. For example, the test on Δc (= $abs(c_{opt} - c_{VF})$) is not redundant with the Δr test of the "σ centering" method. The limiting value of 20° for ΔPA is set from GHASP (Garrido *et al.*, 2002b; Epinat *et al.*, 2009) and CALIFA (Garcia-Lorenzo *et al.*, 2015) and references therein. *Note 1*: Δc = abs($c_{opt}$-$c_{VF}$), and one has to account for the emission line peak location precision convolved with the astrometric precision when relating kinematics and imagery; *Note 2*: the exact value delineating disks from mergers has to be refined and could be lowered in the near future;
*Note 3*: the test on $\epsilon$ can be affected by the presence of a large bulge, which has to be verified from estimations of B/T from imagery; in practice B/T ≤ 0.2 amounts to a negligible impact.

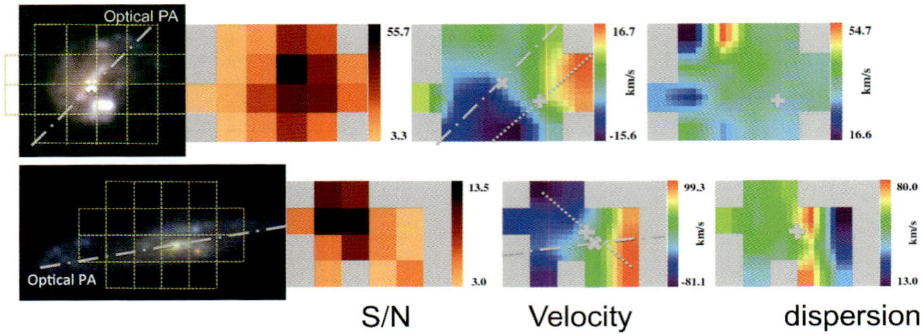


Fig. 4.31: IMAGES galaxies J033239.72-275154.7 (top) at $z = 0.4151$ and J033213.06-274204.8 (bottom) at $z = 0.4215$, with superposed IFU. From left to right: rgb colors using $v, i$, and $z$ filter from HST/ACS, S/N, velocity field, and velocity dispersion maps. $PA_{opt}$ and $PA_{VF}$ are indicated by dot-dash and dot lines, respectively. Cross and plus signs mark the optical center ($c_{opt}$) and the dynamical center ($c_{VF}$), respectively.

kinematic decision tree with "kinemetry" alone would require significant progress in the understanding of the limitation which apply to $v_{asym}$ and $\sigma_{asym}$ (see Sec. 4.5.3).

Figure 4.31 illustrates the efficiency of the decision tree in classifying two galaxies from IMAGES (see also other examples in Sec. 5.2). J033239.72-275154.7 exhibits a young star forming bar and its kinematic center ($c_{VF}$) is 4.5 kpc off from the old star center ($c_{opt}$). $\Delta c$ =0.8 arcsec is larger than the 0.35 arcsec uncertainty, the latter value resulting from the quadratic convolution of the astrometry (0.25 arcsec) and the line peak location (0.25 arcsec) uncertainties, respectively. Interestingly, the galaxy $PA_{VF}$ is almost aligned with $PA_{opt}$, and without high spatial resolution imagery, the system could have been confusedly classified as a rotating disk. However, if using the "$\sigma$ centering" method, it would not have passed the $\Delta$r test (compare the cross and plus sign locations shown in the top panels of Fig. 4.31). J033239.72-275154.7 has been fully analyzed and modeled as a 3:1, near completion merger by Peirani *et al.* (2009), which satisfactorily explains its kinematics as well as its barred morphology. In Fig. 4.31 (bottom panel), J033213.06-274204.8 resembles an almost edge-on spiral galaxy. Kinematics is essential here because after examining the S/N map, the most illuminated spaxels are well outside the edge-on disk. Moreover, $PA_{VF}$ is at $\sim 50°$ from the $PA_{opt}$, which definitively discards a rotating spiral classification for this object. Both examples illustrate how kinematics and imagery are insufficient when they are separately used, while they turn out to be a very powerful technique when they are combined into a morpho-kinematics classification.

### 4.5.5 *The morpho-kinematic classification*

Neichel *et al.* (2008) found an excellent agreement between the above kinematic classification and that based on pure morphological analysis (see Sec. 2.4). The similarities between the large scale motions of the ionized gas and the spatial distribution of stars further supports the validity of the kinematic method when applied comprehensively through a decision tree such as in Fig. 4.30. It is indeed expected that strong events such as major mergers have strong impact and then affect more than a single property of the assumed pre-existing disks. To account for this, we define three different morpho-kinematical classes:

- rotating spiral disks are galaxies showing a rotating velocity field with a dispersion peak at the dynamical center (see, e.g., Flores *et al.*, 2006), and with the morphology of a spiral galaxy;
- nonvirialized systems are galaxies that have a strong discrepancy from a rotational velocity field and whose morphology is peculiar;
- and semivirialized systems that possess either a rotational velocity field and a peculiar morphology or a velocity field discrepant from a rotation but a spiral morphology.

This degree of virialization scheme has been presented by Hammer *et al.* (2009) and subsequently used to model the properties of IMAGES galaxies.[14] For example, J033239.72-275154.7 and J033213.06-274204.8 are classified nonvirialized and semivirialized, respectively, since both their kinematics indicate discrepancy from

[14]Note that Hammer *et al.* (2009) used the term "relaxed" instead of "virialized", which is preferred here since a relaxed system needs to reach a state of energy equipartition that requires a much longer time compared to virialization.

rotation from their kinematics while the first has peculiar morphology, the second looks like an edge on spiral.

### 4.6 Conclusion

3D spectroscopy is crucial to understand how galaxies have been formed and how they have acquired their angular momentum. When coupled with adaptive optics, it can even detect black holes or reveal internal motions such as nuclear outflows. Such results are presently limited to few of the brightest distant galaxies, while elsewhere 3D spectroscopy suffers from low spatial resolution and low S/N. It is however the only way to interpret the large scale motions of distant galaxies. Rotating disks at high redshifts are particularly easy to identify because the beam smearing effects unavoidably leads to a $\sigma$ peak at the dynamical center of an inclined galaxy.

Methods experimented in the local Universe provide guidance for analyzing velocity fields of distant galaxies, though one has to account for the smaller number of available spatial resolution elements. Accompanying high spatial resolution imagery is essential to derive physical quantities such as $V_{\max}$ or simply to distinguish a rotating disk from a merger. Here, we propose a simple methodology applicable from a kinematics decision tree, which we hope may improve interpretations and analyses of the numerous observations in this rapidly growing field. Beyond this, we propose a full comparison between morphological and kinematic properties of a galaxy, to establish the degree of virialization for a system, a method hereafter called the morpho-kinematic classification.

### Bibliography

Afanasiev V., Vlasiuk V., Dodonov S., Silchenko O. (1990). Multi-pupil fiber spectrograph of the 6-meter telescope, *The Special Astrophysical Observatory of the Russian Academy of Sciences* **54**, pp. 1–12.

Allington-Smith J. (2006). Basic principles of integral field spectroscopy, *New Astronomy Review* **50**, pp. 244–251.

Bacon R., Adam G., Baranne A. *et al.* (1995). 3D spectrography at high spatial resolution. I. Concept and realization of the integral field spectrograph TIGER, *A&AS* **113**, pp. 347–357.

Bellocchi E., Arribas S., Colina L. (2012). Studying the kinematic asymmetries of disks and post-coalescence mergers using a new "kinemetry" criterion, *A&A* **542**, A54.

Bender R. (1990). Unraveling the kinematics of early-type galaxies — Presentation of a new method and its application to NGC4621, *A&A* **229**, pp. 441–451.

Bosma A. (1978). *The distribution and kinematics of neutral hydrogen in spiral galaxies of various morphological types*, Ph.D. thesis, Groningen University.

Bournaud F., Jog C. J., Combes F. (2007). Multiple minor mergers: formation of elliptical galaxies and constraints for the growth of spiral disks, *A&A* **476**, pp. 1179–1190.

Bowen I. S. (1938). The image-slicer a device for reducing loss of light at slit of stellar spectrograph., *ApJ* **88**, p. 113.

Brinchmann J., Abraham R., Schade D. *et al.* (1998). Hubble Space Telescope Imaging of the CFRS and LDSS Redshift Surveys. I. Morphological Properties, *ApJ* **499**, pp. 112–133.

Cappellari M., Copin Y. (2003). Adaptive spatial binning of integral-field spectroscopic data using Voronoi tessellations, *MNRAS* **342**, pp. 345–354.

Chemin L., Carignan C., Foster T. (2009). HI kinematics and dynamics of messier 31, *ApJ* **705**, pp. 1395–1415.

Clare S. (1997). *Functional Magnetic Resonance Imaging: Methods and Applications*, University of Nottingham.

Courteau S. (1997). Optical rotation curves and linewidths for Tully–Fisher applications, *AJ* **114**, p. 2402.

Courtes G. (1982). *IAU Colloq. 67: Instrumentation for Astronomy with Large Optical Telescopes, Astrophysics and Space Science Library*, Vol. 92 (ed.), Humphries, C. M., p. 123.

Dahlen T., Mobasher B., Dickinson M. *et al.* (2007). Evolution of the luminosity function, star formation rate, morphology, and size of star-forming galaxies selected at rest-frame 1500 and 2800 Å, *ApJ* **654**, pp. 172–185.

Davies R., Förster Schreiber N. M., Cresci G. *et al.* (2011). How well can we measure the intrinsic velocity dispersion of distant disk galaxies? *ApJ* **741**, p. 69.

Delgado-Serrano R., Hammer F., Yang Y. B. *et al.* (2010). How was the Hubble sequence 6 Gyr ago? *A&A* **509**, A78.

Dutton A. A. (2009). On the origin of exponential galaxy discs, *MNRAS* **396**, pp. 121–140.

Ealet A., Prieto E., Aumeunier M.-H. *et al.* (2008). An image slicer spectrograph for the snap mission, SPIE Newsroom

Epinat B., Amram P., Balkowski C. *et al.* (2010). Evidence for strong dynamical evolution in disc galaxies through the last 11 Gyr. GHASP VIII — a local reference sample of rotating disc galaxies for high-redshift studies, *MNRAS* **401**, pp. 2113–2147.

Epinat B., Contini T., Le Fèvre O. *et al.* (2009). Integral field spectroscopy with SINFONI of VVDS galaxies. I. Galaxy dynamics and mass assembly at 1.2 < z < 1.6, *A&A* **504**, pp. 789–805.

Flores H., Hammer F., Puech M. *et al.* (2006). 3D spectroscopy with VLT/GIRAFFE. I. The true Tully Fisher relationship at z ∼ 0.6, *A&A* **455**, pp. 107–118.

Förster Schreiber N. M., Genzel R., Bouché N. *et al.* (2009). The SINS survey: SINFONI integral field spectroscopy of $z \sim 2$ star-forming galaxies, *ApJ* **706**, pp. 1364–1428.

Förster Schreiber N. M., Genzel R., Lehnert M. D. *et al.* (2006). SINFONI Integral Field Spectroscopy of z ∼ 2 UV-selected Galaxies: Rotation Curves and Dynamical Evolution, *ApJ* **645**, pp. 1062–1075.

Förster Schreiber N. M., Genzel R., Newman S. F. *et al.* (2014). The Sins/zC-Sinf Survey of z ∼ 2 Galaxy Kinematics: Evidence for Powerful Active Galactic Nucleus-Driven Nuclear Outflows in Massive Star-Forming Galaxies, *ApJ* **787**, p. 38.

Gallagher J. S., Hunter D. A., Bushouse H. (1989). Star-formation rates and forbidden O II emission in blue galaxies, *AJ* **97**, pp. 700–707.

Garcia-Lorenzo B., Marquez I., Barrera-Ballesteros J. K. *et al.* (2015). Ionized gas kinematics of galaxies in the CALIFA survey I: Velocity fields, kinematic parameters of the dominant component, and presence of kinematically distinct gaseous systems, *A&A* **573**, A59, doi:10.1051/0004-6361/201423485.

Garrido O., Marcelin M., Amram P. *et al.* (2002). GHASP: An H$\alpha$ kinematic survey of spiral and irregular galaxies. I. Velocity fields and rotation curves of 23 galaxies, *A&A* **387**, pp. 821–829.

Gunn J. E. (1982). *Astrophysical Cosmology Proceedings*, (eds.) Brueck, H. A. and Coyne, G. V. and Longair, M. S., pp. 233–259.

Guzmán R., Gallego J., Koo D. C. *et al.* (1997). The nature of compact galaxies in the hubble deep field. II. Spectroscopic Properties and Implications for the Evolution of the Star Formation Rate Density of the Universe1, *ApJ* **489**, pp. 559–572.

Hammer F., Flores H., Lilly S. J. *et al.* (1997). Canada–France redshift survey. XIV. Spectral Properties of Field Galaxies up to z = 1, *ApJ* **481**, pp. 49–82.

Hammer F., Flores H., Puech M. *et al.* (2009). The Hubble sequence: Just a vestige of merger events? *A&A* **507**, pp. 1313–1326.

Hopkins P. F., Cox T. J., Hernquist L. (2008). Dissipation and the fundamental plane: observational tests, *ApJ* **689**, pp. 17–48.

Hung C.-L., Rich J. A., Yuan T. *et al.* (2015). Kinematic classifications of local interacting galaxies: implications for the merger/disk classifications at high-z, *Astrophys. J* **803**(62), doi:10.1088/0004-637X/803/2/62.

Krajnović D., Cappellari M., de Zeeuw P. T. *et al.* (2006). Kinemetry: a generalization of photometry to the higher moments of the line-of-sight velocity distribution, *MNRAS* **366**, pp. 787–802.

Law D. R., Steidel C. C., Erb D. K. *et al.* (2009). The Kiloparsec-scale Kinematics of High-redshift Star-forming Galaxies, *ApJ* **697**, pp. 2057–2082.

Liang Z.-P. (2007). *ISBI* in IEEE, pp. 988–991.

Mestel L. (1963). On the galactic law of rotation, *MNRAS* **126**, p. 553.

Neichel B., Hammer F., Puech M. *et al.* (2008). IMAGES. II. A surprisingly low fraction of undisturbed rotating spiral disks at z ~ 0.6 The morpho-kinematical relation 6 Gyr ago, *A&A* **484**, pp. 159–172.

Peirani S., Hammer F., Flores H. *et al.* (2009). A giant bar induced by a merger event at z = 0.4? *A&A* **496**, pp. 51–56.

Perret V. (2014). *Mergers of teenage galaxies: from simulations to observations*, Theses, Aix-Marseille Université.

Pizagno J., Prada F., Weinberg D. H. *et al.* (2007). The Tully–Fisher relation and its residuals for a broadly selected sample of galaxies, *ApJ* **134**, pp. 945–972.

Plummer H. C. (1911). On the problem of distribution in globular star clusters, *MNRAS* **71**, pp. 460–470.

Puech M., Flores H., Hammer F. *et al.* (2006). 3D spectroscopy with VLT/GIRAFFE. III. Mapping electron densities in distant galaxies, *A&A* **455**, pp. 131–134.

Puech M., Flores H., Hammer F. *et al.* (2008). IMAGES. III. The evolution of the near-infrared Tully–Fisher relation over the last 6 Gyr, *A&A* **484**, pp. 173–187.

Puech M., Hammer F., Flores H. *et al.* (2010a). The baryonic content and Tully–Fisher relation at z ~ 0.6, *A&A* **510**, A68.

Puech M., Rosati P., Toft S. *et al.* (2010b). Simulating the physics and mass assembly of distant galaxies out to z ~ 6 with the E-ELT, *MNRAS* **402**, pp. 903–922.

Ramsey L. W. (1988). *Fiber Optics in Astronomy, Astronomical Society of the Pacific Conference Series*, Vol. 3 (ed.), Barden, S. C., pp. 26–39.

Rodrigues M., Puech M., Hammer F. *et al.* (2012). A decrease of the gas exchanges between galaxies and the intergalactic medium, from 12 to 6 billion years ago, *MNRAS* **421**, pp. 2888–2903.

Rogstad D. H., Lockhart I. A., Wright M. C. H. (1974). Aperture-synthesis observations of H I in the galaxy M83. *ApJ* **193**, pp. 309–319.

Salucci P., Lapi A., Tonini C. *et al.* (2007). The universal rotation curve of spiral galaxies — II. The dark matter distribution out to the virial radius, *MNRAS* **378**, pp. 41–47.

Schoenmakers R. H. M., Franx M., de Zeeuw P. T. (1997). Measuring non-axisymmetry in spiral galaxies, *MNRAS* **292**, p. 349.

Shapiro K. L., Genzel R., Förster Schreiber N. M. *et al.* (2008). Kinemetry of SINS high-redshift star-forming galaxies: Distinguishing rotating disks from major mergers, *ApJ* **682**, pp. 231–251.

Sharp R., Parkinson H. (2010). Sky subtraction at the Poisson limit with fibre-optic multiobject spectroscopy, *MNRAS* **408**, pp. 2495–2510.

Slipher V. M. (1914). The radial velocity of the Andromeda Nebula, *Popular Astronomy* **22**, pp. 19–21.

van der Kruit P. C. (1976). Observations of velocity fields in three spiral galaxies using optical emission lines, *A&AS* **25**, p. 527.

Vanderriest C. (1980). A fiber-optics dissector for spectroscopy of nebulosities around quasars and similar objects, *PASP* **92**, pp. 858–862.

Vegte J. V. D. (2001). *Fundamentals of Digital Signal Processing with Cdrom*, 1st edn. (Prentice Hall PTR, Upper Saddle River, NJ, USA), ISBN 0130160776.

Vergani S. D., Flores H., Covino S. *et al.* (2011). GRB 091127/SN 2009nz and the VLT/X-shooter spectroscopy of its host galaxy: probing the faint end of the mass-metallicity relation, *A&A* **535**, A127.

Vetterling W. T., Teukolsky S. A., Press W. H. (1992). *Numerical Recipes: Example Book (C)*, 2nd edn. (Press Syndicate of the University of Cambridge), ISBN 0521437202.

Warner P. J., Wright M. C. H., Baldwin J. E. (1973). High resolution observations of neutral hydrogen in M33 — II. The velocity field. *MNRAS* **163**, pp. 163–182.

Weitzel L., Krabbe A., Kroker H. *et al.* (1996). 3D: The next generation near-infrared imaging spectrometer. *A&AS* **119**, pp. 531–546.

Wisnioski E., Förster Schreiber N. M., Wuyts S. *et al.* (2015). The KMOS$^{3D}$ survey: design, first results, and the evolution of galaxy kinematics from $0.7 < z < 2.7$, *ApJ* **799**, p. 209.

Wolf M. (1914). *Vierteljahresschr Astron. Ges.* **49**, p. 162.

Wong T., Blitz L., Bosma A. (2004). A search for kinematic evidence of radial gas flows in spiral galaxies, *ApJ* **605**, pp. 183–204.

Yang Y., Flores H., Hammer F. *et al.* (2008). IMAGES. I. Strong evolution of galaxy kinematics since $z = 1$, *A&A* **477**, pp. 789–805.

Yang Y., Rodrigues M., Puech M. *et al.* (2013). Accurate sky continuum subtraction with fibre-fed spectrographs, *The Messenger* **151**, pp. 10–13.

Yuan T.-T., Kewley L. J., Swinbank A. M. *et al.* (2011). Metallicity gradient of a lensed face-on spiral galaxy at redshift 1.49, *ApJ* **732**, L14.

Zheng X. Z., Hammer F., Flores H. *et al.* (2004). HST/WFPC2 morphologies and color maps of distant luminous infrared galaxies, *A&A* **421**, pp. 847–862.

# Chapter 5

# Applications and Scaling Relations

## 5.1 Applications to different distant galaxy studies

The quality of observations of individual galaxies at moderate redshifts ($z \leq 1$) has now sufficiently improved to allow one to analyze their properties with the same level of detail as for local galaxies.

Chapters 1–4 describe the overall methodology required to reach this goal, including when and how a particular approach can be applied to more distant galaxies.

This might sound a bit tedious, but it is extremely important to understand and validate the analysis process from the beginning to the end.

Chapter 5 describes a few specific examples where a methodological approach is beneficial. This can be generalized to the estimate of all physical quantities that can be derived from observational data.

The process begins with the analysis of a few individual distant galaxies (Sec. 5.2), demonstrating how much we can already learn about these objects by analyzing their shape, internal motion, SED, chemistry, and gas phase distribution.

Astronomy of distant galaxies is increasingly less about applying a cosmological approach (correlating galaxies as points in various diagrams) and more about understanding physical processed in galaxies. Sec. 5.3 introduces a methodological approach to study the evolution of the most important scaling relations. Sec. 5.3.1 presents the evolution of the Tully–Fisher relationship that relates rotation velocity to baryonic mass and is perhaps one of the most fundamental scaling relation for rotating disks of galaxies. Sec. 5.3.2 summarizes the relationship of O/H and stellar mass, its evolution, and how further work needed to better quantify observational biases. Sec. 5.3.3 describes the distribution of galaxies in the SFR-$M_{\rm stellar}$ plane, which is the most discussed topic in the current literature.

The scaling relations presented in this chapter depend on the mass, and more precisely on the stellar mass, which is discussed in Sec. 2.6.4. A proper estimate of both stellar and gas mass of galaxies is an important challenge in modern cosmology and this is discussed in Sec. 5.4.

## 5.2 Dissecting the properties of distant galaxies

### 5.2.1 *An example of a multi-instrument and multi-wavelength survey*

The goal of IMAGES (see an exhaustive description in Hammer *et al.*, 2012) is to gather a sufficient number of constraints for $z = 0.4 - 0.8$ galaxies and to link those constraints to their descendants, the local galaxies.

Its methodology follows the one described in Chapters 1–4. The galaxy selection criterion is an absolute J band magnitude ($M_J$(AB) $< -20.3$), which is a relatively good tracer of stellar mass, and this leads to a complete sample of galaxies (see Chapter 1) with $M_{\rm stellar} > 1.5 \times 10^{10} M_\odot$ (Delgado-Serrano *et al.*, 2010), and with an average mass similar to that of the MW.

IMAGES includes 63 starburst galaxies[1] for which the well resolved [O II] doublet ($EW_0 > 15$Å) allowed to derive spatially-resolved kinematics. Within the IMAGES redshift range, the fraction of these starbursts is 60% (Hammer *et al.*, 1997), allowing Delgado-Serrano *et al.* (2010) to study their morphological properties as well as that of a complimentary sample of quiescent galaxies with $EW_0$([O II]) $\leq 15$Å (see Fig. 1.9). The resulting sample of galaxies is complete because it is based on a random selection according to a single criterion, and because it matches the luminosity function

[1]This is the official number of IMAGES galaxies (see, e.g., Neichel *et al.*, 2008 and Hammer *et al.*, 2012) from which subsamples have been identified based on further limitations, e.g., the existence of 3 color-bands with HST for morphology classification (see Hammer *et al.*, 2009a; Delgado-Serrano *et al.*, 2010).

in the J band at the corresponding redshift. Moreover, IMAGES galaxies have been observed in four different fields of view to avoid cosmological variance effects.

The set of measurements includes:

- HST imagery (1 to 3 orbits in b, v, i, and z bands) to recover color-morphology comparable to the depth and resolution ($\sim 400$ pc) of local galaxies in the Sloan Digitized Spectroscopic Survey (SDSS, see Sec. 2.4);
- Spatially-resolved kinematics from FLAMES/GIRAFFE at the VLT (integration times from 8 to 24 h) to sample gas motion at 7 kpc resolution scale (see Sec. 4.5);
- Deep VLT/FORS2 observations (integration time of 3h at $R = 1500$) to recover the gas metal abundances (see Sec. 3.6.5);
- Spitzer 24 $\mu$m observations to estimate the extinction-corrected SFR as well as Spitzer IRAC and GALEX deep observations providing photometric points to constraint the SED from the rest-frame UV to IR (see Sec. 2.6.5).

### 5.2.2 *Description of a few individual distant galaxies*

Here, we show how these observations, when combined together, provide robust constraints on the evolutionary status of individual IMAGES galaxies. Due to space constraints, we limit this discussion to only 4 galaxies that have been observed in the HUDF. This is intended to illustrate the need for sufficient depth in imagery and photometry, even at moderate redshifts.

**J033237.54-274838.9** The fraction of starburst, rotating spirals found by the IMAGES survey at $z = 0.4$–$0.8$ is 19%. J033237.54-274838.9 is such an object as evidenced by its morphological appearance and its kinematic properties (see Fig. 5.1). It is a giant spiral with many star forming regions along the arms. The small bulge ($B/T = 0.06$) is redder than the disk though it is not quiescent since its color is comparable to that of an Sbc. It is surrounded by an inner ring. The overall disk is not fully symmetrical in the center and shows distorted arms on the left of the image. The kinematics is that of a

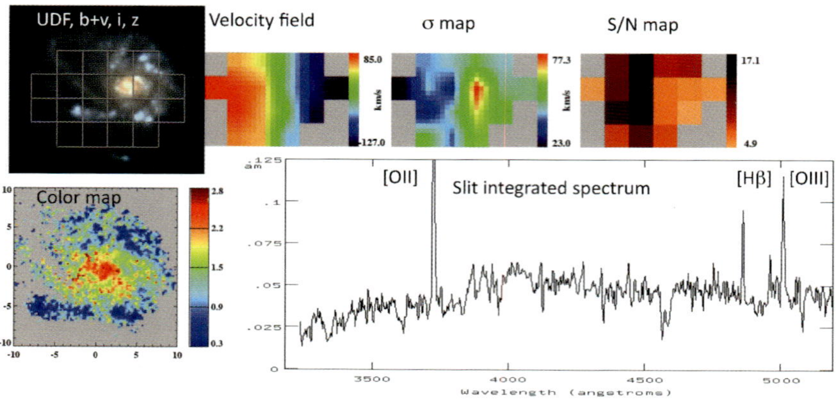


Fig. 5.1: Multiple properties of J033237.54-274838.9, a giant, starburst spiral galaxy with $M_{\rm stellar} = 5 \times 10^{10}\,{\rm M}_\odot$ at $z = 0.6638$. From left to right, top to bottom, the b + v, i, z combined rgb color image, the velocity field, dispersion and S/N maps (see Chapter 4), the v–z color map were determined following Chapter 2 and the integrated spectrum from FORS2 (see Chapter 3).

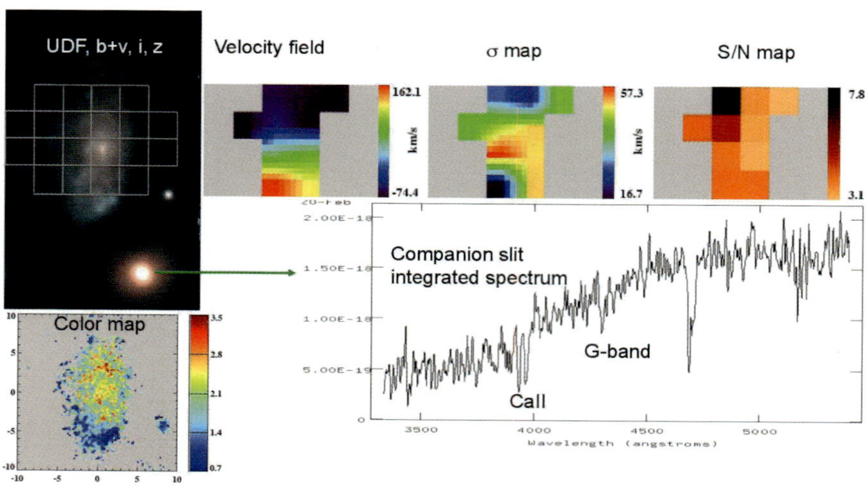


Fig. 5.2: Multiple properties of J033238.60-274631.4, a giant, starburst spiral galaxy with $M_{\rm stellar} = 3.4 \times 10^{10}\,{\rm M}_\odot$ at $z = 0.62066$, which is interacting with an elliptical galaxy at $z = 0.622$. Same panels than in Fig. 5.1, except the spectrum, which is that of the elliptical companion.

regular, rotating disk though it is not very thin given its modest $V/\sigma = 5.6$. It is a LIRG (SFR $= 30\,\mathrm{M}_\odot\,\mathrm{yr}^{-1}$) a property shared by many massive rotating starburst galaxies (see Hammer *et al.*, 2005). The very strong color gradient (see the color map) is evidence of an inside-out growing process in the disk, a property shared by all galaxies of that category (Neichel *et al.*, 2008). The stellar disk is thus young and probably no older than $\sim 1.5$ Gyr according to the SFR/$M_{\mathrm{stellar}}$ ratio. Such a recent formation is supported by the presence of A stars in the spectrum as well as the presence of the Balmer continuum, which almost dilutes the 400nm break.

**J033238.60-274631.4** This is a starburst, rotating disk, with a tiny bulge ($B/T = 0.02$). The disk is diffuse and shows a chain of starburst regions in the bottom edge of the disk, in the direction of the elliptical companion (see Fig. 5.2). The latter has a stellar mass that is nearly half of the mass of J033238.60-274631.4. The spectrum of the elliptical companion reveals a much flatter continuum below 400nm than what would be expected for an elliptical galaxy. It is very likely that the two galaxies are observed after a first interaction enhanced the star formation in the two galaxies ($\Delta$V$=$ 250 kms$^{-1}$, projected distance, 15 kpc). Such an interacting pair of galaxies would not have been identified using only photometric redshift ($z_{\mathrm{phot}} = 0.46$) or even the 3D-HST slitless grism redshift ($z = 0.656$) of the companion. This illustrates how some methods are unable to identify pairs of galaxies, especially interacting pairs where the interaction mainly affects the SED.

**J033245.11-274724.0** This is a distant, luminous compact galaxy ($R_{\mathrm{half}} = 1.7$ kpc), which is also a LIRG (SFR $= 23\,\mathrm{M}_\odot\mathrm{yr}^{-1}$) and has been studied in details by Hammer *et al.* (2009b). It is dominated by a disk which is so dust-enshrouded that it can only be detected in the HUDF (compare it to the GOODS image) despite being a relatively low redshift galaxy (see Fig. 5.3). The spectrum (corrected from reddening) reveals the predominance of young, 0.2–0.8 Gyr old stars. The kinematical axis is well offset from the optical major axis of the disk. The disk rotates extremely fast ($V_{\mathrm{flat}} = 290$ km$s^{-1}$),

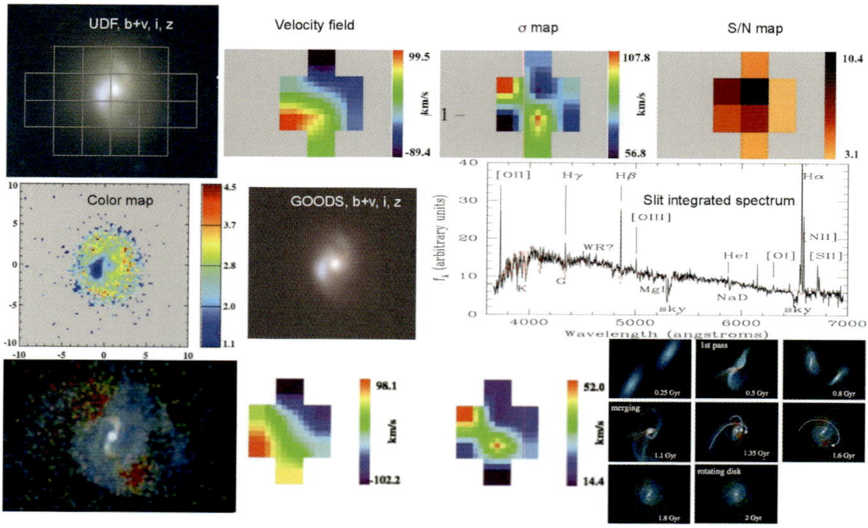


Fig. 5.3: Multiple properties of J033245.11-274724.0, a galaxy rebuilding a disk with $M_{\rm stellar} = 6.3 \times 10^{10}$ M$_\odot$ at $z = 0.43462$. Same panels as in Fig. 5.1, except in the middle in which we have added a GOODS ACS combined image for comparison to the HUDF image. The bottom row shows a simple model of an inclined 1:1 merger ($r_p$= 8 kpc, from Barnes (2002)); from left to right: distribution of gas and young stars 0.5 Gyr after the merger showing a bar and a symmetric-arm system whose motion does not follow the gas rotation, creating shocks, and high velocity dispersions (seen as red particles in the simulation); reproduction of the velocity field and $\sigma$ map; evolution of the merger in which the arrows in the central panels (at 1.35 and 1.6 Gyr) indicate the gas motions and the exchanges of angular momentum after the merger; note that, conversely to the disk, the two armed system does not rotate (compare panels at 1.6 and 1.8 Gyr) and they regulate the final disk rotation.

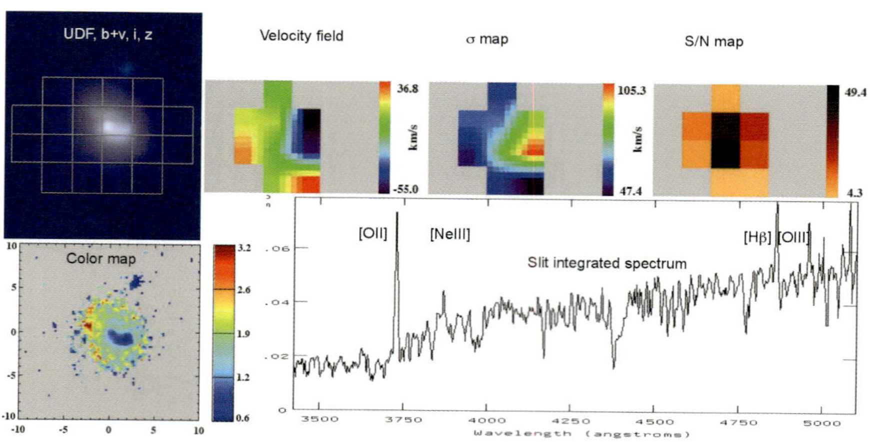

Fig. 5.4: (Figure on facing page) Properties of J033244.20-274733.5, a complex merger with $M_{\rm stellar} = 4.2 \times 10^{10} M_\odot$ at $z = 0.73605$ (same panels than in Fig. 5.1). Notice the presence of [Ne III] and the flat blue continuum, both indicative of a strong starburst.

100 km$s^{-1}$ faster than the prediction from the Tully Fisher relation for its stellar mass. The detailed morphology and kinematics of J033245.11-274724.0 are similar to the predictions of models of a merger of two equal mass, gaseous rich galaxies, 0.5 Gyr after the merging event. The combined system of a bar and of two nonrotating[2] spiral arms regulates the accreting material, induces large velocity dispersions ($\sigma > 100$ kms$^{-1}$), and redistributes the angular momentum. Gas and young stars spiral inwards together rapidly to feed the disk and the central regions.

**J033244.20-274733.5** The morphology of this object is quite similar to J033245.11-274724.0, although perhaps a less evolved system, with a forming bulge with extremely strong star formation in one inner arm, and a counterpart on the other side of the galaxy that is strongly extinguished. This object is a powerful LIRG (SFR = 67.8 $M_\odot$ yr$^{-1}$) and an example of a compact dust enshrouded object. Its outer parts are redder than its center, approaching the color of an S0 galaxy, implying the presence of dust. The kinematics of this object are very complex and show no evidence of rotation (see Fig. 5.4). This, when combined to the large measured velocity dispersion (50 to 100 km/s), suggests a merger. The morphology of this object indeed resemble that of NGC 1614, a well known merger (and also a LIRG), if that object was observed at high redshift.

The modeling of IMAGES galaxies in the GOODS-S field can be found in Hammer *et al.* (2009a). Detailed analysis of individual cases have been also performed for a few of them, although more if not all of the sample could be similarly analyzed.

For instance, J033239.72-275154.7 has been analyzed and modeled by Peirani *et al.* (2009) as a 3:1 merger just before the fusion of the

[2]We also suggest the reader to examine the simulation at `http://www.ifa.hawaii.edu/~barnes/research/gassy_mergers/index.html`, which is a model with 1:1 mass ratio, INClined orbit and pericenter $r_p = 0.2$.

two nuclei. Figure 4.31 of Chapter 4 reveals that the kinematics is dominated by the motion of the secondary, which is harassing the main galaxy and creating a gigantic bar in its center.

A full SED analysis of J033241.88-274853.9 (see Fig. 2.20), Puech *et al.* (2009) showed this object to be a currently forming, young disk galaxy at $z = 0.6679$. Using an inversion of the Schmidt–Kennicutt law,[3] the gas fraction was estimated to be ~80%. It is an asymmetric disk that shows strong gas emission in regions devoid of stars. The galaxy could be the result of a recent galaxy merger or, alternatively, be a protogalaxy, although at a moderate redshift.

Another very large bar was identified in J033227.07-274404.7, and subsequently modeled by Fuentes-Carrera *et al.* (2010). Yang *et al.* (2009) analyzed J033210.76-274234.6 which has two nuclei and a ring. Both of these objects are due to a recent collision of two galaxies and have been fully modeled.

### 5.2.3 *Spiral disk building and merger rate*

The examples above show how individual distant galaxies are likely complex systems and that such a complexity cannot be understood using approximative methods.

Disentangling the physical properties of galaxies requires deep imagery (the redshifted $R_{25}$ with S/N = 4 in two elements of resolution can only be detected at distance up to $z = 0.4$–$0.5$ with 3 HST/ACS orbits), spatially resolved kinematics with more than 2 resolved elements, deep integrated spectroscopy including [O II] to at least [O III], and an SED spanning from the UV to at least 24 $\mu$m. After examining all possible mechanisms, IMAGES revealed that approximately half of MW–mass spirals were in the process of major mergers, 6 Gyr ago, with an equipartition of phases before (two components) and after (one component) the merger, some showing evidences for disk rebuilding (e.g., J033245.11-274724.0).

[3]The Schmidt law relates the gas and star formation surface densities (Kennicutt, 1998). Mapping the SFR by an IFU allows in principle to derive the gas mass. It however assumes that the gas spatial distribution follows that of available tracers of SFR (ionized component or UV and IR light).

Calculation of the merger rates is of considerable importance to understand galaxy formation and cosmology. Merger rates were at first estimated using observations of galaxy pairs (see, e.g., Le Fèvre *et al.*, 2000 and references therein). This however assumes (1) the absence of contamination by foreground or background sources; (2) the subsequent dynamical evolution of galaxies in pairs that may be simply fly-by encounters; and (3) the knowledge of the characteristic time for the merging completion (or visibility timescale of pairs) that directly scales the merger rate calculations.[4]

The average contamination by unrelated sources can be estimated, requiring the study of merger rates using different lines of sight to avoid cosmological variance effects. The merger time scale is fundamentally related to galaxy total masses and determined using hydrodynamical simulations (see, e.g., Lotz *et al.*, 2010a, 2010b). This leads to serious uncertainties that depend on baryonic mass, its fraction to the total mass, and the ingredients of simulations.

The difficulties can be circumvented by performing full modeling of galaxy mergers with a complete sample. Individual merger time scales have been derived from the Hammer *et al.* (2009a) modeling of IMAGES galaxies, and Puech *et al.* (2012) derived a merger rate that agrees particularly well with predictions from the Λ-CDM semi-empirical models of Hopkins *et al.* (2010) out to $z \sim 1.5$.

## 5.3 Evolution of scaling relations

### 5.3.1 *Tully–Fisher Relation*

Galaxies are self-gravitating systems, which can be described simply using their overall radius, mass, and velocity. This is the reason why scaling relations between galaxy mass (or luminosity), size, and velocity (or velocity dispersion for ellipticals) have been extensively studied both in the local and the distant Universe.

[4] $N_{\rm merger}(z) = 0.5 \times n(z) \times f_{\rm merger}(z) \times \tau^{-1}$, where 0.5 converts the number of merging galaxies into the number of merger events (generally 2 to 1), $n(z)$ is the comoving number density of galaxies, $f_{\rm merger}(z)$ is the merger fraction and $\tau$ is the merger time scale (see, e.g., Rawat *et al.*, 2008).

The Tully–Fisher relation (hereafter TFR, Tully and Fisher, 1977), which relates the rotational velocity to the luminosity (as a proxy for stellar mass) of galaxies, is the tightest scaling relation that exists between these fundamental galaxy properties (e.g., Courteau *et al.*, 2007). Using the rotation disk velocity as a proxy for the dynamical mass, this relation has been widely used to compare observations to galaxy formation models (e.g., Eisenstein and Loeb, 1996; Silk, 1997; Navarro and Steinmetz, 2000; Portinari and Sommer-Larsen, 2007; McGaugh *et al.*, 2010; McGaugh, 2012). The evolution of the TFR is therefore a very useful tool to study the evolution of (disk) galaxies and the relation between galaxies and their halos as a function of redshift (e.g., Dutton *et al.*, 2011).

The TFR is generally fitted using a simple linear relation (in log–log space, see Ex. *How to construct a stellar-mass Tully–Fisher relation at high* $z$?). Earlier studies used blue filters (Vogt *et al.*, 1996, 1997; Ziegler *et al.*, 2002; Böhm *et al.*, 2004) but progressively redder bands have been used. Stellar mass (Conselice *et al.*, 2005; Flores *et al.*, 2006; Kassin *et al.*, 2007; Miller *et al.*, 2011) have also been used in order to minimize the impact of dust and variations due to star formation history, facilitating the comparison to models. Also, the interest has progressively shifted from using slits to using IFUs (Flores *et al.*, 2006; Puech *et al.*, 2008; Cresci *et al.*, 2009; Vergani *et al.*, 2012) in order to better separate noncircular motions from the underlying rotation pattern (see Chapter 4).

**Example 5.1: *How to construct a stellar-mass Tully–Fisher relation at high $z$?***

(1) Define a representative sample of emission line galaxies at a given $z$, using a single criterion that is a proxy for stellar mass (e.g., in rest-frame red band, see Chapter 1). Selection effects are known to strongly impact the slope of the TFR (Kannappan *et al.*, 2002; Hammer *et al.*, 2007). For

(*Continued*)

**Example 5.1: *(Continued)***

all galaxies, you will need IFU data, high resolution imaging, and multi-wavelength photometry covering at least the rest-frame NIR;

(2) Obtain IFU data for every galaxy and measure kinematics from which you can estimate the observed rotation velocity (see Chapter 4);

(3) Using high-resolution images, measure the inclination angles of the galaxies (see Chapter 2). This angle should not be derived directly from the kinematic data because limited spatial resolution makes this much less accurate (Epinat *et al.*, 2009);

(4) Use the morpho-kinematic data to identify reliable rotating disks (see Chapter 4);

(5) Correct the observed line-of-sight radial maximal velocity for both beam-smearing and inclination. To estimate the beam smearing correction factor, use maximum likelihood methods based on rotating disk models degraded to the observed seeing (see Chapters 2 and 4);

(6) Using multi-wavelength photometry, estimate the total stellar mass of each galaxy (see Chapter 2). Do no forget to apply statistical $k$-corrections (see Chapter 1) and to correct for extinction (see methods dedicated to rotating disks in Chapter 2);

(7) Construct a detailed uncertainty budget accounting for both random *and* systematic effects (see, e.g., Puech *et al.*, 2008). Take particular care of the possible systematic effects which usually dominate the error budget in the evolution of the TFR;

(8) Construct the mass vs. rotation velocity log–log plot and isolate the rotating disks. Fit the TFR while only considering the rotating disks: all other objects have unreliable measurements of the rotation velocity since they are not

*(Continued)*

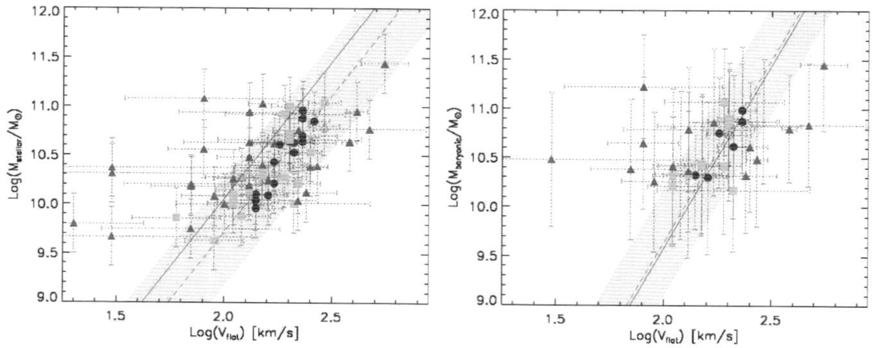


Fig. 5.5: *Left*: Evolution of the stellar-mass TFR between $z \sim 0.6$ (dashed line) and $z = 0$ (solid line) from the IMAGES survey (adapted from Puech *et al.*, 2010). Distant rotating disks are shown using dots, perturbed rotators using squares, and galaxies with complex kinematics are shown using triangles. *Right*: Evolution of the baryonic TFR in the CDFS sub-sample. The solid line shows the local bTFR, while the dashed line shows a fit to the distant relation. The grey areas represent the total systematic uncertainties on the evolution of the zero point (see text).

**Example 5.1: *(Continued)***

rotationally supported or might not be rotating at all. Use preferentially $\log(V_{\rm flat})$ as an independent variable during the fit since associated uncertainties are better controlled than those associated to mass (see Chapter 2). The result at $z \sim 0.6$ is illustrated in Fig. 5.5.

The TFR of local galaxies is very tight when using absolute magnitudes in red bands ($\sigma \sim 0.4$ mag, Pizagno *et al.*, 2007) which are a good proxy for stellar mass. The scatter increases by almost two orders of magnitude[5] from $z = 0$ to $z > 0.5$ (Conselice *et al.*, 2005; Flores *et al.*, 2006, see also Fig. 5.5). This increased scatter appears to

[5]Recent studies (Miller *et al.*, 2011, 2012) based on slit spectroscopy do not find such a large scatter, claiming to have achieved a better sampling of the rotation curve plateau. These studies are however methodologically flawed since they arbitrarily discard compact galaxies (one third of the targets according to Glazebrook, 2013), and they cannot evaluate the origin of the rotation in absence of IFUs (see Chapter 4). For example, J033227.07-274404.7 is assumed to be a rotating disk while it is found to be an almost accomplished merger (see Fuentes-Carrera *et al.*, 2010).

be linked to galaxies with peculiar kinematics and/or morphologies (Flores *et al.*, 2006; Kassin *et al.*, 2007; Puech *et al.*, 2008). Figure 5.5 illustrates this by comparing the distribution of rotating disks to that of perturbed or complex kinematics. A similar effect was also discovered when comparing local spirals to local interacting galaxies, the latter being found by Kannappan and Barton (2004) to be the only TFR outliers. In fact, several order magnitudes deviations (or up to 1 dex in $\log(V_{\rm flat})$, see Covington *et al.*, 2010) can be the result of large scales perturbations caused by major mergers.

Attempts to reduce the scatter in the TFR have been done by using the more general tracer $S$ which is a quadratic combination of velocity dispersion and rotational velocity (Weiner *et al.*, 2006; Kassin *et al.*, 2007; Puech *et al.*, 2010; Vergani *et al.*, 2012). This is however not enough to reproduce the tight TFR observed in local (Pizagno *et al.*, 2007) and distant rotating disks (see Puech *et al.*, 2010). Such a result is not unexpected if mergers are not energy conservative events.

The evolution of the TFR zero point as a function of redshift has been quite debated, perhaps because of the numerous systematic uncertainties that are present (Puech *et al.*, 2008, 2010; Cresci *et al.*, 2009; Miller *et al.*, 2012; Vergani *et al.*, 2012) (see Fig. 5.5). Hammer *et al.* (2007) showed that the commonly used relation from Verheijen (2001) is based on Ursa Major cluster galaxies and hence biased[6] towards low mass emission line galaxies, when compared to field selected galaxies (SDSS, see Pizagno *et al.*, 2007). Instead, one should compare the distant TFR to that of the SDSS, which led Puech *et al.* (2010) to quote for zero point evolutions $\sim$0.3 and 0.6 dex at $z = 0.65$ and 2, respectively. An evolution of 1.3 dex is found at $z \sim 3$ (Gnerucci *et al.*, 2011), although this was estimated using a TFR with a large amount of scatter, which is expected given the difficulty in classifying morphology and kinematics when lacking data with sufficient depth. Estimates of the zero point evolution of the stellar mass TFR remain somewhat uncertain. For instance, the total

[6]as well as that of Bell and de Jong (2001) since their study was based on the same Ursa Major sample than the one used in Verheijen (2001).

systematic uncertainty ~0.55 dex at $z \sim 0.6$ is largely dominated by that on the zero point of the local relation, which is found to be ~0.5 dex by bootstrapping the distribution of $\log(V_{\rm rot}) \geq 2.2$ galaxies (for which the SDSS TFR is representative of the local galaxy population, see Puech *et al.*, 2008).

In the local universe, McGaugh *et al.* (2000) showed that dwarf galaxies with stellar masses $< 10^9\,M_\odot$ fall below the relation defined by more massive, gas-poor disks. Once both gas and stellar masses are combined, one obtains a baryonic Tully–Fisher relation that holds over ~5 decades in mass. Measuring the gas fractions in distant galaxies will remain impossible until the next generation of radio telescopes such as the SKA. However, gas masses can be estimated indirectly by inverting the Schmidt–Kennicutt relation between SFR and gas mass densities, provided that reliable SFR and sizes can be measured (see however footnote in Sec. 5.2).

Puech *et al.* (2010) used this method to estimate the evolution of baryonic TFR at moderate redshifts ($z = 0.65$, 6 Gyr ago) and found no significant evolution (see Fig. 5.5), which might also hold at $z > 1$ (Vergani *et al.*, 2012). This result is based on a very limited number of rotating spiral galaxies at $z \sim 0.65$ and by the unknown nature of galaxies at higher redshifts, and will be considerably improved by future facilities (ELTs, SKA) that will provide the required spatial sampling of rotational curves in distant galaxies (see Chapter 4). The gas fraction is found to evolve mildly between $z \sim 0.6$ and $z = 0$ (~0.1–0.15 dex, Rodrigues *et al.*, 2012), which implies that the zero point evolution of the stellar-mass and baryonic TFRs should be different by that order (Dutton *et al.*, 2011). Given the quite large uncertainties associated to the evolution of the zero points (see above and Fig. 5.5), the evolutions of both relations are found to be broadly consistent one with the other.

The confirmation of a nonevolving baryonic TFR (hereafter bTFR) would be a priceless stepping stone in cosmology. McGaugh (2005) developed a formalism to characterize the stellar mass-to-light ratio leading to a very robust local bTFR, with $M_{\rm vir} \sim (V_{\rm flat})^4$ instead of $\sim (V_{\rm flat})^3$ as simply expected from Λ-CDM (White, 1997). The debate is not settled (Desmond, 2012) and other slopes are generally found ranging between 3 and 4 (see, e.g., Zaritsky *et al.*, 2014). The major uncertainty is associated to the determination of the stellar mass-to-light ratio, which is linked to the IMF and perhaps to its variations at different galaxy mass ranges (Weidner *et al.*, 2013). Deviations from the bTFR are indeed expected, if for example, the baryonic fraction varies considerably from one galaxy to another (McGaugh *et al.*, 2010). Witnessing the evolution of such properties will be one major possible outcome of future ELTs equipped with multiple IFU spectrographs.

### 5.3.2 *The evolution of the M–Z relation*

The stellar mass vs. gas-phase metallicity (M–Z) relation is the main scaling relation of the chemical evolution of galaxies. It is a reflection of the gas enrichment by stars (e.g., supernovae and dying stars) and also of exchanges with the environment (i.e., gas inflow and outflow).

In the local Universe, the M–Z relation (often calibrated using $R_{23}$, see Sec. 3.6.5) is a correlation that spans over 3 decades in mass (Tremonti *et al.*, 2004). In the later study, metallicity increases linearly with stellar mass up to a characteristic mass of $\sim 3 \times 10^{10}$ $M_\odot$ above which the metallicity saturates. It seems that the metallicity saturation problem has been exacerbated by an important Malmquist bias that leads to a considerable over abundance of massive star forming galaxies in the Tremonti *et al.* sample (see e.g., Rodrigues *et al.*, 2016). When corrected for this effect, the local M–Z relation becomes mostly linear, as it would be expected if galaxies were mostly isolated systems, transforming their gas into stars with an efficiency that is linked to their masses.

The different methods used to calibrate O/H can lead to quite different values, resulting in an absolute metallicity scale that is very uncertain (see e.g., Liang *et al.*, 2006b; Yin *et al.*, 2007; Kewley and Ellison, 2008). This does not impede studying the evolution of the M–Z relation as long as consistent O/H calibrations are adopted at different epochs. Such an homogeneity is not always possible, since for example, the $H\alpha$ is redshifted to IR wavelengths at increasing redshifts.

At the very least, available conversion between metallicity calibrations are required (see e.g., Kewley and Ellison, 2008) before studying the M–Z evolution, although it is unclear whether such conversion is valid at higher redshifts. Note that a similar procedure should be adopted to estimate masses at different epochs (see Chapter 2).

Numerous works have studied the M–Z relation at intermediate (e.g., Liang *et al.*, 2006a; Rodrigues *et al.*, 2008; Zahid *et al.*, 2011) and at high redshift (e.g., $z \sim 2$, Erb *et al.*, 2006, $z \sim 3$ Maiolino *et al.*, 2008). Comparing results at different redshifts is a complex task, mainly due to possible selection effects. For example, the fraction of $L^*$ emission line galaxies varies from 17% to 80% between $z = 0$ to $z \geq 1$. Thus, comparing O/H abundances between two different epochs could be an attempt to relate completely different galaxy populations, conversely to the descendent-progenitor problem considered in Sec. 1.7. The problem is exacerbated when considering the S/N limits on emission lines that should be accounted for when estimating metal abundances. Since the S/N depends on the emission line apparent luminosity as well as on the overall instrumental throughput, O/H evolutionary studies may appear to be an unsolvable problem.

It is therefore remarkable that, despite the different methods that are used[7] e.g., IMAGES (Rodrigues *et al.*, 2008) and DEEP2 (see Zahid *et al.*, 2011 see also Moustakas *et al.*, 2011), results are consistent with a redshift evolution by $\sim$0.22$\pm$0.04 dex in metal abundance from $z = 0$ to $z = 0.7$ (see Fig. 5.6). Beyond this redshift range the comparison is not robust as it uses [N II]/H$\alpha$ instead of the $R_{23}$ diagnostic and then the comparison depends on the N/O ratio, which evolution could be not dependent on metallicity (see e.g., Zahid *et al.*, 2013).

Other reasons to limit such comparison to $z \sim 1$ is the probable lack of completeness in samples of higher $z$ galaxies (see Chapter 1).

The following list includes the steps required to perform a careful study of the evolution of the M–Z relation:

[7]Notice that in the two panels of Fig. 5.6 that both O/H and $M_{\mathrm{stellar}}$ calibrations differ, which leads for the later to a change by $\sim$0.4 dex.

**Example 5.2: *How to construct a stellar-mass metallicity relation at high z?***

(1) Pre-select galaxies in two redshift bins using a single criterion that can be used as a proxy for the stellar mass (for example an absolute magnitude in a red band) and verify its representativeness through comparison to luminosity functions at these redshifts (see Chapter 1).

(2) Check the spectral resolution ($R > 1000$ is mandatory) and signal-to-noise (5 is necessary for all useful lines), and verify the spectrophotometry through at least two broad-band imaging filters (see Chapter 3).

(3) Data reduction and analysis (see Chapter 3). Subtract the stellar continuum in each spectrum by using a template spectrum. Measure emission lines in the rest-frame spectrum and correct the line fluxes for dust extinction using the Balmer decrement. Reject galaxies with possible AGN contamination using a diagnostic diagram. Keep galaxies classed as 'intermediate-AGN' in the analysis while identifying them separately from H II, star-forming galaxies. Estimate stellar masses using a simple criterion based on the red-band absolute magnitude and a bluer rest-frame color (see Chapter 2).

(4) O/H may be be estimated using $R_{23}$, [O III]$\lambda$5007/[O II] $\lambda$3727 (for the ionizing potential) and another metallicity estimator if necessary to verify the metallicity branch (see Chapter 3).

(5) Selection effects linked to O/H determination. In each mass bins, keep track of the fraction of galaxies having sufficiently bright emission lines for determining the O/H metallicity. Evolution of metallicity and dust may provide subtle biases. For example, a high metallicity leads to small [O III]$\lambda$5007 lines and a high extinction may limit the detection of high order Balmer lines. Also, keep track of these objects by

(*Continued*)

**Example 5.2: *(Continued)***

identifying these galaxies using flux limits whenever it is possible.

(6) The uncertainty of the final quantities depends on the propagation of the individual lines measurement uncertainties (see Chapter 3). For $R_{23}$, it has to include the errors associated the necessary extinction correction made using the [O II]$\lambda$3727Å/H$\beta$ ratio.

(7) M–Z relation in logarithm can be fitted using either a straight line or, alternatively, any other function (see e.g., Zahid *et al.*, 2013). Any deviation from a linear relation should be checked for possible biases, especially at both ends of the galaxy distribution.

Figure 5.6 (left) shows the distribution of IMAGES galaxies at $z \sim 0.65$ and the scattering of this distribution, which at ~0.18 dex is similar to that at $z = 0$ (~0.15 dex).

Even when it is not affected by completeness biases such as Malmquist bias (see e.g., Rodrigues *et al.*, 2016), the shape of the M–Z relation may depend on the O/H calibration.

For example, the O/H–$R_{23}$ relationship flattens at 12+log(O/H)~ 9.1–9.2 (see Fig. 3.21 of Chapter 3), and low $R_{23}$ values thus lead to fainter and lower S/N [O II]$\lambda$3727Å and [O III]$\lambda$5007 lines. The resulting bias against massive metal rich galaxies is perhaps the cause of the very rapidly decrease of the M–Z scatter with increasing mass/metallicity (see, e.g., Tremonti *et al.*, 2004; Moustakas *et al.*, 2011).

If correct, this implies that the distant M–Z relation should be compared to a complete, volume-limited sample of local galaxies as in [Rodrigues *et al.* (2016)]. Otherwise, the decrease of O/H as a function of redshift would be hidden by the bias against metal rich galaxies in the high mass bin. This would lead to an artificially flattening of the M–Z relation, as well as to underestimating the evolutionary effect in this mass range. Figure 5.6 (right) shows this flattening at $z = 0$, with a less pronounced effect at $z = 0.78$ that

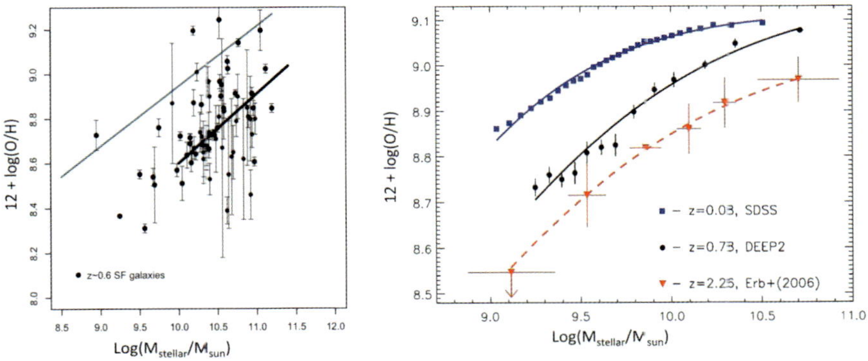


Fig. 5.6: Evolution of the M–Z evolution. *Left*: The local reference established using a complete, volume limited sample from the SDSS, is shown using a solid light blue line (adapted from Rodrigues *et al.*, 2016). Data from the IMAGES survey at ∼0.6 (Rodrigues *et al.*, 2008) are shown using symbols. Galaxies outside the range of stellar mass where the sample is representative are shown using open symbols and a linear fit to the data is shown using a the solid black line. The observed evolution is ∼0.25 dex at $M_{\rm stellar} = 10^{10}\,{\rm M}_\odot$. Metallicities were estimated using the Tremonti *et al.* (2004) calibrator (R23 parameter) and the stellar mass were derived using the method describe in Bell *et al.* (2003) and a Salpeter IMF. *Right*: The local relation derived from the SDSS survey are shown using blue symbols, the relation at $z \sim 0.78$ from DEEP2 survey (Zahid *et al.*, 2011) is shown in black. Metallicities were estimated using the Kobulnicky and Kewley (2004) calibrator ($R_{23}$ and $O_{32}$ parameters) and stellar masses were derived using SED fitting and a Chabrier IMF. The observed evolution is ∼ 0.18 dex in the same mass range. Star-forming galaxies at $z \sim 2.26$ from Erb *et al.* (2006) and for which metallicities were estimated using the $N_2$ parameter are shown using red triangles.

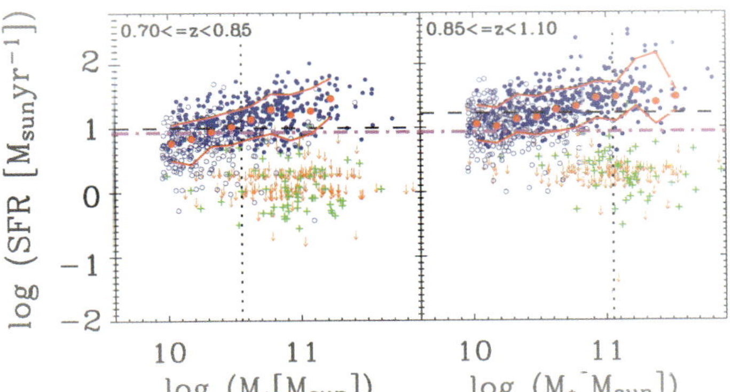


Fig. 5.7: SFR–$M_{\rm stellar}$ relation of star forming galaxies in the $z = 0.7{-}1.1$ range (from Noeske *et al.*, 2007 ©AAS, reproduced with permission). See text for the significance of symbols.

could be well fitted by a single line (this done by Moustakas *et al.*, 2011 in their Fig. 18, for example).

Consequences of these biases are not insignificant: a considerable number of studies have considered a small evolution of O/H as a proof that these objects have already synthesized the bulk of their metals at early epochs, more than 8 Gyr ago. Conversely, if it evolves by ~0.25 dex over the last 6 Gyr, it favors a scenario in which the population galaxies evolved as closed-systems, converting their own gas reservoirs into stars (Rodrigues *et al.*, 2012). Further studies are certainly necessary to compare large and complete samples at different redshifts. Although the evolution of the M–Z relationship is established, the amplitude of this relation remains to be determined better.

### 5.3.3 *SFR-$M_{\rm stellar}$ diagram*

Analyses of distant galaxies have revealed a SFR-$M_{\rm stellar}$ sequence[8] that appears as a relatively tight correlation (~0.3 dex). It was first observed by Noeske *et al.* (2007) and then Elbaz *et al.* (2007), on the basis of targets selected from the DEEP2 redshift survey and from a redshift catalog of the GOODS-S field, respectively. Figure 5.7 shows some of the selection effects described in Chapter 1, including an increase of the low-mass limit as a function of redshift, (95% completeness, see vertical dotted lines) and of the expected limits on SFR (24 $\mu$m 80% completeness limits, see horizontal dashed lines). Interpreting Fig. 5.7 requires accounting for further biases:

- Less than 50% of the galaxies included in Fig. 5.7 (filled blue dots) have their SFRs derived from 24 $\mu$m luminosity. The others are either from [O II]λ3727Å (open blue dots and green crosses, see Chapter 3 and example *Can we estimate SFR from the forbidden* [O II] *line?*), or are limits (orange arrow);
- In the $z = 0.7-1.1$ range, the DEEP2 redshift survey is limited in the R band ($R \leq 24.1$), implying a rest-frame selection ranging

[8]also dubbed the "Main Sequence" of star formation or of star forming galaxies, although its physical grounds cannot be compared to that of Main Sequence stars.

from 382 to 295nm, i.e., providing a considerable bias against many intrinsically red galaxies (see Sec. 1.3.6), which are thus missing in Fig. 5.7.

It is thus unclear whether the SFR–$M_{\rm stellar}$ sequence includes a significant fraction (i.e., >50%) of galaxies, even amongst the star forming ones. The rest-frame selection is however slightly improved in Elbaz *et al.* (2007) who used the B band, but at the cost of an large diversity in selection criteria for the spectroscopy (see, e.g., Ravikumar *et al.*, 2007). The following is a list of steps that are required for a careful study of the SFR–$M_{\rm stellar}$ diagram:

**Example 5.3: *How to establish the SFR-$M_{stellar}$ diagram at $z \sim 0.9$?***

(1) Pre-select galaxies using a single criterion avoiding biases against red galaxies at $z \sim 0.9$, for example a selection in observed J or H bands samples the rest-frame continuum between 600nm and 800nm. Verify its representativeness through comparison to luminosity functions at these redshifts (see Chapter 1).

(2) The spectroscopy has to be performed in the observed J band to avoid biases against red sources and includes the H$\alpha$ emission. Allow for sufficient spectral resolution ($R >$ 3000 is mandatory in the NIR, see Chapter 3) and depth. Complimentary spectroscopy should be done in the $z$ band in order to measured H$\beta$ and estimate extinction values. Verify the spectrophotometry through at least two broadband imaging filters per spectrum and correct for aperture effects accordingly (see Chapter 3).

(3) Subtract the stellar continuum in each spectrum by using a template spectrum. Measure emission lines in the rest-frame spectrum and correct the line fluxes from dust extinction using the Balmer decrement.

(4) Calculate stellar masses using a red-band absolute magnitude and a bluer rest-frame color (see Chapter 2) and SFRs

(*Continued*)

**Example 5.3: *(Continued)***

from extinction corrected H$\alpha$ luminosity. Calculate the mass limit for objects with no or with S/N $\leq$ 3 H$\alpha$ and H$\beta$ fluxes. Compare SFR values and limits to those provided by deep UV and IR photometry, if it exists.

(5) Limit the redshift range as much as possible to avoid selection effects for the redshift determination and for the H$\alpha$ flux measurement, Finally, draw the SFR–$M_{\rm stellar}$ diagram, after accounting for the incompleteness in the redshift measurements (see Chapter 1).

Alternatively the SFR can be measured using both UV and IR luminosities (see Chapter 2), if extremely deep IR photometry is available. The validity of the "Main Sequence" could then be estimated, including its scatter. The exact shape of the SFR–$M_{\rm stellar}$ distribution is still uncertain, especially at high masses. For example, Brinchmann *et al.* (2004) noticed[9] that at $z = 0$, "the distribution of SFRs broadens significantly for $M_{\rm stellar} \geq 10^{10}\,{\rm M}_{\odot}$ and that the correlation between stellar mass and SFR breaks down" (see their Fig. 17 and also Drory and Alvarez, 2008).

Further methodological studies are required to establish the validity of the "Main Sequence", and alternatively whether it is just a maximal envelop of star formation for "normal" star forming galaxies (i.e., excluding mergers in their fusion phase, see e.g., Puech *et al.* 2014). The later uncertainty is true at all redshifts, because the SFR calibration of quiescent galaxies is always an important issue (for example at $z = 0$ see, e.g., Abramson *et al.*, 2014 and also their Fig. 2).

Independently of the nature of the SFR–$M_{\rm stellar}$ relation, its redshift evolution to higher SFR/$M_{\rm stellar}$ is beyond doubt. This could

[9]Even at $z = 0$, the SFR–$M_{\rm stellar}$ relation from Brinchmann *et al.* (2004) could be affected by numerous galaxies showing no H$\alpha$ lines and for which the SFR/$M_{\rm stellar}$ has been estimated through a correlation with the 400nm break. This correlation does not provide any increase in precision while bringing an artificial SFR–$M_{\rm stellar}$ correlation for the quiescent galaxies (see e.g., Fig. 1 of Renzini and Peng, 2015).

be caused by a significant evolution in the gas supply (through gas exhaustion) as suggested by Noeske *et al.* (2007), perhaps simply linked to the gas fraction evolution (Rodrigues *et al.*, 2012).

Finally the SFR–$M_{\text{stellar}}$ relation cannot be taken as being as robust as the Tully Fisher or the M–Z relationships. Estimating its scatter is not a trivial exercise because there is no reason why the galaxies forming more stars than those on the "Main Sequence" should be distributed symmetrically relatively to those which are forming fewer stars. The dominant mode of star formation can therefore not the predicted, e.g., the presence of major mergers cannot be excluded (Puech *et al.*, 2014), and Drory and Alvarez (2008) found the relation to be consistent with ~1 major merger per massive object at $1.5 > z > 0$. Perhaps ISM masses, evaluated through measurements of the dust continuum using ALMA, could provide important clues about the dominant star formation mode (Scoville *et al.*, 2015).

## 5.4 An important challenge: baryonic mass estimate

As we discussed above, mass is the essential parameter to understand galaxy physics and evolution. It is however one of the most difficult parameter to estimate using observations, because most of the starlight is coming from stars that do not constitute the bulk of a galaxy mass and also because gas may be under different phases, including warm and hot phases, which could be diffuse within the halo scales and therefore almost undetectable.

### 5.4.1 *Stellar mass estimates*

As described in Chapter 2, stellar mass estimates are affected by numerous biases and systematic effects. Furthermore, there are considerable uncertainties associated to whatever assumptions are made about the star formation history. Figure 5.8 (top) shows how different approximations of the stellar mass lead to systematic shifts. This is caused by the different adopted IMF (see, e.g., Bell *et al.*, 2003), and also by differences depending on the stellar mass range. Most determinations of stellar masses assume a constant Initial Mass Function (IMF) that fixes the converting rate of light into mass ($M_{\text{stellar}}/L$).

Kroupa *et al.* (2013) reviewed the main properties of the IMF, concluding that the IMF depends on the SFR density and on the metal abundance. The upper mass limit of stars in a cluster is a strong function of the stellar mass of the embedded cluster (see Fig. 5.8, left), which necessarily impacts the IMF since most stars took their birth into clusters.

Kroupa *et al.* (2013) quoted: "the existence of the physical $m_{\rm max} - M_{\rm ecl}$ relation implies, for example, that $10^5$ clusters, each weighing $10\,{\rm M}_\odot$, cannot have the same composite (i.e., combined) IMF as one cluster with $10^6\,{\rm M}_\odot$ because such small clusters can never make stars more massive than about $2.5\,{\rm M}_\odot$". The IMF therefore depends on the star formation history of each galaxy (see Fig. 5.8, right). For example, it explains pretty well why the Salpeter IMF (slope $\alpha = 2.35$) that applies to many star clusters is less steep (or more top heavy) than the Scalo IMF (slope $\alpha = 2.7$) since the latter is an average over the whole Galaxy and it therefore includes small clusters.

Another example is the SFR calculated from the $H\alpha$ line that depends on massive stars and that is likely underestimated for dwarf galaxies that lack massive stars (see Fig. 5.8). Calculations of SFR, $M_{\rm stellar}/L$ and $M_{\rm stellar}$ would become more accurate for galaxies if one could account for the physical variations of the IMF, that is expected to be more top-heavy in e.g., major mergers in the fusion processes than in more quiescent stages.

One has to be cautious when calculating $M_{\rm stellar}$ and SFR and assuming a constant IMF. This common assumption is probably not valid when comparing properties over a large range of galaxy scales, and systematic effects can be very important. To account for these, we further advice to use the most simplistic evaluation, e.g., an analytical form to calculate $M_{\rm stellar}/L$ (see Chapter 2), since it would be then easier to correct the result for possible systematic effects. Notice that the Bell *et al.* (2003) method appear also more appropriate than that used in the SDSS (Kauffmann *et al.*, 2003) because there could be less luminous stars per mass unit formed in dwarf than in giant galaxies. The latter would cause a constant IMF to potentially lead to an underestimate of the dwarf stellar masses.

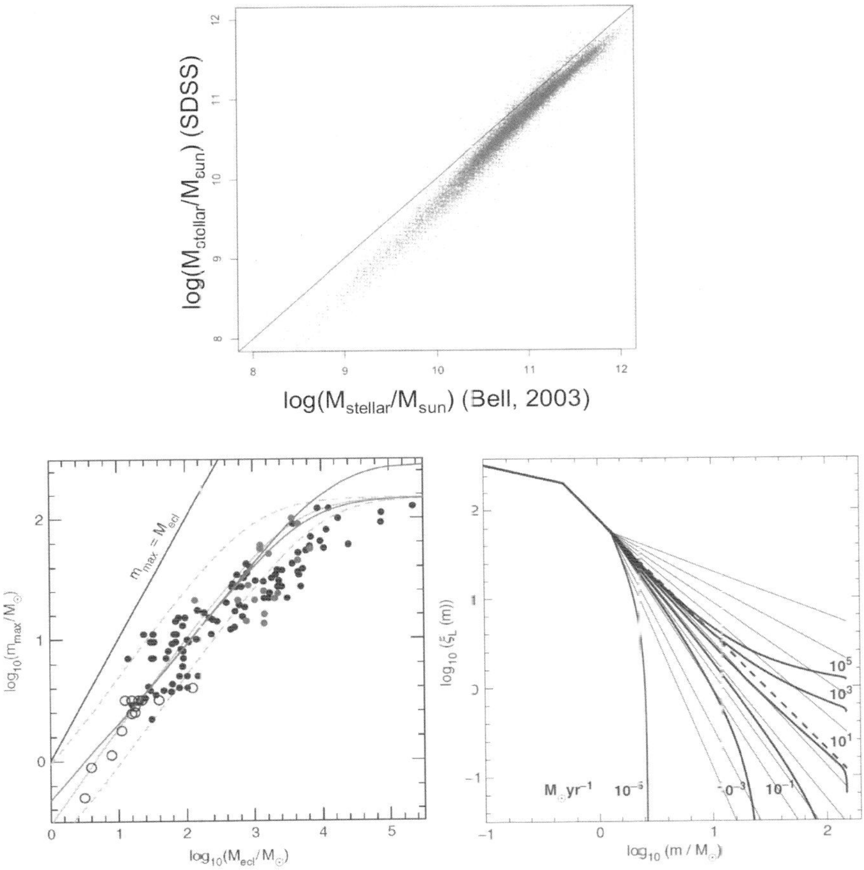


Fig. 5.8: *Top*: Comparison between stellar mass derived using the Bell *et al.* (2003) method and from the SDSS (Kauffmann *et al.*, 2003). Aside from a systematic shift that is caused by using a different IMF, the second method underestimates dwarf masses, when compared to the first method. *Bottom-left*: the strong relationship between the most massive star and the embedded cluster mass ($M_{\rm ecl}$, open and full dots) and *bottom-right*, the IMF of galaxies and the variation of the high stellar mass branch with SFR (adapted from Figs. 5 and 35 of Kroupa *et al.*, 2013, with kind permission from Springer Science + Business Media, and references therein).

### 5.4.2 *Gas mass estimates*

According to the cosmic baryon inventory made by Fukugita *et al.* (1998) and by Fukugita and Peebles (2004) only 6.2% and 1.7% of

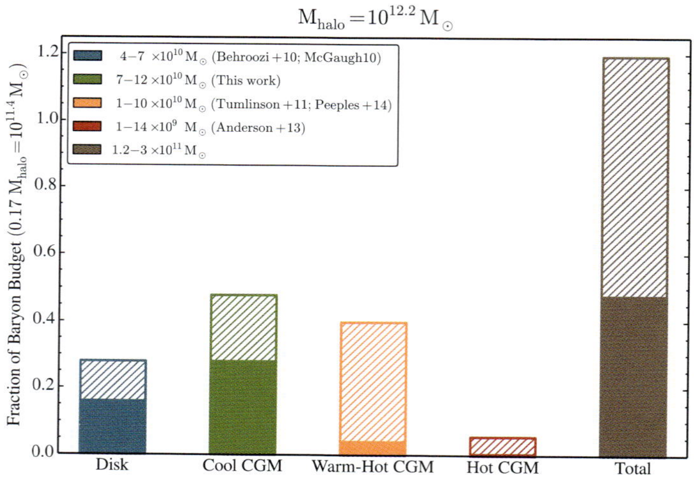


Fig. 5.9: Baryon inventory within ~160 kpc of typical $z \sim 0.2$ L* galaxies based on Werk *et al.* (2014) and references therein. The first column represents stellar mass and neutral gas associated with the disk, the next three columns show the circumgalactic gas (CGM) at three different temperatures. Filled histograms show the preferred lower limit of Werk *et al.* (2014), and the open histograms include corrections for damped absorptions. The last column shows the total amount of baryons that may populate a galaxy and its halo (adapted from Werk *et al.*, 2014, ©AAS, reproduced with permission).

them are made of stars or of neutral gas. Accounting for star residuals, it let ~90% of the baryons in the shape of ionized gas, from cool ($2\ 10^4 < T < 10^5$ K), warm-hot ($10^5 < T < 10^7$) to hot gas ($T > 10^7$), respectively. This gas resides galaxy halos or the intergalactic medium and is too diffuse to be directly detected from X-ray and UV observations.

Figure 5.9 summarizes the most recent inventory of baryons using absorption lines at different ionization levels in spectra of background QSOs. This Figure shows that only a small fraction of the baryons is in the form of stars or of neutral gas, and, remarkably, confirms the earlier expectations of Fukugita and Peebles (2004). This is sufficient to resolve the "missing baryons" problem.[10] This Figure also shows

[10]The "missing baryons" refer to the baryon content derived from the modeling of the $z \sim 3$ Ly forest significantly exceeding the amount of baryon estimated to be in stars and in neutral gas at $z = 0$ (see, e.g., Rauch *et al.*, 1997; Weinberg *et al.*, 1997; Kim *et al.*, 2001).

how large the uncertainties (and systematic effects) are and how these could affect various scaling relations such as the baryonic Tully–Fisher relation shown in Fig. 5.5.

At higher redshifts, investigations of the gaseous component are only beginning, apart from (about neutral) gas mass estimates through the inversion of the Schmidt–Kennicutt law (see, e.g., Erb *et al.*, 2006 and footnote in Sec. 5.2). It is expected (and marginally observed) that large quantities of ionized gas are produced in the high-$z$ progenitors of local massive galaxies since high star formation and mergers were more common in the past. Understanding the cooling/heating/interactions between ionized/neutral/molecular gas that constitute the large majority of baryons is crucial to the understanding galaxy evolution and formation. Perhaps the evolution of the Tully–Fisher relation, as well as of other scaling relations, is simply a reflection of a decrease, as a function of redshift, of the amount of gas that is available and cold enough to allow for stars to form.

## Bibliography


Abramson L. E., Kelson D. D., Dressler A. *et al.* (2014). The mass-independence of specific star formation rates in galactic disks, *ApJ* **785**, L36.

Barnes J. E. (2002). Formation of gas discs in merging galaxies, *MNRAS* **333**, pp. 481–494.

Bell E. F., de Jong R. S. (2001). Stellar mass-to-light ratios and the Tully–Fisher relation, *ApJ* **550**, pp. 212–229.

Bell E. F., McIntosh D. H., Katz N. *et al.* (2003). The optical and near-infrared properties of galaxies. I. Luminosity and stellar mass functions, *ApJS* **149**, pp. 289–312.

Böhm A., Ziegler B. L., Saglia R. P. *et al.* (2004). The Tully–Fisher relation at intermediate redshift, *A&A* **420**, pp. 97–114.

Brinchmann J., Charlot S., White S. D. M. *et al.* (2004). The physical properties of star-forming galaxies in the low-redshift universe, *MNRAS* **351**, pp. 1151–1179.

Conselice C. J., Bundy K., Ellis R. S. *et al.* (2005). Evolution of the near-infrared tully-fisher relation: Constraints on the relationship between the stellar and total masses of disk galaxies since $z \sim 1$, *ApJ* **628**, pp. 160–168.

Courteau S., Dutton A. A., van den Bosch F. C. *et al.* (2007). Scaling relations of spiral galaxies, *ApJ* **671**, pp. 203–225.

Covington M. D., Kassin S. A., Dutton A. A. *et al.* (2010). Evolution of the stellar mass Tully–Fisher relation in disk galaxy merger simulations, *ApJ* **710**, pp. 279–288.

Cresci G., Hicks E. K. S., Genzel R. *et al.* (2009). The SINS survey: Modeling the dynamics of $z \sim 2$ galaxies and the high-z Tully–Fisher relation, *ApJ* **697**, pp. 115–132.

Delgado-Serrano R., Hammer F., Yang Y. B. *et al.* (2010). How was the Hubble sequence 6 Gyr ago? *A&A* **509**, A78.

Desmond H. (2012). The baryonic Tully–Fisher relation predicted by cold dark matter cosmogony, arXiv:124.1497 [astro-ph.CO].

Drory N., Alvarez M. (2008). The contribution of star formation and merging to stellar mass buildup in galaxies, *ApJ* **680**, pp. 41–53.

Dutton A. A., van den Bosch F. C., Faber S. M. *et al.* (2011). On the evolution of the velocity-mass-size relations of disc-dominated galaxies over the past 10 billion years, *MNRAS* **410**, pp. 1660–1676.

Eisenstein D. J., Loeb A. (1996). Can the Tully–Fisher relation be the result of initial conditions? *ApJ* **459**, p. 432.

Elbaz D., Daddi E., Le Borgne D. *et al.* (2007). The reversal of the star formation-density relation in the distant universe, *A&A* **468**, pp. 33–48.

Epinat B., Contini T., Le Fèvre O. *et al.* (2009). Integral field spectroscopy with sinfoni of vvds galaxies. i. Galaxy dynamics and mass assembly at $1.2 < z < 1.6$, *A&A* **504**, pp. 789–805.

Erb D. K., Shapley A. E., Pettini M. *et al.* (2006). The mass-metallicity relation at $z > 2$, *ApJ* **644**, pp. 813–828.

Flores H., Hammer F., Puech M. *et al.* (2006). 3d spectroscopy with vlt/giraffe. i. The true Tully–Fisher relationship at $z \sim 0.6$, *A&A* **455**, pp. 107–118.

Fuentes-Carrera I., Flores H., Yang Y. *et al.* (2010). Determining the morpho-kinematic properties of a face-on merger at $z \sim 0.7$, *A&A* **513**, A43.

Fukugita M., Hogan C. J., Peebles P. J. E. (1998). The cosmic baryon budget, *ApJ* **503**, pp. 518–530.

Fukugita M., Peebles P. J. E. (2004). The cosmic energy inventory, *ApJ* **616**, pp. 643–668.

Glazebrook K. (2013). The dawes review 1: Kinematic studies of star-forming galaxies across cosmic time, *PASA* **30**, e056.

Gnerucci A., Marconi A., Cresci G. *et al.* (2011). Dynamical properties of amaze and lsd galaxies from gas kinematics and the Tully–Fisher relation at $z \sim 3$, *A&A* **528**, A88.

Hammer F., Flores H., Elbaz D. *et al.* (2005). Did most present-day spirals form during the last 8 gyr? A formation history with violent episodes revealed by panchromatic observations, *A&A* **430**, pp. 115–128.

Hammer F., Flores H., Lilly S. J. *et al.* (1997). Canada–France Redshift Survey. XIV. Spectral Properties of Field Galaxies up to $z = 1$, *ApJ* **481**, pp. 49–82.

Hammer F., Flores H., Puech M. *et al.* (2009a). The hubble sequence: Just a vestige of merger events? *A&A* **507**, pp. 1313–1326.

Hammer F., Flores H., Yang Y. B. *et al.* (2009b). A forming, dust-enshrouded disk at $z = 0.43$: the first example of a massive, late-type spiral rebuilt after a major merger? *A&A* **496**, pp. 381–387.

Hammer F., Puech M., Chemin L. *et al.* (2007). The milky way, an exceptionally quiet galaxy: Implications for the formation of spiral galaxies, *ApJ* **662**, pp. 322–334.

Hammer F., Yang Y., Flores H. *et al.* (2012). The formation of large galactic disks through the hierarchical scenario: Further consequences, *MPLA* **27**, 1230034.

Hopkins P. F., Croton D., Bundy K. *et al.* (2010). Mergers in ΛCDM: Uncertainties in theoretical predictions and interpretations of the merger rate, *ApJ* **724**, pp. 915–945.

Kannappan S. J., Barton E. J. (2004). Tools for identifying spurious luminosity offsets in Tully–Fisher studies: Application at low redshift and implications for high redshift, *AJ* **127**, pp. 2694–2710.

Kannappan S. J., Fabricant D. G., Franx M. (2002). Physical sources of scatter in the Tully–Fisher relation, *AJ* **123**, pp. 2358–2386.

Kassin S. A., Weiner B. J., Faber S. M. *et al.* (2007). The stellar mass Tully–Fisher relation to $z = 1.2$ from aegis, *ApJ* **660**, pp. L35–L38.

Kauffmann G., Heckman T. M., White S. D. M. *et al.* (2003). Stellar masses and star formation histories for $10^5$ galaxies from the sloan digital sky survey, *MNRAS* **341**, pp. 33–53.

Kennicutt, Jr. R. C. (1998). The global schmidt law in star-forming galaxies, *ApJ* **498**, pp. 541–552.

Kewley L. J., Ellison S. L. (2008). Metallicity calibrations and the mass–metallicity relation for star-forming galaxies, *ApJ* **681**, pp. 1183–1204.

Kim T.-S., Cristiani S., D'Odorico S. (2001). The Lyalpha forest at $1.5 < z < 4$, *A&A* **373**, pp. 757–781.

Kobulnicky H. A., Kewley L. J. (2004). Metallicities of $0.3 < z < 1.0$ Galaxies in the GOODS-north field, *ApJ* **617**, pp. 240–261.

Kroupa P., Weidner C., Pflamm-Altenburg J. *et al.* (2013). The stellar and substellar initial mass function of simple and composite populations, *Planets, Stars and Stellar Systems. Volume 5: Galactic Structure and Stellar Populations*, (eds.) Oswalt T. D. and Gilmore G., p. 115, doi:10.1007/978-94-007-5612-0 4.

Le Fèvre O., Abraham R., Lilly S. J. *et al.* (2000). Hubble Space Telescope imaging of the CFRS and LDSS redshift surveys — IV. Influence of mergers in the evolution of faint field galaxies from $z \sim 1$, *MNRAS* **311**, pp. 565–575.

Liang Y. C., Hammer F., Flores H. (2006a). Significant evolution of the stellar mass–metallicity relation since $z = 0.65$, *A&A* **447**, pp. 113–119.

Liang Y. C., Yin S. Y., Hammer F. *et al.* (2006b). The oxygen abundance calibrations and N/O abundance ratios of ~40,000 SDSS star-forming galaxies, *ApJ* **652**, pp. 257–269.

Lotz J. M., Jonsson P., Cox T. J. *et al.* (2010a). The effect of gas fraction on the morphology and time-scales of disc galaxy mergers, *MNRAS* **404**, pp. 590–603.

Lotz J. M., Jonsson P., Cox T. J. *et al.* (2010b). The effect of mass ratio on the morphology and time-scales of disc galaxy mergers, *MNRAS* **404**, pp. 575–589.

Maiolino R., Nagao T., Grazian A. *et al.* (2008). AMAZE. I. The evolution of the mass–metallicity relation at z > 3, *A&A* **488**, pp. 463–479.

McGaugh S. S. (2005). The Baryonic Tully–Fisher relation of galaxies with extended rotation curves and the stellar mass of rotating galaxies, *ApJ* **632**, pp. 859–871.

McGaugh S. S. (2012). The baryonic Tully–Fisher relation of gas-rich galaxies as a test of ΛCDM and MOND, *AJ* **143**, 40.

McGaugh S. S., Schombert J. M., Bothun G. D. *et al.* (2000). The baryonic Tully–Fisher relation, *ApJ* **533**, pp. L99–L102.

McGaugh S. S., Schombert J. M., de Blok W. J. G. *et al.* (2010). The baryon content of cosmic structures, *ApJ* **708**, pp. L14–L17.

Miller S. H., Bundy K., Sullivan M. *et al.* (2011). The assembly history of disk galaxies. i. The Tully–Fisher relation to $z \sim= 1.3$ from deep exposures with deimos, *ApJ* **741**, 115.

Miller S. H., Ellis R. S., Sullivan M. *et al.* (2012). The assembly history of disk galaxies. ii. Probing the emerging Tully–Fisher relation during 1 < z < 1.7, *ApJ* **753**, 74.

Moustakas J., Zaritsky D., Brown M. *et al.* (2011). Evolution of the stellar mass–metallicity relation since $z = 0.75$, arXiv:1112.3300 [astro-ph.CO].

Navarro J. F., Steinmetz M. (2000). Dark halo and disk galaxy scaling laws in hierarchical universes, *ApJ* **538**, pp. 477–488.

Neichel B., Hammer F., Puech M. *et al.* (2008). Images. ii. A surprisingly low fraction of undisturbed rotating spiral disks at $z$ ˜ 0.6 the morpho-kinematical relation 6 gyr ago, *A&A* **484**, pp. 159–172.

Noeske K. G., Weiner B. J., Faber S. M. *et al.* (2007). Star formation in aegis field galaxies since $z = 1.1$: The dominance of gradually declining star formation, and the main sequence of star-forming galaxies, *ApJ* **660**, pp. L43–L46.

Peirani S., Hammer F., Flores H. *et al.* (2009). A giant bar induced by a merger event at $z = 0.4$? *A&A* **496**, pp. 51–56.

Pizagno J., Prada F., Weinberg D. H. *et al.* (2007). The Tully–Fisher relation and its residuals for a broadly selected sample of galaxies, *AJ* **134**, pp. 945–972.

Portinari L., Sommer-Larsen J. (2007). The Tully–Fisher relation and its evolution with redshift in cosmological simulations of disc galaxy formation, *MNRAS* **375**, pp. 913–924.

Puech M., Flores H., Hammer F. *et al.* (2008). Images. iii. The evolution of the near-infrared Tully–Fisher relation over the last 6 gyr, *A&A* **484**, pp. 173–187.

Puech M., Hammer F., Flores H. *et al.* (2010). The baryonic content and Tully–Fisher relation at $z \sim 0.6$, *A&A* **510**, A68.

Puech M., Hammer F., Flores H. *et al.* (2009). A forming disk at $z \sim 0.6$: collapse of a gaseous disk or major merger remnant? *A&A* **493**, pp. 899–906.

Puech M., Hammer F., Hopkins P. F. *et al.* (2012). Galaxy disks do not need to survive in the Λcdm paradigm: The galaxy merger rate out to $z \sim 1.5$ from morpho-kinematic data, *ApJ* **753**, 128.

Puech M., Hammer F., Rodrigues M. *et al.* (2014). The main sequence of star-forming galaxies at z ∼ 0.6: reinstating major mergers, *MNRAS* **443**, pp. L49–L53.

Rauch M., Miralda-Escudé J., Sargent W. L. W. *et al.* (1997). The opacity of the Ly$\alpha$ forest and implications for $\Omega_b$ and the ionizing background, *ApJ* **489**, pp. 7–20.

Ravikumar C. D., Puech M., Flores H. *et al.* (2007). New spectroscopic redshifts from the CDFS and a test of the cosmological relevance of the GOODS-South field, *A&A* **465**, pp. 1099–1108.

Rawat A., Hammer F., Kembhavi A. K. *et al.* (2008). Toward a robust estimate of the merger rate evolution using near-IR photometry, *ApJ* **681**, pp. 1089–1098.

Renzini A., Peng Y.-j. (2015). An objective definition for the main sequence of star-forming galaxies, *ApJ* **801**, L29.

Rodrigues M., Hammer F., Flores H. *et al.* (2016). The severely biased mass-metallicity relation in the SDSS. The representative relation from 17,000 galaxies and its origin. *MNRAS, submitted.*

Rodrigues M., Hammer F., Flores H. *et al.* (2008). IMAGES IV: strong evolution of the oxygen abundance in gaseous phases of intermediate mass galaxies from $z \sim 0.8$, *A&A* **492**, pp. 371–388.

Rodrigues M., Puech M., Hammer F. *et al.* (2012). A decrease of the gas exchanges between galaxies and the intergalactic medium, from 12 to 6 billion years ago, *MNRAS* **421**, pp. 2888–2903.

Scoville N., Sheth K., Aussel H. *et al.* (2015). ISM masses and star formation at $z = 1$ to 3 ALMA observations of dust continuum in 180 Galaxies in COSMOS, *ArXiv e-prints*.

Silk J. (1997). Feedback, disk self-regulation, and galaxy formation, *ApJ* **481**, pp. 703–709.

Tremonti C. A., Heckman T. M., Kauffmann G. *et al.* (2004). The Origin of the Mass-Metallicity Relation: Insights from 53,000 star-forming galaxies in the sloan digital sky survey, *ApJ* **613**, pp. 898–913.

Tully R. B., Fisher J. R. (1977). A new method of determining distances to galaxies, *A&A* **54**, pp. 661–673.

Vergani D., Epinat B., Contini T. *et al.* (2012). Massivse: Mass assembly survey with sinfoni in vvds. iv. Fundamental relations of star-forming galaxies at 1 < z < 1.6, *A&A* **546**, A118.

Verheijen M. A. W. (2001). The ursa major cluster of galaxies. V. H I Rotation curve shapes and the Tully–Fisher relations, *ApJ* **563**, pp. 694–715.

Vogt N. P., Forbes D. A., Phillips A. C. *et al.* (1996). Optical rotation curves of distant field galaxies: Keck results at reshifts to *Z* approximately 1, *ApJ* **465**, p. L15.

Vogt N. P., Phillips A. C., Faber S. M. *et al.* (1997). Optical rotation curves of distant field galaxies: Sub-L* systems, *ApJ* **479**, pp. L121–L124.

Weidner C., Kroupa P., Pflamm-Altenburg J. *et al.* (2013). The galaxy-wide initial mass function of dwarf late-type to massive early-type galaxies, *MNRAS* **436**, pp. 3309–3320.

Weinberg D. H., Miralda-Escudé J., Hernquist L. *et al.* (1997). A lower bound on the cosmic baryon density, *ApJ* **490**, pp. 564–570.

Weiner B. J., Willmer C. N. A., Faber S. M. *et al.* (2006). A survey of galaxy kinematics to $z \sim 1$ in the tkrs/goods-n field. ii. Evolution in the Tully–Fisher relation, *ApJ* **653**, pp. 1049–1069.

Werk J. K., Prochaska J. X., Tumlinson J. *et al.* (2014). The COS–Halos survey: Physical conditions and baryonic mass in the low-redshift circumgalactic medium, *ApJ* **792**, 8.

White S. D. M. (1997). *Galaxy Scaling Relations: Origins, Evolution and Applications*, (eds.) da Costa, L. N. and Renzini, A., p. 3.

Yang Y., Hammer F., Flores H. *et al.* (2009). A surviving disk from a galaxy collision at z = 0.4, *A&A* **501**, pp. 437–443.

Yin S. Y., Liang Y. C., Hammer F. *et al.* (2007). Empirical strong-line oxygen abundance calibrations from galaxies with electron-temperature measurements, *A&A* **462**, pp. 535–546.

Zahid H. J., Geller M. J., Kewley L. J. *et al.* (2013). The chemical evolution of star-forming galaxies over the Last 11 billion years, *ApJ* **771**, L19.

Zahid H. J., Kewley L. J., Bresolin F. (2011). The mass–metallicity and luminosity–metallicity relations from DEEP2 at z = 0.8, *ApJ* **730**, 137.

Zaritsky D., Courtois H., Muñoz-Mateos J.-C. *et al.* (2014). The baryonic Tully–Fisher relationship for $S^4G$ galaxies and the "Condensed" baryon fraction of galaxies, *AJ* **147**, 134.

Ziegler B. L., Böhm A., Fricke K. J. *et al.* (2002). The evolution of the Tully–Fisher relation of spiral galaxies, *ApJ* **564**, pp. L69–L72.

# Appendix A

# Most Intense Emission and Absorption Lines

Table A.1: Most intense emission lines found in AGNs, H II regions, and planetary nebulae.

| Line | $\lambda$ [Å] | Line | $\lambda$ [Å] |
|---|---|---|---|
| Ly $\epsilon$ | 937.8 | [O III] | 4958.9 |
| Ly $\delta$ | 949.75 | [O III] | 5006.85 |
| Ly $\gamma$ | 972.5 | [N I] | 5198.0 |
| Ly $\beta$ | 1025.7 | [N I] | 5200.4 |
| [O VI] | 1031.9 | [Fe XIV] | 5302.9 |
| [O VI] | 1037.6 | [Fe XIV] | 5303.6 |
| Ly $\alpha$ | 1215.65 | [O I] | 5577.35 |
| [N V] | 1238.8 | [N II] | 5754.6 |
| [N V] | 1242.8 | He I | 5875.6 |
| Si IV | 1393.75 | [O I] | 6300.3 |
| Si IV | 1402.75 | [S III] | 6312.1 |
| C IV | 1548.2 | [O I] | 6363.8 |
| C IV | 1550.8 | [Fe X] | 6374.5 |
| He II | 1640.0 | [N II] | 6548.1 |
| [C III] | 1909.0 | H $\alpha$ | 6562.8 |
| N II | 2142.0 | [N II] | 6583.4 |
| C II | 2326.0 | He I | 6678.1 |
| [Ne IV] | 2424.0 | [S II] | 6716.4 |
| Mg II | 2795.5 | [S II] | 6730.8 |
| Mg II | 2802.7 | [Ar V] | 7005.7 |
| [O III] | 3132.9 | He I | 7065.2 |
| He I | 3187.8 | [Ar III] | 7135.8 |
| [Ne V] | 3345.85 | [Ar IV] | 7170.6 |
| [Ne V] | 3425.9 | [Ar IV] | 7237.3 |
| [O II] | 3726.05 | [Ar IV] | 7262.8 |
| [O II] | 3728.8 | He I | 7281.35 |

(*Continued*)

Table A.1: (*Continued*)

| Line | λ [Å] | Line | λ [Å] |
|---|---|---|---|
| H 12 | 3750.1 | [O II] | 7319.9 |
| H 11 | 3770.6 | [O II] | 7330.2 |
| H 10 | 3797.8 | [Ar III] | 7751.1 |
| H 9 | 3835.3 | He I | 7816.2 |
| [Ne III] | 3868.75 | [Cl II] | 8046.1 |
| He I | 3888.6 | He II | 8236.6 |
| H 8 | 3889.0 | O I | 8446.4 |
| [Ne III] | 3967.5 | H P10 | 9014.9 |
| H $\epsilon$ | 3970.1 | [S III] | 9069.0 |
| [S II] | 4068.5 | [Fe XIII] | 10776.8 |
| [S II] | 4076.2 | [Fe XIII] | 10797.9 |
| H $\delta$ | 4101.7 | He I | 10830.0 |
| H $\gamma$ | 4340.5 | | |
| [O III] | 4363.2 | | |
| He I | 4471.5 | | |
| He II | 4685.75 | | |
| H $\beta$ | 4861.3 | | |

Wavelengths are given in the vacuum up to [Ne IV] and in the air red-ward.

Table A.2: Most intense absorption lines found in QSOs and galaxies.

| Line | λ [Å] | Line | λ [Å] |
|---|---|---|---|
| (Ly-limit | 912.0) | Ca II (K) | 3933.65 |
| N II | 915.6 | Ca II (H) | 3968.45 |
| Ly $\epsilon$ | 937.8 | H $\epsilon$ | 3970.1 |
| Ly $\delta$ | 949.75 | Blend Fe–Mn | 4030. |
| Ly $\gamma$ | 972.5 | H $\delta$ | 4101.7 |
| C III | 977.0 | Ca I | 4226.7 |
| Ly $\beta$ | 1025.7 | CH + metals(G) | (4280–4320) |
| [O VI] | 1031.9 | | $\lambda_{eff} \sim 4304$ |
| C II | 1036.3 | H $\gamma$ | 4340.5 |
| [O VI] | 1037.6 | Fe I | 4383.6 |
| Ly $\alpha$ | 1215.65 | H $\beta$ | 4861.3 |
| [N V] | 1238.8 | Mg I | 5167.3 |
| [N V] | 1242.8 | Mg I | 5172.7 |
| Si II | 1260.4 | Mg I | 5183.6 |
| O I | 1302.15 | | |
| Si II | 1304.4 | | |
| C II | 1334.5 | Ca + Fe | 5269.0 |
| Si IV | 1393.75 | (Fe) | (5406) |

(*Continued*)

Table A.2: (*Continued*)

| Line | $\lambda$ [Å] | Line | $\lambda$ [Å] |
|---|---|---|---|
| Si IV | 1402.75 | Na I (D2) | 5889.95 |
| Si II | 1526.7 | Na I (D1) | 5895.9 |
| C IV | 1548.2 | H $\alpha$ | 6562.8 |
| C IV | 1550.8 | | |
| Fe II | 1608.45 | | |
| Al II | 1670.8 | | |
| Al III | 1854.7 | | |
| Al III | 1862.8 | | |
| Fe II | 2343.5 | | |
| Fe II | 2373.75 | | |
| Fe II | 2382.0 | | |
| Fe II | 2585.9 | | |
| Fe II | 2599.4 | | |
| Mg II | 2795.5 | | |
| Mg II | 2802.7 | | |
| Mg I | 2852.1 | | |
| He I | 3187.75 | | |
| H 12 | 3750.1 | | |
| H 11 | 3770.6 | | |
| H 10 | 3797.8 | | |
| Blend Mg I | 3830. | | |
| CN (L) | (3783–3883) | | |
| | $\lambda_{eff}$ ~3880 | | |
| H 9 | 3835.3 | | |
| He I | 3888.6 | | |
| H 8 | 3888.9 | | |

Wavelengths are given in the vacuum up to [Al III] and in the air red-ward.

# Index

**K**



**L**



**M**

## V



## W



## Z